STUDIES IN APPLIED MECHANICS 12

Local Effects in the Analysis of Structures

STUDIES IN APPLIED MECHANICS

1. Mechanics and Strength of Materials (Skalmierski)

2. Nonlinear Differential Equations (Fučík and Kufner)

3. Mathematical Theory of Elastic and Elastico-Plastic Bodies
An Introduction (Necas and Hlaváček)

4. Variational, Incremental and Energy Methods in Solid Mechanics
and Shell Theory (Mason)

5. Mechanics of Structured Media, Parts A and B (Selvadurai, Editor)

6. Mechanics of Material Behavior (Dvorak and Shield, Editors)

7. Mechanics of Granular Materials: New Models and Constitutive Relations
(Jenkins and Satake, Editors)

8. Probabilistic Approach to Mechanisms (Sandler)

9. Methods of Functional Analysis for Application in Solid Mechanics (Mason)

10. Boundary Integral Equation Methods in Eigenvalue Problems of Elastodynamics
and Thin Plates (Kitahara)

11. Mechanics of Material Interfaces (Selvadurai and Voyiadjis, Editors)

12. Local Effects in the Analysis of Structures (Ladevèze, Editor)

STUDIES IN APPLIED MECHANICS 12

Local Effects in the Analysis of Structures

Edited by

Pierre Ladevèze

Laboratoire de Mécanique et Technologie (E.N.S.E.T./Université Paris 6/C.N.R.S.), Cachan, France

ELSEVIER

Amsterdam — Oxford — New York — Tokyo 1985

ELSEVIER SCIENCE PUBLISHERS B.V.
1 Molenwerf,
P.O. Box 211, 1000 AE Amsterdam, The Netherlands

Distributors for the United States and Canada:

ELSEVIER SCIENCE PUBLISHING COMPANY INC.
52, Vanderbilt Avenue
New York, NY 10017, U.S.A.

Library of Congress Cataloging-in-Publication Data
Main entry under title:

Local effects in the analysis of structures.

(Studies in applied mechanics ; 12)
Selection of papers presented at the EUROMECH Colloquium "Inclusion of Local Effects in the Analysis of Structures," held Sept. 11-14, 1984 at Laboratoire de mécanique et technologie, Cachan, France.
Bibliography: p.
1. Structures, Theory of--Congresses. 2. Stress concentration--Congresses. I. Ladevèze, Pierre, 1945- . II. EUROMECH Colloquium "Inclusion of Local Effects in the Analysis of Structures" (1984 : Laboratoire de mécanique et technologie, Cachan, France) III. Series.

TA645.L63 1985 624.1'71 85-13150
ISBN 0-444-42520-9 (U.S.)

ISBN 0-444-42520-9 (Vol. 12)
ISBN 0-444-41758-3 (Series)

Printed in The Netherlands

F O R E W O R D

At the present time, the Inclusion of Local Effects in the Analysis of Structures is undoubtedly a question of prime importance for Engineering Design. The classical computational approaches are not readily adapted to take into account the local effects - appropriate treatments are necessary. This book attempts to provide an introduction to and a survey of the specific computational methods. It begins with the various theories which allow to separate and then to determine the local and global effects. Chapter 2 discusses edge effects for composite structures. Chapter 4 deals with general numerical methods, especially for effects due to large local variations of geometry. Chapter 3 concerns some dynamic problems - it is an opening towards non-conventional local effects in Structural Mechanics.

The papers, for a part, have been presented at the EUROMECH Colloquium "Inclusion of Local Effects in the Analysis of Structures" which has been held on September 11-14, 1984 in CACHAN (France) - LABORATOIRE DE MECANIQUE ET TECHNOLOGIE.

The Scientific Committee included :

P. LADEVEZE
R. OHAYON
M. PREDELEANU
E. SANCHEZ-PALENCIA
N.Q. SON.

CONTENTS

FOREWORD V

CHAPTER 1 : FUNDAMENTAL APPROACHES ON LOCAL EFFECTS

P. LADEVEZE
"On Saint-Venant's Principle in Elasticity" 3

R.D. GREGORY - F.Y.M. WAN
"Edge Effects in the Stretching of Plates" 35

N. NGUETSENG - E. SANCHEZ-PALENCIA
"Stress Concentration for Defects Distributed near a Surface" 55

A.M.A. VAN DER HEIJDEN
"On the Influence of Free Edges in Plates and Shells" 75

F. PECASTAINGS
"On a Method to Evaluate Edge Effects in Elastic Plates" 101

V. BERDICHEVSKII - L. TRUSKINOVSKII
"Energy Structure of Localization" 127

CHAPTER 2 : EDGE EFFECTS IN COMPOSITE STRUCTURES

M. SAYIR
"Edge Effects in Rotationally Symmetric Composite Shells"

J.L. DAVET - Ph. DESTUYNDER - Th. NEVERS
"Some Theoretical Aspects in the Modelling of Delamination for Multilayered Plates" 181

D. ENGRAND
"Local Effects Calculations in Composite Plates by a Boundary Layer Method" 199

H. DUMONTET
"Boundary Layers Stresses in Elastic Composites" 215

CHAPTER 3 : LOCAL EFFECTS IN DYNAMICS

J. BALLMANN - H.J. RAATSCHEN - M. STAAT
"High Stress Intensities in Focussing Zones of Waves" 235

C.H. SOIZE
"The Local Effects in the Linear Dynamic Analysis of Structures in the Medium Frequency Range" 253

CHAPTER 4 : GENERAL NUMERICAL APPROACHES OF LOCAL EFFECTS

J. JIROUSEK
"Implementation of Local Effects into Conventional and Non-Conventional Finite Element Formulations" 279

R. PILTNER
"Special Finite Elements for an Appropriate Treatment of Local Effects" 299

W. DIRSCHMID
"On the Consideration of Local Effects in the Finite Element Analysis of Large Structures" 315

E. SCHNACK
"Local Effects of Geometry Variation in the Analysis of Structures". 325

CHAPTER 1:

FUNDAMENTAL APPROACHES ON LOCAL EFFECTS

ON SAINT-VENANT'S PRINCIPLE IN ELASTICITY

P. LADEVEZE

Laboratoire de Mécanique et Technologie, E.N.S.E.T./Université PARIS 6/C.N.R.S., 61, Avenue du Président Wilson - 94230 CACHAN (France)

The Saint-Venant's Principle is considered from the angle of its present practical interest in Structural Mechanics, especially for composite structures. The interior large wavelength effect must be separated from the edge or extremity effects with a small wavelength in order to be computed. In a more precise way, it is a theorem which expresses conditions ensuring localization of displacements and stresses. This point of view is built on certain characteristic properties of the solutions and not on the properties of zero resultant-moment loadings. Moreover, and this is an important point, the diameter Ø of the beam or the thickness 2h of the plate are not considered as small parameters. The approach has nothing to do with asymptotic methods.

Beams and plates are studied as far as possible from a unitary point of view. The problem is restricted to the so-called Saint-Venant Problem for which the non-zero loadings and displacements are only prescribed on the edges of the structure. Additional loadings on the lateral surface of the beam or on the faces of the plate essentially only modify the interior effect. This effect is now well-known [6] [15] [16] [17] [18] [19] [20] [32].

This paper starts by proposing several major properties for the solutions which are localized or not. They are inferred from the particularities of the geometry. The localization concepts are stated precisely. The corresponding solutions decrease exponentially as a function of the distance to the edges such that the decaying length is O(Ø) for beams and O(h) for plates. In fact, the Saint-Venant Principle which characterizes such solutions expresses the orthogonality to the interior large wavelength solutions. Some auxiliary problems are used to write this condition in terms of the data on the edges for any boundary conditions.

The splitting up of the solution into the interior effect and the edge effects is a central problem. Several results concerning the existence and the unicity of such a splitting up are presented.

1. GENERALITIES - HYPOTHESES

1.1. Beams

Cylindrical beams are considered (figure 1). The domain Ω is defined by

$\Omega = \{M = X + t.N \; ; \; t \in]0,L[\; ; X \in S\}$

where S denotes the cross-section. t is the N- coordinate. The boundary of the cross-section is supposed as having the usual regularity, namely corner points can occur.

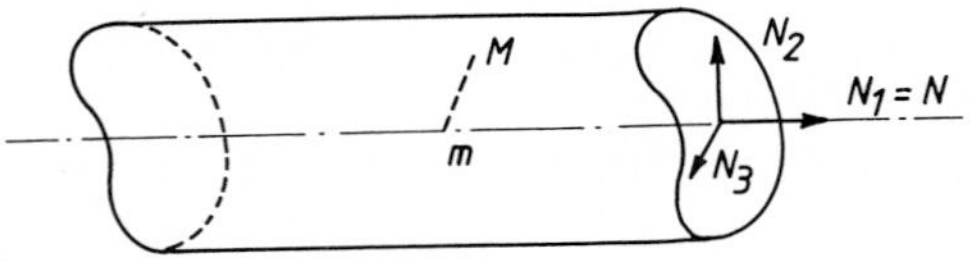

Figure 1

The end cross-sections are denoted by S_0 and S_L. The Hooke tensor $\mathbb{K}$ is taken to be constant on lines parallel to the center line.

1.2. Plates

The thickness is constant. It is supposed that the material is homogeneous on planes parallel to the middle surface Σ. The domain is defined by

$\Omega = \{M = m + Nz \quad , \quad m \in \Sigma \, , \, z \in]-h,h[\}$

and the boundary of the middle surface is supposed as having the usual regularity, namely corner points can occur.

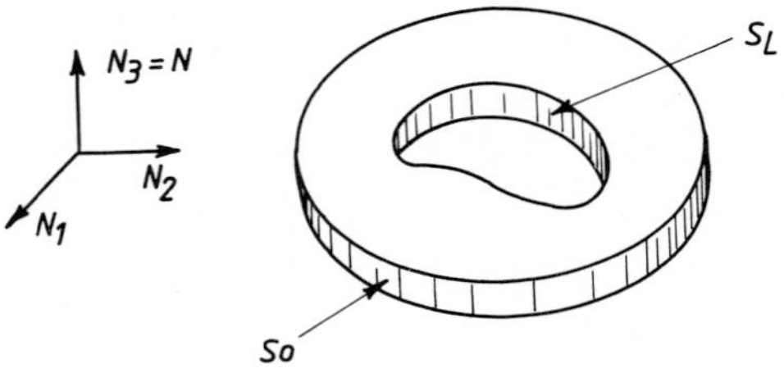

Figure 2

1.3. Basic problem

Our aim is to study end effects, so we shall restrict the paper to this specific problem which, by analogy with the beam theory, will be designated as the Saint-Venant Problem. It is characterized by the following assumptions :

- the body-forces are zero
- the lateral surfaces of the beam and the upper and lower surfaces of the plate are free.

In other words, the non-zero forces and displacements are only prescribed on the end surfaces S_0 and S_L.

Moreover, we will consider only physically admissible solutions namely those with finite energy.

Remarks

- This framework contains two important particular cases :
  . composite beams with homogeneous layers parallel to the center line
  . composite plates with homogeneous layers parallel to the middle surface.
- The boundary conditions on the end cross-section can be of any type.

Notations

The scalar product of the vectors V,W will be written $\overline{V}W$. π denotes both the orthogonal projection on the cross-section for the beams and the orthogonal projection on the middle surface for the plates.

1.4. Saint-Venant's solutions

They are large wavelength "solutions" of the Saint-Venant Problem. They verify all the equations except for the boundary conditions on S_0 and S_L.

- Beams

The Saint-Venant solutions can be written as follows :

$$U_t^* = \mathbb{A}\, T^* + \mathbb{B}\, M_t^* + V_t + \Omega_t\, X$$

$$\sigma^* N\big|_t = \mathbb{A}^0\, T^* + \mathbb{B}^0 M^*$$

where

- T^*, M^* : resultant-moment of normal stress vector
- V, Ω : vectors which are constant related to X-coordinates
- $\mathbb{A}, \mathbb{B}, \mathbb{A}^0, \mathbb{B}^0$: linear operators which are constant related to t.

- Plates

For the sake of simplicity, the Saint-Venant solution is written for homogeneous isotropic materials. It is the classical Kirchhoff-Love solution :

$$\pi U^* = \text{grad}\left[-z\, W + \frac{3\lambda + 4\mu}{6(\lambda+2\mu)}\, z^3\, \Delta_m W\right] + u + \frac{\lambda}{8(\lambda+\mu)}\, z^2\, \text{grad}_m\, \omega$$

$$\overline{N}\,U^{*} = W + \frac{1}{\lambda+2\mu}\left[\frac{\lambda}{2}z^{2} - 2h^{2}(\lambda+\mu)\right]\Delta_m W - \lambda z\,\frac{\omega}{4(\lambda+\mu)}$$

with $W(m)$, $u(m)$, $w(m)$ such as :

$$\Delta_m \Delta_m W = 0 \quad \frac{2\lambda\mu}{\lambda+2\mu}\,\mathrm{grad}_m(\mathrm{div}_m u) + 2\mu\,\mathrm{div}_m(\pi\varepsilon(u)\pi) = 0$$

$$\mathrm{div}_m u = \frac{\lambda+2\mu}{4(\lambda+\mu)}\,\omega$$

1.5. Interior and exterior problems

The beam and the plate described by the figures (1) and (2) have in fact two edges. Therefore, it is natural to introduce :

- the interior problem related to an end section S_0

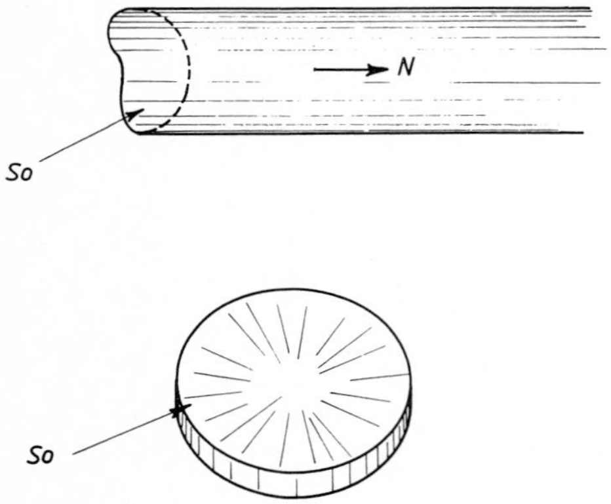

Figure 3

- the exterior problem related to an end section S_0

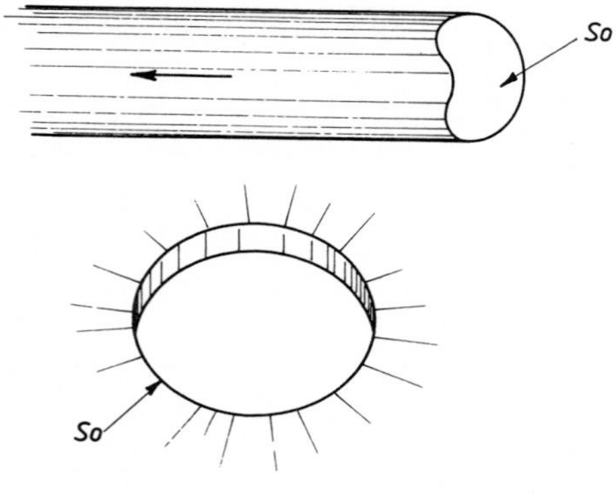

Figure 4

These problems are particular Saint-Venant Problems. It should be recalled that the solutions have to lead to finite energy. "+" shall denote quantities connected with the interior problem, "-" those connected with the exterior one.

For plates, the following coordinates system is introduced :

$C_t, t \in]-\infty,+\infty[$ is a family of closed curves such that :

- $C_t\big|_{t=0} = C_0$
- the domain interior to C_t decreases with t and tends to zero with $t \to \infty$
- the domain exterior to $C_{t'}$ decreases with (-t'). The interior diameter of $C_{t'}$ tends to infinity with (-t').

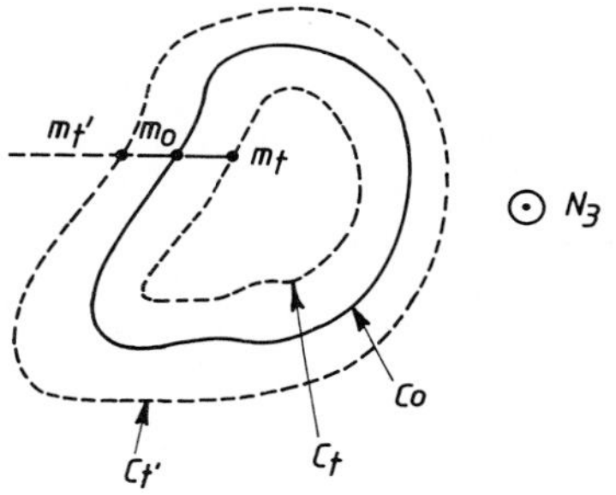

Figure 5

For starshaped domains, it is possible to use

$$t = -\log \frac{|Om_t|}{|Om_0|}$$

where m_t, m_0 are homothetic points.
Moreover, the t-section is defined by $S_t = C_t \times]-h,h[$. So, a plate is mapped on an abstract beam with a variable "cross-section". t is the coordinate generating this different "cross-sections".
With these notations, beams and plates can be studied using the same terminology.

2. INTERIOR AND EXTERIOR PROBLEMS - BASIC PROPERTIES

2.1. Semi-groups $\mathbb{R}^+, \mathbb{R}^-$

- Beams

Let D be the space of displacement values at the cross-section S ($D = [H^{1/2}(S)]^3$). S_0 is the chosen reference cross-section and U_0 a given displacement belonging to D.

U_t^+ designates the solution of the interior problem for the displacement boundary conditions :

$$U_t^+\Big|_{t=0} = U_0$$

$U_{t'}^-$ is defined in the same way for $t' < 0$. It is clear that :

$$U_t^+ = \mathbb{R}_t^+ . U_0 \qquad t \geq 0$$

$$U_{t'}^- = \mathbb{R}_{t'}^- U_0 \qquad t' \leq 0$$

where $\mathbb{R}_t^+$, $\mathbb{R}_{t'}^-$ are operators on D which verify :

$$\mathbb{R}_0^+ = Id$$

$$\mathbb{R}_{t_1}^+ . \mathbb{R}_{t_2}^+ = \mathbb{R}_{t_1+t_2}^+ \qquad \forall\, t_1 \geq 0 \;\; \forall\, t_2 \geq 0$$

$$\mathbb{R}_0^- = Id$$

$$\mathbb{R}_{t_1'}^- . \mathbb{R}_{t_2'}^- = \mathbb{R}_{t_1'+t_2'}^- \qquad \forall\, t_1' \leq 0 \;\; \forall\, t_2' \leq 0$$

Property 2-1 $\{\mathbb{R}_t^+ \quad t \geq 0\}$ and $\{\mathbb{R}_{t'}^- \quad t' \leq 0\}$ are one parameter semi-groups.

- Plates

For plates, the same kind of properties exist but the semi-groups involve two parameters, namely

$$U_t^+ = \mathbb{R}^+(t,t')U_{t'}^+ \qquad t \geq t'$$

$$U_{t'}^- = \mathbb{R}^-(t',t)U_t^- \qquad t' \leq t$$

2.2. Scalar products on D

Let U,V be two displacements of D. The following energy scalar product can be defined on D (in fact on the quotient of D by the space of rigid body displacements) :

$$< U,V >_{+|S_0} = \int_{\Omega_0^+} Tr\left[\mathbb{K} . \varepsilon(U_t^+)\varepsilon(V_t^+)\right] d\Omega$$

$$< U,V >_{-|S_0} = \int_{\Omega_0^-} Tr\left[\mathbb{K} . \varepsilon(U_{t'}^-)\varepsilon(V_{t'}^-)\right] d\Omega$$

where U_t^+, V_t^+, $U_{t'}^-$, $V_{t'}^-$ are prolongations of U,V. Ω_0^+ is the interior domain to the cross-section S_0 , Ω_0^- is the complementary part.

From [25] [27], it can be seen that :

$$< U,V >_{+|S_0} = + \int_{S_0} \overline{V}\, \sigma(U_t^+)n\Big|_{t=0}\, dS$$

$$< U,V >_{-|S_0} = - \int_{S_0} \overline{V}\, \sigma(U_{t'}^-)n\Big|_{t'=0}\, dS$$

where n is the unit outward normal to S_0^+ . and $\sigma(.)$ the stress obtained through the constitutive relation.

For beams, $< U,V >_{+|S_0}$ and $< U,V >_{-|S_0}$ are independent of the reference cross-section. For plates, however the scalar products

$$<.,.>_t^+ \qquad <.,.>_t^-$$

related to the reference cross-section S_t have to be introduced.

The following property expresses a relation between $\mathbb{R}^+$ and $\mathbb{R}^-$, that is $\mathbb{R}^-$ can be defined through $\mathbb{R}^+$.

Property 2-2

$\forall$ t,t' $\quad t \geqslant t'$

$\forall$ U,V $\in$ D

$$< U,\mathbb{R}^-(t',t)V >_{t'}^+ + < U,\mathbb{R}^-(t',t)V >_{t'}^- =$$

$$= < V,\mathbb{R}^+(t,t')U >_t^+ + < V,\mathbb{R}^+(t,t')U >_t^-$$

The proof is obtained by applying Stokes' formula to the quantity

$$A_{t,t'} = \int_{\Omega_{t,t'}} Tr\left[\mathbb{K}.\, \varepsilon(\mathbb{R}^+(t,t')U)\varepsilon(\mathbb{R}^-(t',t)V)\right] d\Omega$$

Remarks

Supplementary properties have been obtained in the case of beams where the cross-section is a symmetric plane for the material [26] [27].

- Let us introduce the scalar product

$$<< U,V >>_t = < U,V >_t^+ + < U,V >_t^-$$

The previous property can be expressed in the form

$$<< U,\mathbb{R}^-(t',t)V >>_{t'} = << V,\mathbb{R}^+(t,t')U >>_t$$

Let U_t^+, V_t^- be solutions of the interior and exterior problems related to different end cross-section S_{t_1}, S_{t_2}. It can be deduced from (2.2) that the quantity

$$<< U_t^+,V_t^- >>_t$$

is a constant $\underline{\gamma}$ independent of t on the common domain, i.e.,

$t_1 < t < t_2$

It is easy to see that this property is still true for the arbitrary cross-section S, even for the curved cross-section S which cuts the beam or the plate into two parts :

$$\underline{\gamma} = \ll U^+_{|S}, V^-_{|S} \gg_{|S}$$

Another extension can be made by considering solutions of the Saint-Venant problem related to the domain Ω_{t_1,t_2}. Let U_t, V_t be such solutions. Using the Stokes formula, the following quantity is obtained :

$$\gamma\ (U_t,V_t) = \int_S \{\overline{V}_t\sigma(U_t)n\}_{|S}\ dS - \int_S \{\overline{U}_t\sigma(V_t)n\}_{|S}\ dS$$

It is a constant independent of the cross-section S (possibly curved).

In the particular case where U_t,V_t are solutions of the interior and exterior problem :

$$\gamma\ (U^+_t,V^-_t) = \ll U^+_t,V^-_t \gg_t$$

So, γ defines invariant quantities which play, as will be seen below, an essential role in the understanding of edge effects.

2.3. New formulations

- Beams

In [27], we have proved that $\mathbb{R}^+,\mathbb{R}^-$ are contraction semi-groups of class C^0. The norms are associated to the scalar products $<\ >_+$, $<\ >_-$. Traditionally, the generators are defined by :

$$\underline{\mathbb{A}}^+U = \lim_{t\to 0^+} \frac{(Id - \mathbb{R}^+_t)\,U}{t} \qquad \underline{\mathbb{A}}^-U = \lim_{t'\to 0^-} \frac{(Id - \mathbb{R}^-_{t'})\,U}{t'}$$

and using the Hille-Yoshida-Phillips theorem, the initial problems will be equivalent to the following differential equations :

$$\frac{d}{dt}U^+ + \underline{\mathbb{A}}^+ U^+ = 0 \qquad \frac{d}{dt'}U^- + \underline{\mathbb{A}}^- U^- = 0$$

$$U^+_{|t=0} = U_0 \qquad U^-_{|t'=0} = U_0$$

- Plates.

Similar properties are easily derived, but the generators depend on t. Moreover, the equivalence between the initial problems and the differential equations have not been proved.

Remarks

- For beams, the initial problems are equivalent to differential equations with "constant coefficients". A natural method to solve them is to use the spectral decomposition of the generators $\mathbb{A}^+$, $\mathbb{A}^-$. The eingenvalues which are complex numbers are the same for $\mathbb{A}^+$ and $\mathbb{A}^-$. The computation of $(\lambda_i, \Psi_i^+(0), \Psi_i^-(0))$ must solve an eingenvalue problem on the cross-section [27]. In the particular case of the semi-infinite strip, these eingenfunctions can be stated. They are the famous Papkovitch functions [7]. Through the differential equations, solutions $\Psi_i^+(t)\Psi_i^-(t')$ are associated to $\Psi_i^+(0)$, $\Psi_i^-(0)$ such that the behavior on t,t' is the product of an exponential term by a polynomial.
- For plates, the differential equations do not have "constants coefficients". Nethertheless, a similar technique can be used (see paper by F. Pecastaings) resulting in the functions derived in [34] by Lure.
- For the semi-infinite strip, Gregory has proved convergence results for the expansions using $\{\Psi_i^+ \;\; i > 1\}$ or $\{\Psi_i^- \;\; i > 1\}$ [13]. Similar results have been derived for plates by Pecastaings [25].
- $\{\Psi_i^+(t)\}$ and $\{\Psi_j^-(t)\}$ have orthogonality properties which have already been noted for many particular cases. It is easy to prove without additional hypothesis that :

$$\langle\langle \Psi_i^+(t), \Psi_j^-(t) \rangle\rangle_t = \delta_{ij}$$

3. LOCALIZED SOLUTIONS OF THE INTERIOR AND EXTERIOR PROBLEMS

Let us consider the interior and exterior problems associated to the reference end cross-section S_0.

3.1. Definitions

The Saint-Venant solutions which locally have finite energy are denoted by U_t^*, σ_t^*. L_S^* is the space of the traces of U_t^*, σ_t^* on the cross-section S.

Now, let us suppose that U_t^+ is orthogonal to the Saint-Venant solutions, i.e. :

$$\gamma(U_t^+, U_t^*) = 0 \qquad \forall\, U^*$$

This property is equivalent to the verification on one cross-section S belonging to Ω_0^+ of the relation :

$$\int_S \overline{U}^{*} \sigma(U^{+}) n \, dS - \int_S \overline{U}^{+} \sigma(U^{*}) n \, dS = 0 \qquad \forall \; U^{*}, \sigma^{*} \in L_S^{*}$$

and then in particular

$$\int_{S_0} \overline{U}^{*} \sigma(U^{+}) n \, dS - \int_{S_0} \overline{U}^{+} \sigma(U^{*}) n \, dS = 0 \qquad \forall \; U^{*}, \sigma^{*} \in L_{S_0}^{*}$$

This last relation is in fact a condition on the data prescribed on the end cross-section S_0. If these conditions are verified on S_0 , they are also verified by U_t^{+} on any cross-section, possibly even curved, belonging to the domain Ω_0^{+} . Moreover, these conditions hold for any cross-section if, and only if, they are verified for one cross-section.

Definitions

. U^{+} is localized if

$$\gamma(U_t^{+}, U_t^{*}) = 0 \qquad \forall \; U^{*}$$

. U^{-} is localized if

$$\gamma(U_{t'}^{-}, U_{t'}^{*}) = 0 \qquad \forall \; U^{*}$$

3.2. Characterization of localized solutions

- Beams

Let us consider the interior problem. From the expression of the Saint-Venant solutions and the relation

$$\int_S \overline{U}^{+} \sigma^{*} n \, dS - \int_S \overline{U}^{*} \sigma(U^{+}) n \, dS = 0 \qquad \forall \; U^{*}, \sigma^{*} \in L_S^{*}$$

the following results :

Property 3-1

The solution of the interior problem is localized if, and only if, the following relations hold for a cross-section S :

. $\int_S \sigma^{+} N \, dS = \int_S X \wedge \sigma^{+} N \, dS = 0$

. $\int_S \overline{\mathbb{A}}^{0} U^{+} \, dS - \int_S \overline{\mathbb{A}} \, \sigma^{+} N \, dS = 0$

. $\int_S \overline{\mathbb{B}^{0}} \, U^{+} \, dS - \int_S \overline{\mathbb{B}} \, \sigma^{+} N \, dS = 0$

They are 12 scalar conditions. For U^-, similar conditions apply :

Property 3-2

The solution of the exterior problem is localized if, and only if, the following relations hold for a cross-section S :

. $\int_S \sigma^- N \, dS = \int_S X \wedge \sigma^- N \, dS = 0$

. $\int_S \overline{\mathbb{A}}^0 U^- dS - \int_S \overline{\mathbb{A}} \, \sigma^- N \, dS = 0$

. $\int_S \overline{\mathbb{B}}^0 U^- \, dS - \int_S \overline{\mathbb{B}} \, \sigma^- N \, dS = 0$

- Plates

The fact that the conditions are independent of the cross-section means that the conditions can be written in a local form. For homogeneous isotropic materials, the following applies :

Property 3-3

The solutions of the interior and exterior problems are localized if, and only if :

- the global equilibrium is satisfied,
- the following relations hold for a cross-section S

$$\int_{-h}^{h} a^+ \, dz = 0$$

$$\int_{-h}^{h} a^+ z \, dz = 0$$

$$\int_{-h}^{h} (3\lambda+4\mu) z^3 a^+ dz - 12\mu(\lambda+\mu) \int_{-h}^{h} (z^2-h^2) \, \overline{N} \, U^+ \, dz = 0$$

$$\lambda \int_{-h}^{h} (z^2-h^2) \, \mathrm{grad}_m \, a^+ \, dz + 8\mu(\lambda+\mu) \int_{-h}^{h} \pi \, U^+ \, dz = 0$$

where $a^+ = \lambda \, \mathrm{Tr}\, [\varepsilon(U^+)] + 2\mu \, \mathrm{Tr}[\pi \varepsilon(U^+) \, \pi]$

$$\int_{-h}^{h} a^- dz = 0$$

$$\int_{-h}^{h} a^- z \, dz = 0$$

$$(3\lambda+ 4\mu) \int_{-h}^{h} z^3 a^+ \, dz - 12\mu(\lambda+\mu) \int_{-h}^{h} (z^2-h^2) \, \overline{N} \, U^- \, dz = 0$$

$$\lambda \int_{-h}^{h} (z^2-h^2) \, \mathrm{grad}_m \, a^- \, dz + 8\mu(\lambda+\mu) \int_{-h}^{h} \pi \, U^- \, dz = 0$$

where $a^- = \lambda \mathrm{Tr}[\varepsilon(U^-)] + 2\mu \, \mathrm{Tr}[\pi \varepsilon(U^-) \pi]$

There is no particular problem in the first part of the proof. The orthogonality condition to the Saint-Venant solutions applies for any regular cross-section. The relations are then reduced using the equations verified by any solution, localized or not, giving the previous relations.

Let us now suppose that the previous relations are satisfied on an arbitrary cross-section S(*). The question is to prove that they are also verified everywhere. Let us consider the interior problem and let C be the trace of S' on the middle surface Σ^+. From the equations verified by any solutions, the following equation can be obtained :

$$\Delta \left[\int_{-h}^{h} a^+ dz\right] = 0$$

$$\Delta \left[\int_{-h}^{h} za^+ \, dz\right] = 0$$

$$\Delta \left[(3\lambda+4\mu) \int_{-h}^{h} z^3 a^+ \, dz - 12\mu(\lambda+\mu) \int_{-h}^{h} (z^2-h^2) \, \overline{N}U^+ \, dz\right] = -6(\lambda+2\mu) \int_{-h}^{h} a^+ z dz$$

on Σ^+. As the quantities between brackets are zero on C, they are equally zero on the interior part of the middle surface bounded by C. Moreover, these quantities are analytical functions of m on Σ^+ and consequently they are zero everywhere.

For the last relation, the following is derived from $\int_{-h}^{h} a^+ \, dz = 0$ and from the local equations verified by any solution :

$$\mathbf{div}_m \left[\lambda \int_{-h}^{h} (z^2-h^2) \, \mathrm{grad}_m \, a^+ \, dz + 8\mu(\lambda+\mu) \int_{-h}^{h} \pi U^+ dz\right] = 0$$

$$\mathrm{div}_m \left[N \pi \int_{-h}^{h} \pi U^+ \, dz\right] = \text{constant} = C$$

on the middle surface Σ^+. Then, the mean value of πU^+ can be written :

$$\int_{-h}^{h} \pi U^+ = \mathrm{grad}_m \psi - N \wedge Om. \frac{C}{2}$$

and then

$$\Delta \left[\int_{-h}^{h} (z^2-h^2) \, a^+ \, dz + 8\mu(\lambda+\mu) \int_{-h}^{h} \psi dz\right] = 0 \qquad (1)$$

on Σ^+.

Let us denote by Z the quantity

$$Z = \lambda \int_{-h}^{h} (z^2-h^2) \, \mathrm{grad}_m \, a^+ dz + 8\mu(\lambda+\mu) \int_{-h}^{h} \pi . U^+ dz \qquad \text{giving :}$$

(*) can be non-regular.

div $Z = 0$ on Σ^+.

So, Ψ exists such that :

$$Z = N\wedge \text{grad}_m \Psi \quad \text{on} \quad \Sigma^+$$

with $\Delta\Psi = C$.

The condition Z equal to zero on C leads to the following relations :

$\Psi,_n = 0$ (n : unit outward normal to C)

Ψ = constant

on C. Consequently, Ψ has to be constant on Σ^+ and $C = 0$. This is merely to point out that the proof only uses the condition $\Psi,_n = 0$ on C, i.e., $\bar{n}(N\wedge Z) = 0$ on C. So, the quantity

$$\lambda \int_{-h}^{h} (z^2-h^2)a^+ \, dz + 8\mu(\lambda+\mu) \int_{-h}^{h} \Psi dz = 0$$

is constant on C and from (1) it is equally constant on the domain bounded by C. Being an analytical function of m, it remains constant on the whole domain Σ^+ and the relation

$$Z = 0$$

is satisfied everywhere on Σ^+.

Remark

From the previous proof, it appears that the condition

$$\lambda \int_{-h}^{h} (z^2+h^2) \, \text{grad}_m \, a^+ \, dz + 8\mu(\lambda+\mu) \int_{-h}^{h} \pi U^+ \, dz = 0$$

represents in fact a scalar relation. Thus, we obtain four scalar conditions in all at every point of C.

3.4. Decaying properties

- Beams

Property 3-4

The localized solutions verify :

$$||U_t^+||_+ < \exp(-t/\ell^+) \, ||U_0||_+$$

$$||U_{t'}^-||_- < \exp(t'/\ell^-) \, ||U_0||_-$$

with

$$0 < \frac{1}{\ell^+} = \inf_{W \in D} \frac{<A^+W,W>_+}{<W,W>_+} \qquad 0 < \frac{1}{\ell^-} = \inf_{W \in D} \frac{<A^-W,W>_-}{<W,W>_-}$$

Proof :

for $U_t^+ \in D$, one has

$$< A^+ U_t^+,U_t^+>_+ = \frac{1}{2} \int_{S_t} \mathrm{Tr}\left[\mathbb{K}\,\varepsilon(\mathbb{R}_\eta^+ U_t)\,\varepsilon(\mathbb{R}_\eta^+ U_t)\right] dS$$

and then

$$\frac{d}{dt} <U_t^+,U_t^+>_+ = -2 < A^+ U_t^+,U_t^+>_+ \qquad t > 0$$

So,

$$\frac{d}{dt} <U_t^+,U_t^+>_+ + \frac{2}{\ell^+} < U_t^+,U_t^+>_+ < 0$$

The function $t \longmapsto <U_t^+,U_t^+>_+ \exp(2t/\ell^+)$ is then a decaying function on $]0,\infty[$. A similar proof holds for U_t^-.

Remarks

- The previous result can be written

$$E_t^+ < \exp(-2t/\ell^+)\ E_0^+ \tag{2}$$

where E_t^+ denotes the strain energy contained in the part of beam situated beyond the cross-section S_t.

- The norm $L^2(S)$ of the displacement follows a similar decaying property. Using the norm associated to the scalar product $<<,>>$, the same kind of results for U^+,U^- are obtained. The decaying length is the same :

$$\frac{1}{\ell} = \inf_{W \in D} \frac{<<A^+W,W>>}{<<W,W>>} = - \inf_{W \in D} \frac{<<A^-W,W>>}{<<W,W>>} \tag{3}$$

- The above improves Toupin's famous results [2] on two points. The decaying is purely exponential near the end cross-section, a property which has already been proved by another method in [5]. Moreover, the lengths ℓ^+, ℓ^- which characterize the cross-section geometry and the material are optimal.

- Plates

The decaying properties are the same for the interior and the exterior problem.

So, we will only deal with the interior problem. The index + will be omitted.

. A preliminary property

If the conditions (3-3) are satisfied on the end cross-section S , we have seen that they are equally verified on the whole middle surface Σ. Taking into account the local equations verified by any solution, it is easy to show the following relations which hold on the interior of Σ:

$$
\begin{aligned}
&(3\lambda+4\mu) \int_{-h}^{h} z^3 a \, dz - 12\mu(\lambda+\mu) \int_{-h}^{h} (z^2-h^2)\, \overline{N}\, U \, dz = 0 \\
&\lambda \int_{-h}^{h} (z^2-h^2)\, \mathrm{grad}\, a + 8\mu(\lambda+\mu) \int_{-h}^{h} \pi U \, dz = 0 \\
&\int_{-h}^{h} \mathrm{div} \left[N \wedge \pi U \right] dz = 0 \\
&\int_{-h}^{h} \pi\, U z \, dz + {}_{-h}^{\;h} \pi\sigma N . \frac{z^2-h^2}{2\mu}\, dz + \frac{\lambda+2\mu}{4\mu(\lambda+\mu)} \int_{-h}^{h} z . \mathrm{grad}\, (\overline{N\sigma N})\, dz = 0
\end{aligned}
\qquad (4)
$$

These six scalar conditions link certain mean values and moments of the displacement to the stresses.

. Infinite plate submitted to periodic data on the edge

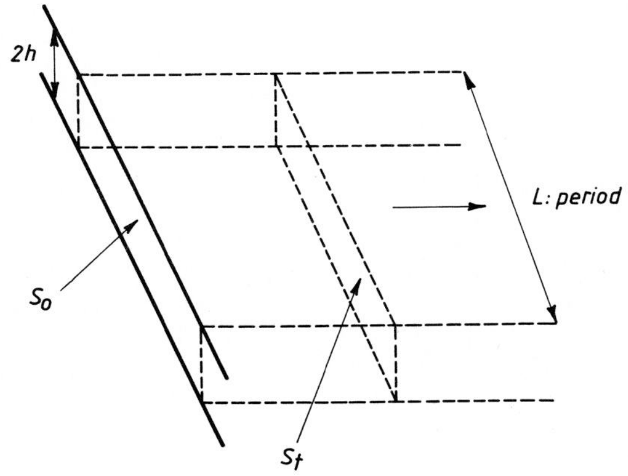

Figure 6

Let us consider the domain restricted to a period. It can be classed as a beam with a rectangular cross-section, and thus the semi-group $\mathbb{R}$ involves only one parameter.

Property 3-5

The localized solutions verify

- $||U_t|| < \exp(-t/kh)\, ||U_0||$

$$- ||U_t||_{L^2(S)} < k' \, k^{1/2} \, \exp(-t/kh) \, ||U_0||$$

where k,k' are dimensionless constants O(1). The optimal value of k is defined by

$$\frac{1}{k} = \inf_{W \text{ localized}} \frac{\langle AW,W\rangle}{\langle W,W\rangle} \cdot h$$

Proof :

By applying the proof as above for beams, it is sufficient to derive a lower bound of (1/kh). Let us consider the strain energy

$$2E_t = \int_{S_t} - \overline{U}.\sigma \, N \, dS$$

which can be written :

$$2E_t = - \int_{S_t} \overline{N\sigma N}. \overline{NU} \, dS - \int_{S_t} \overline{\pi U} \, \pi\sigma \, N \, dS$$

From

$$\int_{-h}^{h} \overline{NU}.(z^2-h^2)dz = \frac{3\lambda+4\mu}{12\mu(\lambda+\mu)} \int_{-h}^{h} z^3 a \, dz$$

one gets

$$\int_{S_t} - \overline{N\sigma N}. \overline{NU} \, dS \leqslant k_1 h \int_{S_t} Tr\left[\sigma \mathbb{K}^{-1} \sigma\right] dS$$

where the dimensionless constant k_1 is O(1).

The difficulty is to bound $||\pi U||_{L^2(S_t)}$. For that, recovering by parallelepipeds of a neighborhood of the cross-section S_t is used.

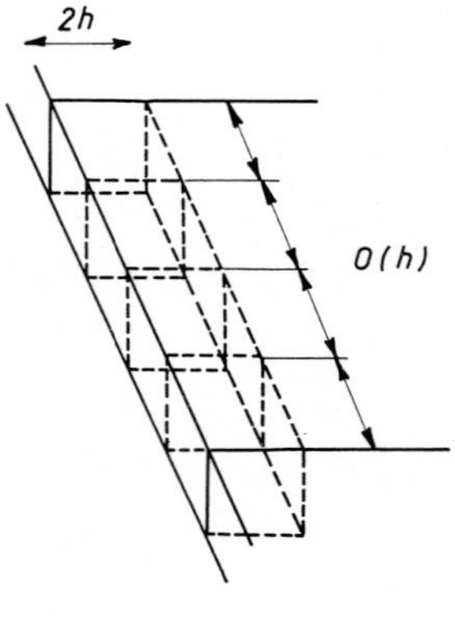

Figure 7

The height of each elementary domain is equal to 2h and the transversal dimensions are of the order of the magnitude of 2h. For each elementary parallelipiped P, the result proved in the Appendix which comes from the preliminary properties is applied :

$$||U||_{L^2(S_P)} = \left[\int_{(P)} \mathrm{Tr}\,[\sigma \mathbb{K}^{-1} \sigma]\, d\Omega \right]^{1/2} \underline{k}\, h^{1/2} \qquad (5)$$

By considering them all, the following is derived

$$2E_t \leqslant k_1 h \int_{S_t} \mathrm{Tr}\left[\sigma \mathbb{K}^{-1} \sigma\right] dS + h^{1/2} k_2 \left[\int_{S_t} \mathrm{Tr}[\sigma \mathbb{K}^{-1} \sigma]\, dS \right]^{1/2} [2E_t]^{1/2}$$

then,

$$4E_t \leqslant k_3 h \int_{S_t} \mathrm{Tr}\left[\sigma \mathbb{K}^{-1} \sigma\right] dS \qquad \text{with } k_3 = O(1)$$

Therefore, (1/kh) is bounded by :

$$\frac{1}{k_3 h} \leqslant \frac{1}{kh} \leqslant \frac{\int_{S_t} \mathrm{Tr}[\sigma K^{-1} \sigma]\, dS}{4E_t} = \frac{\langle AU_t, U_t \rangle}{\langle U_t, U_t \rangle}$$

The decaying property of $||U_t||_{L^2(S_t)}$ comes from the inequality (5).

. <u>Others examples</u>

Instead of dealing with the general case which requires the introduction of many additional notations, we will examine some typical situations.

A)

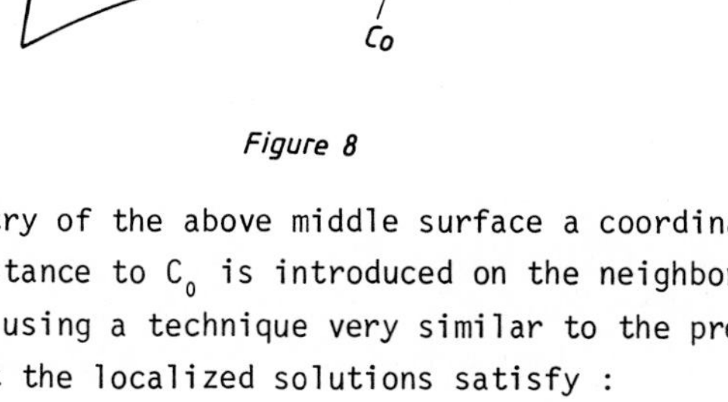

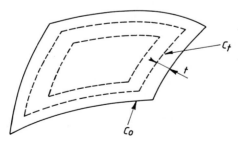

Figure 8

For the geometry of the above middle surface a coordinate system such as t represents the distance to C_0 is introduced on the neighborhood of the end cross-section. By using a technique very similar to the previous one, it is easy to prove that the localized solutions satisfy :

$$||U_t||_t \leqslant \exp(-t/kh) \quad ||U_0||_0$$
$$||U_t||_{L^2(S_t)} \leqslant k' \, h^{1/2} \quad \exp(-t/kh) \quad ||U_0||_0 \qquad (6)$$

where k,k' are O(1).

B)

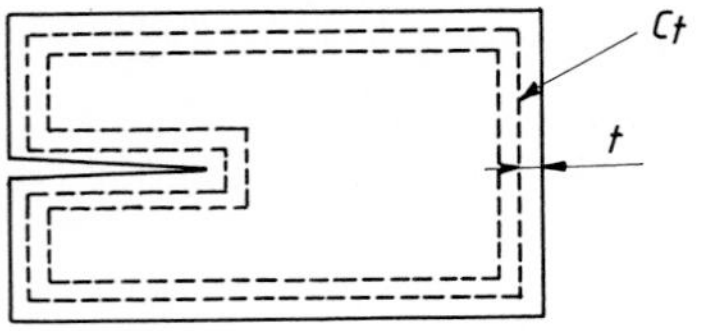

Figure 9

In the case where a crack appears, the previous result is still valid. It is clear that a same kind of re-covering by parallelepipeds can be used.

From this result, it follows that most of the singularity cannot be taken into account by the edge effects, namely by the localized solutions. This point confirms [35] where it is proved that the energy release rate can be computed through the Kirchhoff-Love plate theory for rather large cracks.

C)

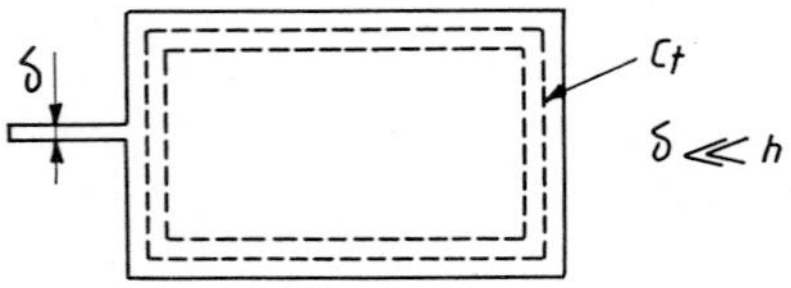

Figure 10

The detail defined by δ has to be included in the edge effects. The previous results remain valid by only considering the rectangular zone of the middle surface. In this way, the considered cross-sections do not contain rapid geometry variations.

4. SAINT-VENANT PRINCIPLE

The Saint-Venant Principle is in fact a theorem which defines necessary and sufficient conditions ensuring the localization of the displacements and the stresses, namely the rapidly decaying properties presented previously. Two principles have to be distinguished because the localization of the stresses does not involve the localization of the displacements (a rigid body displacement has zero stress values). This distinction in only essential for beams.

4.1. Interior and exterior problems

For both interior and exterior problems, the Saint-Venant Principles can be written in the same manner.

- Beams

Saint-Venant Principle for stresses

The stresses are localized if, and only if :

$$\int_{S_0} \sigma N \, dS = \int_{S_0} X \wedge \sigma N \, dS = 0$$

Saint-Venant Principle for displacements

The displacements are localized if, and only if :

$$\int_{S_0} \sigma N \, dS = \int_{S_0} X \wedge \sigma N \, dS = 0$$

$$\int_{S_0} \overline{\mathbb{A}^0} \, U \, dS - \int_{S_0} \overline{\mathbb{A}} \, \sigma N \, dS = 0$$

$$\int_{S_0} \overline{\mathbb{B}^0} \, U \, dS - \int_{S_0} \overline{\mathbb{B}} \, \sigma N \, dS = 0$$

The localization is characterized by the property (3-4). The previous conditions can be written easily for any boundary conditions by solving auxiliary problems. In [27] these conditions will be found for beams and in [8] [30] for the semi-infinite strip. For example, in the case where the displacements are prescribed, the stresses are always localized. The displacements are localized for the interior problem if, and only if :

$$\int_{S_0} \overline{\hat{\mathbb{A}}-\mathbb{A}^0} \, . \, U \, dS = \int_{S_0} \overline{\hat{\mathbb{B}} - \mathbb{B}^0} \, U \, dS = 0$$

where $\hat{\mathbb{A}} T + \hat{\mathbb{B}} M$ is the normal stress vector at $t = 0$, solution of the auxiliary interior problem related to the following value of the prescribed displacements at $t = 0$:

$$\mathbb{A} T + \mathbb{B} M$$

- Plates

Saint-Venant Principles

The displacements and stresses are localized if, and only if :

- the global equilibrium is satisfied,
- the following relations hold on the end cross-section

$$\int_{-h}^{h} a \, dz = \int_{-h}^{h} a \, z \, dz = 0$$

$$(3\lambda+4\mu) \int_{-h}^{h} z^3 a \, dz - 12\mu(\lambda+\mu) \int_{-h}^{h} (z^2-h^2) \, \overline{N} \, U \, dz = 0$$

$$\lambda \int_{-h}^{h} (z^2-h^2) \, \mathrm{grad} \, a \, dz + 8\mu(\lambda+\mu) \int_{-h}^{h} \pi \, U \, dz = 0$$

Remarks

For the last relation, it is possible to take only the tangential component at every point of C_0 which is the trace of the end cross-section on the middle surface.

4.2. General case - General properties of the solutions

Let us consider the finite beam and the plate with a hole described by the figures (1) and (2).

- Beams

In [27], the following main result is proved.

Property 4-1

The displacements can be written in a unique way :

$$U_t = U_t^* + U_t^+ + U_{t-L}^-$$

where U^* : Saint-Venant solution

U^+ : Solution of the interior problem related to S_0

U^- : Solution of the exterior problem related to S_L

- Plates

The existence of such a decomposition is not evident but the uniqueness can be easily proved.

Let us consider an isotropic homogeneous material and let us suppose that the exact solution is known. We denote by Γ_i $i = 1,..,4$ the following quantities computed from the exact solution at the end cross-section S_0 , S_L :

$$\Gamma_1 = \int_{-h}^{h} azdz$$

$$\Gamma_2 = (3\lambda+4\mu) \int_{-h}^{h} z^3 a \, dz - 12\mu(\lambda+\mu) \int_{-h}^{h} (z^2-h^2) \, \overline{N}U \, dz$$

$$\Gamma_3 = \int_{-h}^{h} a \, dz$$

$$\Gamma_4 = \bar{t}\left[\lambda \int_{-h}^{h} (z^2-h^2) \text{ grad a } dz + 8\mu(\lambda+\mu) \int_{-h}^{h} \pi \, U \, dz \right]$$

where t is the tangent unit vector to the traces C_0, C_L of the end cross-sections S_0, S_L on the middle surface.

For the Saint-Venant solutions, these quantities are equal to :

$$\Gamma_1^* = -\frac{8}{3} h^3 \frac{\mu(\lambda+\mu)}{\lambda+2\mu} \Delta W \qquad \Gamma_2^* = 16\mu(\lambda+\mu) h^3 \left[W - \frac{h^2}{5} \frac{11\lambda+12\mu}{\lambda+2\mu} \Delta W \right]$$

$$\Gamma_3^* = \frac{8(\lambda+\mu)}{\lambda+2\mu} \mu h . \text{div } u$$

$$\Gamma_4^* = \bar{t}\left[16h\mu(\lambda+\mu) \, u - \frac{8}{3} h^3 \lambda\mu \frac{(\lambda+\mu)}{\lambda+2\mu} \text{ grad(div } u) \right]$$

For the sake of shortness, let us examine only the case of bending. The Saint-Venant solution will then be defined by :

$$\Delta\Delta W = 0 \quad \text{on } \Sigma$$

$$\Delta W = -\frac{3}{8h^2} \frac{\lambda+2\mu}{\mu(\lambda+\mu)} \Gamma_1 \quad \text{on } C_0 \text{ and } C_L$$

$$W = \frac{\Gamma_2}{16\mu(\lambda+\mu)h^3} - \frac{6}{5h} \frac{11\lambda+12\mu}{\mu(\lambda+\mu)} \Gamma_1 \quad \text{on } C_0 \text{ and } C_L .$$

This boundary value problem admits no more than one solution and then the decomposition between Saint-Venant solution and edge effects is unique. Nethertheless, it exists only if Γ_1, Γ_2 are "continuous" on C_0 and C_L, namely if the data of the problem are more regular than they can be in the classical framework. This point is important. It means that for several situations the decomposition in a finite energy Kirchhoff-Love solution and edge effects is not valid on the whole domain even if it remains true for every interior subdomain. Some matching modification or additional terms have to be introduced. They are associated to the discontinuity points of Γ_i i = 1,.,4 on C_0 and C_L.

4.3. General case - Saint-Venant Principle

- Beams

The generalized quantities can be defined as follows :

$$T_t = \int_{S_t} \sigma N \, dS \qquad M_t = \int_{S_t} X \wedge \sigma N \, dS$$

$$\mu_t = \int_{S_t} (\bar{\mathbb{A}}^0 U - \bar{\mathbb{A}} \sigma N) \, dS \qquad \omega_t = \int_{S_t} (\bar{\mathbb{B}}^0 U - \bar{\mathbb{B}} \sigma N) \, dS$$

where U, σ denote the exact solution. Using auxiliary problems, it is possible to write them in terms of the data [27].

Saint-Venant Principle for stresses

The stresses are localized if, and only if, the stresses due to the Saint-Venant solution are zero, namely

- the prescribed components at the ends of the beam of the generalized stresses are zero,
- the prescribed components at the ends of the beam of the generalized displacements are compatible with a rigid body displacement .

Saint-Venant Principle for displacements

The displacements are localized if, and only if, the Saint-Venant solution is zero, namely

- the prescribed components of the generalized quantities at the ends are zero.

- Plates

  Saint-Venant Principle

  The displacements and stresses are localised if, and only if, the quantities Γ_i i = 1,.,4 are zero on the traces of the different end cross-sections.

Remarks

. Particularly for the plates, this statement of the Saint-Venant Principle is not easy to use, even numerically. Nethertheless, it is possible to assume that the interaction between the different edge effects on the edges is weak, this is very satisfactory in most practical cases. This brings us back to the interior and exterior problems.

. This view of the Saint-Venant Principle can be extended to more complex edges. From the paragraph 2.2., it follows that all the results can easily be extended for curved edges. For beams, the end cross-sections need not be straight. For plates, the edges can be curved in the z-direction.

5. AN APPROXIMATION OF THE SAINT-VENANT PRINCIPLE FOR PLATES

We will consider interior and exterior problems in the light of the last remark. The difficulty for plates is to write in terms of the data the localization conditions. To solve in a simple way this question, further assumptions are needed.

For the sake of simplicity, let us consider the bending of an isotropic semi-infinite plate submitted to periodic data on the edge (interior problem).

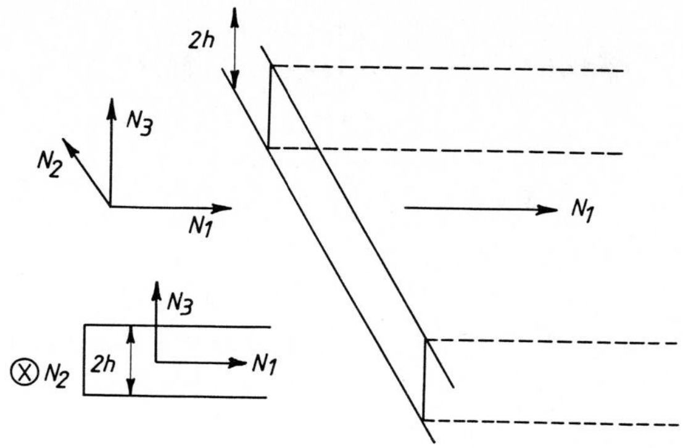

Figure 11

The data are either odd or even z-functions in order to ensure a bending behavior. They are supposed to be very regular in the X_2-coordinate. This is the main assumption.

For bending, the localization conditions are :
- the global equilibrium is satisfied

- $\int_{-h}^{h} a\ z\ dz = 0$ on S_0

- $\int_{-h}^{h} (3\lambda+4\mu)\ az^3 dz - 12\mu(\lambda+\mu) \int_{-h}^{h} \bar{N}U(z^2-h^2)\ dz = 0$ on S_0

In [17], we have proved that the edge effect can be described within an error $O(h^2/L^2)$ in terms of solutions of the plane and anti-plane problems. These problems can be defined as follows :

. Anti-plane problem

Displacement : $N \wedge \text{grad}_m \Psi$

$$\Psi,_{33} + \Psi,_{11} = 0 \quad \text{on } \{X_1, X_3;\ X_1 > 0\ -h < X_3 < h\}$$

$$\Psi,_3 = 0 \quad \text{for} \quad X_3 = \pm h$$

Plane problem

Displacement : $V = V_1 N_1 + V_3 N$

Normal stress vector σ_{13} σ_{11}

$$\text{div}\ (\mathbb{H}\,\varepsilon(V)) = 0 \quad \text{on} \quad \{X_1, X_3;\ X_1 > 0\ \ -h < X_3 < h\}$$

$$\mathbb{H}\,\varepsilon(V).N = 0 \quad \text{for} \quad X_3 = \pm h$$

$\mathbb{H}$: Hooke tensor for plane strain

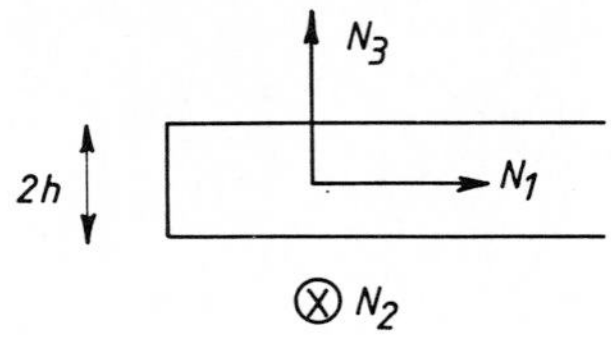

Figure 12

The edge effect is described by the superposition of the both solutions. This is to point out that the localization conditions are related to the plane problem :

$$\int_{-h}^{h} \sigma_{11}\, z\, dz = 0 \qquad \text{on } S_0$$

$$(3\lambda+4\mu) \int_{-h}^{h} \sigma_{11}\, z^3 - 12\mu(\lambda+\mu) \int_{-h}^{h} V_3 (z^2-h^2) dz = 0 \qquad \text{on } S_0 \qquad (7)$$

Furthemore, all the values at the edge namely at $X_1 = 0$, will be underlined and let $(\underline{U}_1, \underline{U}_2, \underline{U}_3)$ and $(\underline{F}_1, \underline{F}_2, \underline{F}_3)$ be displacements and forces at the edge :

$$\underline{U}_1 = -\underline{\Psi}_{,2} + \underline{V}_1 \qquad \underline{F}_1 = -2\mu\, \underline{\Psi}_{,12} + \underline{\sigma}_{11}$$

$$\underline{U}_2 = \underline{\Psi}_{,1} \qquad \underline{F}_2 = \mu \left[\underline{\Psi}_{,11} - \underline{\Psi}_{,22}\right] + \mu\, \underline{V}_{1,2}$$

$$\underline{U}_3 = \underline{V}_3 \qquad \underline{F}_3 = -\mu\, \underline{\Psi}_{,23} + \underline{\sigma}_{13}$$

We have to add to the localization conditions those conditions making it possible to build the solutions of the plane and anti-plane problems. Ψ is an even z-function and so, the Ψ -problem does not involve any condition. At the contrary , for the plane problem, conditions are needed. Two of them are (7) and the others are the following :

$$\int_{-h}^{h} \underline{\sigma}_{13}\, dz = 0$$

$$\int_{-h}^{h} \frac{3\lambda}{24\mu(\lambda+\mu)}\, \underline{\sigma}_{13}\, z^2\, dz + \int_{-h}^{h} z\, \underline{V}_1\, dz = 0$$

All these conditions come from the Saint-Venant Principle for the semi-infinite strip which is a particular beam.

<u>Property 5-1</u>

The second order approximation of the Saint-Venant Principle is defined through the following conditions :

$$\int_{-h}^{h} (\underline{F}_1 + 2\mu\underline{\Psi},_{21})\, z\, dz = 0$$

$$(3\lambda+4\mu) \int_{-h}^{h}(\underline{F}_1 + 2\mu\, \underline{\Psi},_{21})\, z^3 dz - 12\mu(\lambda+\mu) \int_{-h}^{h} \underline{U}_3\, (z^2-h^2)\, dz = 0$$

$$\int_{-h}^{h} (\underline{F}_3 + \mu\underline{\Psi},_{23})\, dz = 0$$

$$\int_{-h}^{h} \frac{3\lambda}{24\mu(\lambda+\mu)} (\underline{F}_3 + \mu\, \underline{\Psi},_{23})\, z^2 dz + \int_{-h}^{h} z(\underline{U}_1 + \underline{\Psi},_2)\, dz = 0$$

$$\underline{\Psi} = \int_0^z d\eta \int_0^\eta d\eta' \left(\underline{U}_{1,2} - \frac{F_2}{\mu}\right)$$

$$\underline{U}_1 = -\underline{\Psi},_2 + \underline{V}_1$$

$$\underline{U}_2 = \underline{\Psi},_1$$

Proof :

From the relations

$$\underline{F}_2 = \mu \left[\underline{\Psi},_{11} - \underline{\Psi},_{22}\right] + \mu\, \underline{V}_{1,2}$$

$$\Psi,_{11} + \Psi,_{33} = 0$$

$$\underline{U}_1 = - \underline{\Psi},_2 + \underline{V}_1$$

it is easy to determine that the value of Ψ at the edge is equal to :

$$\underline{\Psi} = \int_0^z d\eta \int_{-h}^\eta d\eta' \left[\underline{U}_{1,2} - \underline{F}_2 /\mu\right]$$

These conditions can be easily written in terms of the data for any boundary conditions. To illustrate this point, we will consider two cases : given loadings and prescribed displacements.

. Given loadings

The only conditions to take into account are

$$\int_{-h}^{h} \underline{\sigma}_{11}\, z\, dz = 0$$

$$\int_{-h}^{h} \underline{\sigma}_{13}\, dz = 0 \qquad (8)$$

One introduces the function $G(X_1,X_2)$, solution of the anti-plane problem for $G|_{X_1=0} = (z/h)$. Then,

$$0 = \int_{-h}^{h} \underline{\sigma}_{11}\, z\, dz = \int_{-h}^{h} \underline{F}_1\, z\, dz + 2\mu \int_{-h}^{h} z\, \underline{\Psi}_{,21}\, dz$$

$$0 = \int_{-h}^{h} \underline{F}_1\, z\, dz + 2\mu \int_{-h}^{h} h\, \underline{G}_{,2}\, \underline{\Psi}_{,2}\, dz$$

Let us consider the value of Ψ at the edge

$$\underline{\Psi} = \int_0^z d\eta \int_{-h}^{\eta} d\eta' \left[\underline{U}_{1,2} - \underline{F}_2/\mu\right] = \int_0^z d\eta \int_{-h}^{\eta} d\eta' \left[\underline{V}_{1,2} - \underline{F}_2/_{\mu} - \underline{\Psi}_{,22}\right]$$

So, with an error of (h^2/L^2), Ψ can be approximated by :

$$\underline{\Psi} = \int_0^z d\eta \int_0^{\eta} d\eta' \left[\underline{V}_{1,2} - \underline{F}_2/\mu\right]$$

and we obtain :

$$0 = \int_{-h}^{h} \underline{F}_1\, z\, dz + 2\mu \int_{-h}^{h} dz\, \underline{G}_{,1}\, h \int_0^z \int_{-h}^{\eta} d\eta d\eta' (\underline{V}_{1,2} - \underline{F}_2/\mu)_{,2}$$

As $\underline{\sigma}_{11} = 0(\underline{V}_{1,1})$ and then $\underline{V}_{1,22}\, h = 0((h^2/L^2)\, \underline{\sigma}_{11})$, it follows that the previous condition can be written : $(\int_{-h}^{h} \underline{F}_1\, z\, dz \neq 0)$

$$0 = \int_{-h}^{h} \underline{F}_1\, z\, dz - 2 \int_{-h}^{h} dz\, \{\underline{G}_{,1} \int_0^z \int_{-h}^{\eta} d\eta d\eta'\, \underline{F}_{2,2}\, \}h$$

The other condition is

$$0 = \int_{-h}^{h} \underline{\sigma}_{13}\, dz = \int_{-h}^{h} (\underline{F}_3 + \mu\, \underline{\Psi}_{,23}) dz = \int_{-h}^{h} \{\underline{F}_3 + \mu \int_0^z d\eta (\underline{V}_{1,22} - \underline{F}_{2,2}/\mu)\}\, dz$$

$$0 = \int_{-h}^{h} \underline{F}_3\, dz - \int_{-h}^{h} z\, dz\, \{\mu \underline{V}_{1,22} - \underline{F}_{2,2}\}$$

From

$$\int_{-h}^{h} z\, \underline{V}_1\, dz = -\frac{3\lambda}{24\mu(\lambda+\mu)} \int_{-h}^{h} \underline{\sigma}_{13}\, z^2 dz$$

it can be determined that

$$\int_{-h}^{h} z\, \mu\, \underline{V}_{1,22}\, dz = h\, 0(\underline{\sigma}_{13} (h^2/L^2))$$

and then

$$\int_{-h}^{h} (\underline{F}_3 + z\, \underline{F}_{2,2})\, dz = 0$$

Property 5-2

The second order approximation of the Saint-Venant Principle is defined for given loadings by

- $\int_{-h}^{h} \underline{F}_1 \, z \, dz - 2 \int_{-h}^{h} h\{\underline{G},_1 \int_0^z \int_{-h}^{\eta} d\eta d\eta' \, \underline{F}_{2,2}\} \, dz = 0$
- $\int_{-h}^{h} (\underline{F}_3 + z \, \underline{F}_{2,2}) \, dz = 0$
- Prescribed displacements

The conditions to take into account are

$$(3\lambda+4\mu) \int_{-h}^{h} \underline{\sigma}_{11} \, z^3 - 12\mu(\lambda+\mu) \int_{-h}^{h} \underline{V}_3 \, (z^2-h^2) \, dz = 0$$

$$\frac{\lambda}{8\mu(\lambda+\mu)} \int_{-h}^{h} \underline{\sigma}_{13} \, z^2 \, dz + \int_{-h}^{h} z \, \underline{V}_1 \, dz = 0 \qquad (9)$$

One introduces two auxiliary plane problems defined by the following boundary displacement conditions :

$$\hat{U}\big|_{X_1=0} = z^3 N_1 \qquad\qquad \hat{\hat{U}}\big|_{X_1=0} = z^2 \, N_3$$

The corresponding stresses are localized. Let us denote by $\hat{\underline{\sigma}}_{11}, \hat{\underline{\sigma}}_{13}$ and $\hat{\hat{\underline{\sigma}}}_{11}, \hat{\hat{\underline{\sigma}}}_{13}$ the normal stress vectors at the edge.

It follows that :

$$\int_{-h}^{h} \underline{\sigma}_{11} \, z^3 dz = \int_{-h}^{h} \left[\hat{\underline{\sigma}}_{11} \, \underline{V}_1 + \hat{\underline{\sigma}}_{13} \underline{V}_3 \right] dz$$

$$\int_{-h}^{h} \underline{\sigma}_{13} \, z^2 dz = \int_{-h}^{h} \left[\hat{\hat{\underline{\sigma}}}_{11} \, \underline{V}_1 + \hat{\hat{\underline{\sigma}}}_{13} \, \underline{V}_3 \right] dz$$

Consequently, the previous conditions can be written :

$$\int_{-h}^{h} \hat{\underline{\sigma}}_{11} \, \underline{V}_1 + \int_{-h}^{h} \underline{V}_3 \left[\hat{\underline{\sigma}}_{13} - \frac{12\mu(\lambda+\mu)}{3\lambda+4\mu} (z^2-h^2) \right] dz = 0$$

$$\frac{\lambda}{8\mu(\lambda+\mu)} \int_{-h}^{h} \hat{\hat{\underline{\sigma}}}_{13} \underline{V}_3 \, dz + \int_{-h}^{h} \underline{V}_1 \left[\hat{\hat{\underline{\sigma}}}_{11} \frac{\lambda}{8\mu(\lambda+\mu)} + z \right] = 0$$

with

$$\underline{V}_1 = \underline{U}_1 + \underline{\Psi},_2$$

$$\underline{U}_2 = \underline{\Psi},_1$$

$$\underline{U}_3 = \underline{V}_3$$

Let us consider two auxiliary anti-plane problems defined by

$$\hat{\Psi}_{,1}\Big|_{X_1=0} = \hat{\underline{\sigma}}_{11}$$

$$\hat{\hat{\Psi}}_{,}\Big|_{X_1=0} = \hat{\hat{\underline{\sigma}}}_{11}\,\frac{\lambda}{8\mu(\lambda+\mu)} + z$$

giving the following properties :

$$\int_{-h}^{h} \hat{\underline{\sigma}}_{11}\,\underline{\Psi}\,dz = \int_{-h}^{h} \hat{\underline{\Psi}}\,\underline{\Psi}_{,1}\,dz$$

$$\int_{-h}^{h} \left(\hat{\hat{\underline{\sigma}}}_{11}\,\frac{\lambda}{8\mu(\lambda+\mu)} - z\right)\underline{\Psi}\,dz = \int_{-h}^{h} \hat{\hat{\underline{\Psi}}}\,\underline{\Psi}_{,1}\,dz$$

which imply :

$$\int_{-h}^{h} \hat{\underline{\sigma}}_{11}\,\underline{U}_1\,dz + \int_{-h}^{h} \hat{\underline{\Psi}}\,\underline{U}_{2,2}\,dz + \int_{-h}^{h} \left[\hat{\underline{\sigma}}_{13} - \frac{12\mu(\lambda+\mu)}{3\lambda+4\mu}(z^2-h^2)\right]\underline{U}_3\,dz = 0 \qquad (10)$$

$$\int_{-h}^{h} \frac{\lambda}{8\mu(\lambda+\mu)}\,\hat{\hat{\underline{\sigma}}}_{13}\,\underline{U}_3\,dz + \int_{-h}^{h} \hat{\hat{\underline{\Psi}}}_{,1}\cdot\underline{U}_1\,dz + \int_{-h}^{h} \hat{\hat{\underline{\Psi}}}\,\underline{U}_{2,2}\,dz = 0 \qquad (11)$$

Property 5-3

The second order approximation of the Saint-Venant Principle is defined for prescribed displacements by (10) and (11).

Remarks

. The localization conditions given here define the second order approximation of the Saint-Venant Principle for plates. They are very simple to use because the auxiliary problems can be solved once and for all. The extensions to curved edges and to composite materials does not present any fundamental difficulty but does involve a certain degree of technical complication.

. The application of theses conditions makes it possible to build the boundary conditions of the Kirchhoff-Love theory. One finds again certain results obtained by different authors. For example in [29], they are written in the case of an infinite plate with edge wise uniform data for any kind of boundary conditions. In [37], one can find second order conditions for prescribed loadings. These "modified boundary conditions" lead to a second order Kirchhoff-Love plate theory. Nethertheless, this theory is not mathematically well-posed and the authors propose an iterative method for its solution.

. We have studied in [17] the Reissner-Mindlin plate theory. Replacing the shear constant by :

$$\frac{5}{6}\frac{16}{10s^2} \left(s = \sum_{k=1}^{\infty} (2k-1)^{-5}\,\frac{384}{\pi^5} \right)$$

the usual values being 5/6 or $\pi^2/12$, we have proved this theory is a second order theory for free edges. Using the previous Saint-Venant Principle, one obtains the best boundary conditions. In this case, it has been also proved that is a second order theory [17]. This is to point out that these modified boundary conditions are expressed through modifications of the prescribed displacements and loadings which can be determined only in terms of the given data. It follows the problem is mathematically well-posed.

APPENDIX

Let U be a localized solution. It verifies the property (6). One consider a parallelepiped (P) such as the tranverse dimensions are O(2h) and the thickness is 2h. The idea is to separate U into two terms :

- $V = U_0 + \Omega_0 \wedge OM$ (rigid body displacement)
- $\underset{\sim}{U}$

where $\int_M \underset{\sim}{U}\, d\Omega = \int_M OM \wedge \underset{\sim}{U}\, d\Omega = 0$

From [29], one gets that

$$\left[\int_{S_P} (\pi\underset{\sim}{U})^2\, dS\right]^{1/2} < \alpha\, h^{1/2} \left[\int_{(P)} \mathrm{Tr}[\mathbb{K}\, \varepsilon(\underset{\sim}{U})\, \varepsilon(\underset{\sim}{U})]\, d\Omega\right]^{1/2} \qquad (12)$$

$$\left[\int_{S_P} (\pi\underset{\sim}{U})^2\, dS\right]^{1/2} < \beta\, h \left[\int_{(P)} \mathrm{Tr}[\mathbb{K}\, \varepsilon(\underset{\sim}{U})\, \varepsilon(\underset{\sim}{U})]\, d\Omega\right]^{1/2} \qquad (13)$$

where S_P is a face of (P). The constants α,β which are connected respectively to the Steklov and Poincaré constants are O(1). These results are extended to much more complicated domains in [29], [28]. α,β depend on the ratio of the exterior diameter on the interior diameter of the star-shaped domain.

To evaluate V, it is first determined from

$$\int_{-h}^{h} \mathrm{div}_m\, [N \wedge U]\, dz = 0$$

that πV can be written : $\pi V = \pi U_0 + \pi\Omega_0 \wedge Nz$

Let us introduce the function $g(X_1, X_2)$ such that

$$g_{,11} + g_{,22} = -2 \qquad \text{on } \Sigma_P$$

$g = 0$ on the boundary of the middle surface Σ_P

$$\int_{(P)} g(\pi V + \pi \underset{\sim}{U})\, d_{\Omega} = \int_{(P)} g \pi U\, d\Omega$$

$$\int_{(P)} gz(\pi V + \pi \underset{\sim}{U})\, d\Omega = \int_{(P)} gz \pi U\, d\Omega$$

Taking into account the property (6), the following is derived

$$\pi U_0 = \frac{1}{2h \int_{\Sigma_P} g d\Sigma} \left[- \int_{(P)} g \pi \underset{\sim}{U}\, d\Omega - \frac{\lambda}{8\mu(\lambda+\mu)} \int_{(P)} \text{grad a.g.}(z^2-h^2)\, d\Omega \right]$$

$$\pi \Omega_0 \Omega N = \frac{3}{2h^3 \int_{\Sigma_P} g d\Sigma} \left[- \int_{(P)} g.z\pi \underset{\sim}{U}\, d\Omega - \int_{(P)} \pi \sigma N.\frac{z^2-h^2}{2\mu} gdz - \frac{\lambda+2\mu}{4\mu(\lambda+\mu)} \int_{(P)} zg\, \text{grad}(\overline{N}\sigma \underset{\sim}{N}) dz \right] = 0$$

It should be noted that :

$$\int_{(P)} \text{grad}_m\, a\, g(z^2-h^2)\, d\Omega = - \int_{(P)} a\, \text{grad}_m\, g\, (z^2-h^2)\, d\Omega$$

$$\int_{(P)} zg\, \text{grad}_m\, (\overline{N}\sigma N)\, d\Omega = - \int_{(P)} z\, \overline{N}\sigma N\, \text{grad}_m\, g\, d\Omega$$

Using (12) and the previous relations, it is found that

$$|| \pi V ||_{L^2(S_P)} < \delta h^{1/2} \left[\int_{(P)} Tr[\varepsilon(U)\, \mathbb{K}\, \varepsilon(U)]\, d\Omega \right]^{1/2}$$

where $\delta = O(1)$. Finally, the following result holds :

$$|| \pi V ||_{L^2(S_P)} < \underline{k}\, h^{1/2} \left[\int_{(P)} Tr[\varepsilon(U)\, \mathbb{K}\, \varepsilon(U)]\, d\Omega \right]^{1/2}$$

with $\underline{k} = O(1)$.

REFERENCES

1 Saint-Venant A, "De la torsion des prismes", C.R.A.S., Paris, Vol. 14, 1853, p. 253-560.

2 Toupin R, "Saint-Venant's Principle", Arch. Rat. Mech. and Anal., Vol. 18, n°2, 1965, p. 83-96.

3 Koiter W.T, "On the foundations of the linear theory of thin elastic shells", Koninkl, Nederl, Akademie van Wetenschappen, Amsterdam (Reprinted from proceedings serie B-73, n°3, 1970).

4 Roseman J.J, "A point-vise estimate for the stresses in a cylinder and its application of Saint-Venant's Principle", Arch. Rat. Mech. and Anal., Vol, 21, 1966, p. 23-48.

5 Berdichevskii V.L, "On the proof of the Saint-Venant Principle for bodies of arbitrary shape", Appl. Math. and Mech., Vol. 38, n°5, 1974, p. 799-813.

6 Choi I, and Horgan, C.O, "Saint-Venant's Principle and end effects in anisotropic elasticity", J. Appl. Mech., Vol. 44, n°33, 1977, p. 424-430.

7 Papkovich P.F, "Uber eine Form der Lösung des Byharmonische Problems für des Rechteck", Dokl. Aka. Nauk., SSSR, Vol. 27, 1940, p.334-338.

8 Gusein-Zade M.I, "On necessary and sufficient conditions for the existence of decaying solutions of the plane problem of the theory of elasticity for a semi-strip", Appl. Math. and Mech., Vol. 29, n°4, 1965, p. 752-760.

9 Johnson M.W, and Little R.W, "The semi-infinite elastic strip", Quart. Appl., Vol. 22, 1964, p. 335-344.

10 Truesdell C, "History of classical mechanics", Vol. 63, part I : to 1800, p. 53-62 ; part II : the 19th and 20th Centuries, p. 119-130, 1976, Springer-Verlag.

11 Knowles J.K, "On Saint-Venant's Principle in the two dimensionnal linear theory of elasticity", Arch. Rat. Mech. Anal., Vol 21, 1966, p. 1-22.

12 Duvaut G, and Lions J.L, "Les inéquations en mécanique et en physique", Dunod, Paris, 1972.

13 Gregory R.D, "The traction boundary problem for the elastostatic semi-infinite strip ; Existence of solution and completness of the Papkovich-Fadle eigenfunctions", J. of Elasticity, Vol. 10, n°3, 1980, p. 295-327.

14 Brezis H, "Opérateurs maximaux monotones et semi-groupes de contraction dans les espaces de Hilbert", North-Holland, Mathematics Studies, 1973.

15 Ladeveze P, Ladeveze J, Pecastaings F, and Pelle J.P, "Sur les fondements de la théorie linéaire des poutres élastiques", Partie I : J. Mec., Vol. 18, n°1, 1979, p. 129-173 ; Partie II : J. Mec., Vol. 19, n°1, 1980, p. 1-66.

16 Ladeveze P, "Justification de la théorie linéaire des coques élastiques", J. Mec., Vol. 15, n°5, 1976, p. 813-856.

17 Ladeveze P, "On the validity of linear shell theories" Theory of shells, North-Holland, 1980, p. 367-391.

18 Danielson D.A, "Improved error estimates in the linear theory of thin elastic shells", Proc. Kon. Ned. Ak. Wet., B, 74, 1970, p. 294-300.

19 Koiter W.T, and Simmonds J.C, "Foundations of shell theory", Report n° 473, August 1972, Laboratory of Engineering Mechanics, Melkelveg 2, Delft.

20 Simmonds J.G, "Pointwise displacement errors in linear shell theory resulting from errors in the stress-strain relations", ZAMP, Vol. 23, 1972, p. 265-269.

21 Ho C.L, and Knowles J.L, "Energy inequalities and error estimates for torsion of elastic shells of revolution", ZAMP, Vol. 21, 1970, p. 352-377.

22 Gol'Denveizer "Theory of elastic thin shells", Pergamon Press, 1961.

23 Maisonneuve O, "Sur le principe de Saint-Venant", Thèse d'Etat, Poitiers, 1971.

24 Weeler L.T, and Horgan C.O, "A two-dimensional Saint-Venant Principle for second order linear elliptic equations", Quart. Appl. Math.,

Vol. XXXIV, n°3, 1976, p. 257-270.

25 Pecastaings F, "Principe de Saint-Venant pour les plaques", Thèse d'Etat, Paris, 1985.

26 Ladevèze P, "Principes de Saint-Venant en contrainte et en déplacement pour les poutres droites demi-infinies", J. Math. et Phys. Appl. (ZAMP), Vol. 33, 1982, p. 132-139.

27 Ladevèze P, "Sur le principe de Saint-Venant en élasticité", J. de Méca., Théo. et Appl., Vol. 1, n°2, 1983, p. 161-184.

28 Ladevèze J, and Ladevèze P, "Majorations de constantes de Steklov", J. Math. et Phys. Appl. (ZAMP), Vol. 29, 1978, p. 684-692.

29 Ladevèze J, and Ladevèze P, "Majorations de la constante de Poincaré", Rendiconti dei Lincei, série VIII, Vol. LXIV, fasc 6, 1979, p. 548-556.

30 Gregory D, and Wan F, "Decaying states of plane strain in a semi-infinite strip and boundary conditions for plate theory", J. of Elas., Vol. 14, 1984, p. 27-64.

31 Reissner E, "On the derivation of boundary conditions for plate theory", Proc. Royal Soc., A276, 1963, p. 178-186.

32 Nair W, and Reissner E, "On asymptotic expansions and error bounds in the derivation of the two-dimensional shell theory", Studies in Appl. Math., Vol. 56, 1977, p. 189-217.

33 Horgan C, "Saint-Venant end effects in composites", J. of Composites Mat., Vol. 12, 1982.

34 Lure A.I, "Three dimensional problems of the theory of elasticity", Vol. 1, Dunod, Paris, 1968.

35 Simmonds J.G, and Duva J, "Thickness effects are minor in the energy release rate integral for bent plates containing elliptic holes or cracks", J. of Appl. Mech., Vol. 48, 1981, p. 320-326.

36 Fredrichs K.O, and Dressler R.F, "A boundary layer theory for elastic plates", Comm. on Pure and Applied Math., Vol. XIV, 1961, p. 1-33.

37 Van der Heijden A.M.A, "On modified boundary conditions for the free edge of a shell", Thesis, Delft, 1976.

EDGE EFFECTS IN THE STRETCHING OF PLATES*

R.D. GREGORY[1] and F.Y.M. WAN[2]

[1]Department of Mathematics, University of Manchester, Manchester M13 9PL, (England)

[2]Applied Mathematics Program, FS-20, University of Washington, Seattle, Washington 98195 (U.S.A)

ABSTRACT

The stretching of flat plates is investigated by methods first introduced by the authors in the context of plate bending. The elastic reciprocal theorem is used to generate necessary conditions which the prescribed data at the edge of the plate must satisfy in order that it should generate a decaying state within the plate; these decaying state conditions are obtained explicitly for the case of axisymmetric stretching (and torsion) of a circular plate when stress or mixed conditions are prescribed at the plate edge. The conditions which any interior solution must satisfy at the plate edge are then deduced. As an example we obtain the complete interior solution (correct to within exponentially small error) for the problem of a simply supported thick circular plate under a concentrated load. It is shown conclusively that applications of Saint-Venant's principle lead to wrong corrections to the Kirchhoff thin plate theory.

1. INTRODUCTION

The exact solution of linear elastostatic problems for flat plates is known to consist of an <u>interior</u> component and <u>layer</u> components. The interior solution is significant throughout the plate, while a layer solution has only a localized effect in a region with typical linear dimensions of the order of the plate thickness. Boundary conditions at an edge of the plate are generally satisfied only by a combination of the two types of solution components. However, the layer solutions are difficult to obtain and the solution behavior near the plate edges is often not needed for design purposes. Hence, there has been a continual effort over the years to determine the interior solution of plate problems without any reference to the layer solutions. Saint-Venant's principle [1] has been used for this purpose (see [2] and [3] for examples). As it is well known, the principle makes it possible to derive an appropriate set of stress boundary conditions for the interior solution. Strictly speak-

* The research was partly supported by U.S. - NSF Grant No. MCS 830-6592 and by Canadian NSERC Individual Operating Grant No.A9259.

ing, Saint-Venant's principle does not apply to plate problems as one linear dimension of the loaded area, namely the circumference of the plate's midplane, is not small compared to a representative plate span. Another interesting approach to obtaining the interior solution which also invokes (the unproven) Saint-Venant's principle was proposed in [4] but was not entirely satisfactory for reasons previously discussed in [5] and [6].

In [5] and [6], we developed a new and completely different method determining the interior solution without any reference to the layer solutions. The method effectively obtains also an appropriate set of boundary conditions for the interior solution (which includes both the classical and related thick plate theories) corresponding to the given set of admissible stress, displacement or mixed edge data for the three-dimensional elasticity problem. In contrast to all previous attempts, the key to our method lies in a novel application of the Betti-Raleigh reciprocal theorem. The special case of a semi-infinite plate in a state of plane strain induced by edgewise uniform edge data was first analyzed by this method in [5]. For stress edge data, the result obtained there rigorously justifies the application of Saint-Venant's principle. More importantly, correct boundary conditions for the interior solution for plates with displacement and mixed edge data were obtained for the first time. Next, we obtained in [6] the corresponding results for axisymmetric bending of a circular plate. For the stress edge data, the results show that indiscriminate use of Saint-Venant's principle for plate bending problems may lead to quantitatively and qualitatively incorrect solutions even in the plate interior; they in effect delimit the range of applicability of Saint-Venant's principle for axisymmetric bending problems. Results were also obtained in [6] for edgewise nonuniform data along the straight edge of a semi-infinite plate in bending.

In this paper, we complement the discussion in [6] for circular plates with an analysis of the problem of axisymmetric stretching and torsion. The stress boundary conditions obtained for the interior solution again delimit the applicability of Saint-Venant's principle in the case of stress edge data. The stress boundary conditions for axisymmetric extension along with the corresponding conditions for axisymmetric bending from [6] enable us to obtain the first correct interior solution for a circular plate subject to a point force at the center of its upper face and "simply-supported" at its edge. A number of results for this problem are of general interest, including the following:

(1) It is demonstrated conclusively that the results in [2], [7] and [8] for the same problem, obtained by incorrect applications of Saint-Venant or other ad hoc methods, are in significant error for thick plates and may be inferior to the Kirchhoff solution.

(2) The correction to the classical Kirchhoff thin plate solution for the "exact" interior solution is in fact singular at the point of load application as predicted by Reissner's plate theory [9] (see also Appendix IV of the present paper).

(3) There is generally an extensional component in the interior solution of the same order of magnitude as the corrections to the Kirchhoff solution in the bending part of the problem, a fact not known previously.

(4) The effect in the interior solution of the different interpretation of a "concentrated load" and a "simply supported edge" in a three-dimensional setting are of the same order as the corrections to the Kirchhoff solution.

From these results for the point load problem and other results in [6], a conclusion which is of considerable importance in engineering analysis and design may be inferred. In the framework of a thick plate theory, it is not possible to determine the interior solution when only stress resultants and stress couples are prescribed at the edge.

2. DECAYING STATES IN A CIRCULAR PLATE

The central step in our method of approach to obtaining the interior solution is to seek the answer to the following question: What (necessary) condiitions must an admissible set of edge data satisfy in order that the resulting solution in the plate should be a decaying elasto-static state? The notion of a decaying (elasto-static) state for a plate of general shape is defined in [6]. Briefly, the stress and displacement components, σ_{ij} and u_j, of a decay-state decay exponentially away from the plate edge:

$$\{u_j,\sigma_{ij}\} = O(e^{-\gamma d/h}) \qquad \text{as} \quad h \to 0$$

where $2h$ is the plate thickness, d is the distance of the observation point from the plate edge and γ is a positive constant. Once we have a sufficient number of the necessary conditions sought above, they can be translated into boundary conditions for the interior solution of plate problems.

Necessary conditions for various types of admissible edge data have been derived in [6] for plates of general shape by way of the elastic reciprocal theorem. In cylindrical coordinates (r,θ,z) a necessary condition for the stress data $\bar{\sigma}_r^d$, $\bar{\sigma}_{r\theta}^d$ and $\bar{\sigma}_{rz}^d$ (prescribed at the edge of a circular plate of radius a and thickness $2h$) to induce only a decaying state in the plate was found to be

$$\int_{-h}^{h}\int_{0}^{2\pi} \{\bar{\sigma}_{rr}^d u_r^{(2)} + \bar{\sigma}_{r\theta}^d u_\theta^{(2)} + \bar{\sigma}_{rz}^d u_z^{(2)}\}_{r=a} \, a\, d\theta\, dz = 0 \qquad (2.1)$$

for any suffix (2) elasto-static state $\{u_j^{(2)}, \sigma_{jk}^{(2)}\}$ which satisfies the stress-free conditions on the faces $z = \pm h$ and along the edge $r = a$, and has at worst an algebraic growth as $h \to 0$*. With different suffix (2) states, the condition (2.1) gives a whole class of necessary conditions for the stress data to induce a decaying elastostatic state.

It was also shown in [6] that the corresponding necessary condition for displacement edge data $\bar{u}_r^d$, $\bar{u}_\theta^d$, and $\bar{u}_z^d$ to induce only a decaying elastostatic state is

$$\int_{-h}^{h}\int_{0}^{2\pi} \{u_r^d\sigma_{rr}^{(2)} + \bar{u}_\theta^d\sigma_{r\theta}^{(2)} + \bar{u}_z^d\sigma_{rz}^{(2)}\}_{r=a} a\,d\theta dz = 0 \tag{2.2}$$

for any regular suffix state (2) which satisfies the traction-free condition on the faces $z = \pm h$ and the conditions of no displacement at $r = a$. Similarly, a necessary condition for the typical mixed edge data $\bar{u}_r^d$, $\bar{u}_\theta^d$ and $\bar{\sigma}_{rz}^d$ was found to be

$$\int_{-h}^{h}\int_{0}^{2\pi} \{u_r^d\sigma_{rr}^{(2)} + \bar{u}_\theta^d\sigma_{r\theta}^{(2)} - \bar{\sigma}_{rz}^d u_z^{(2)}\}_{r=a} a\,d\theta dz = 0 \tag{2.3}$$

where the regular suffix (2) state satisfies the traction-free conditions on $z = \pm h$ and the homogeneous edge condition $u_r^{(2)} = 0$ and $u_\theta^{(2)} = 0$ and $\sigma_{rz}^{(2)} = 0$ on $r = a$. (In all cases, the edge conditions for the suffix (2) state are the homogeneous counterparts of the actual edge conditions for the decaying state.) The practical difficulty in deriving from (2.1)-(2.3) the desired boundary conditions for the interior solution lies in the determination of suitable suffix (2) states which satisfy the required boundary conditions. This task simplifies considerably for problems with axisymmetry.

For axisymmetric edge data and an axisymmetric suffix (2) state, the integration with respect to θ can be carried out, leaving us with a condition involving only integration across the plate thickness. For example, the condition for stress data becomes

$$\int_{-h}^{h} \{\bar{\sigma}_{rr}^d(z)u_r^{(2)}(a,z) + \bar{\sigma}_{r\theta}^d(z)u_\theta^{(2)}(a,z) + \bar{\sigma}_{rz}^d(z)u_z^{(2)}(a,z)\}dz = 0 \tag{2.4}$$

where we have retained the torsion action by allowing for a nonvanishing $u_\theta^{(2)}$. For axisymmetric bending, one suffix (2) state for stress data is a rigid body translation in the z direction. With $u_r^{(2)} \equiv 0$, $u_\theta^{(2)} \equiv 0$, and $u_z^{(2)} = 1$, the condition (2.4) becomes

---

* An elastostatic state $\{u_j, \sigma_{jk}\}$ which has at worst an algebraic growth as $h \to 0$ is called a regular (elastostatic) state.

$$\int_{-h}^{h} \bar{\sigma}^{d}_{rz} dz = 0 \; . \tag{2.5}$$

A second suffix (2) state was found to be

$$u_r^{(2)} = [(1+\nu)\,\frac{a}{r} + (1-\nu)\,\frac{r}{a}]z \; , \qquad u_\theta^{(2)} \equiv 0$$
$$u_z^{(2)} = -(1+\nu)a\,\ln(\frac{r}{a}) - (1-\nu)\,\frac{r^2}{2a} - \nu\frac{z^2}{a} \tag{2.6}$$

for which (2.4) becomes

$$\int_{-h}^{h} \{\bar{\sigma}^{d}_{rr} - \frac{\nu}{2a}\, z^2 \bar{\sigma}^{d}_{rz}\} dz = 0 \tag{2.7}$$

While (2.5) is the expected requirement of no resultant axial force, the condition (2.7) is not the usual condition ([2], [3])

$$\int_{-h}^{h} z\bar{\sigma}^{d}_{rr} dz = 0 \tag{2.8}$$

resulting from a direct application of Saint-Venant's principle to plate problems with prescribed edge tractions. A thorough discussion of this difference can be found in [6] and will not be repeated here.

We now take $\bar{\sigma}^{d}_{rz} = \sigma^{I}_{rz}(a,z) - \bar{\sigma}_{rz}(z)$ and $\bar{\sigma}^{d}_{rr} = \sigma^{I}_{rr}(a,z) - \bar{\sigma}_{rr}(z)$ where the state with a superscript I is the interior solution of the plate problem and $\bar{\sigma}_{rr}$ and $\bar{\sigma}_{rz}$ are the actual edge traction of the plate. Then (2.5) and (2.7) give two conditions for the determination of the two unknown constants in σ^{I}_{rz} and σ^{I}_{rr}, and hence the interior solution (see examples in [6] and later sections of this paper).

3. AXISYMMETRIC EXTENSION AND TORSION OF A CIRCULAR PLATE

Additional suitable suffix (2) states have to be found to obtain boundary conditions for the generalized plane stress problem which determines the in-plane extension and torsion of the plate. As in the axisymmetric bending problem, simple suffix (2) states can be found for the stress data case. For the stress data to induce only a decaying state, they must satisfy the necessary condition (2.1) for any admissible suffix (2) state. One admissible suffix (2) state is the rigid body displacement field

$$u_r^{(2)} \equiv u_z^{(2)} \equiv 0 \; , \qquad u_\theta^{(2)} = r \tag{3.1}$$

for which (2.1) or (2.4) becomes

$$\int_{-h}^{h} \bar{\sigma}^{d}_{r\theta}(z)dz = 0 \tag{3.2}$$

A second admissible suffix (2) state is

$$u_r^{(2)} = (1-\nu)r + (1+\nu)\frac{a^2}{r}, \quad u_z^{(2)} = -2\nu z, \quad u_\theta^{(2)} \equiv 0$$
$$\sigma_{rr}^{(2)} = E(1 - \frac{a^2}{r^2}), \quad \sigma_{rz}^{(2)} \equiv \sigma_{zz}^{(2)} \equiv \sigma_{z\theta}^{(2)} \equiv \sigma_{r\theta}^{(2)} \equiv 0 \tag{3.3}$$

for which (2.4) becomes

$$\int_{-h}^{h} [\bar{\sigma}^{d}_{rr} - \frac{\nu}{a} z\sigma^{d}_{rz}]dz = 0 \tag{3.4}$$

The interior solution for axisymmetric in-plane stress components, even in z, is expressed in terms of a biharmonic stress function F by (see Appendix I)

$$\sigma^{I}_{rr}(r,z) = \frac{1}{r}[F_{,r} + \frac{\nu(h^2-3z^2)}{6(1+\nu)}(\nabla^2 F)_{,r}] \tag{3.5a}$$

$$\sigma^{I}_{r\theta}(r,z) = -\{\frac{1}{r}[F_{,\theta} + \frac{\nu(h^2-3z^2)}{6(1+\nu)}(\nabla^2 F)_{,\theta}]\}_{r} . \tag{3.5b}$$

With $\bar{\sigma}^{d}_{ij} = \sigma^{I}_{ij}(a,z) - \bar{\sigma}_{ij}$, the necessary conditions (3.2) and (3.4) may be written as

$$-2h[(r^{-1}F_{,\theta})_{,r}]_{r=a} = \int_{-h}^{h} \bar{\sigma}_{r\theta}dz, \quad 2h[r^{-1}F_{,r}]_{r=a} = \int_{-h}^{h} [\bar{\sigma}_{rr} - \frac{\nu z}{a}\bar{\sigma}_{rz}]dz \tag{3.2',4'}$$

For axisymmetric extension and torsion, we have

$$F(r,\theta) = c_0\theta + [B'\ell n(r) + C'r^2 + A'r^2\ell nr + d'] \equiv F_T(\theta) + F_E(r) \tag{3.5d}$$

It is evident from (3.2'), (3.4'), and (3.5) that there is no coupling between torsion and extension in axisymmetric problems. The condition (3.2') determines c_0 and therefore the solution $F_T(\theta)$ for the torsion problem in terms of the actual edge data $\bar{\sigma}_{r\theta}(z)$:

$$2h\frac{c_0}{a^2} = \int_{-h}^{h} \bar{\sigma}_{r\theta}(z)dz . \tag{3.6}$$

For in-plane extension, we set $d' = 0$ as this term gives rise to no stress fields. We must set $A' = 0$ in order to have single valued displacement fields (see p. 68 of [3]). For a circular plate with only one edge at $r = a$, B' may be set equal to zero for bounded stress fields or determined by some other prescribed condition at the origin as we shall see later (in Appendix III). Application of the condition (3.4'), which now takes the form

$$2h\left[\frac{1}{r} F_{E,r}\right]_{r=a} = \int_{-h}^{h} [\bar{\sigma}_{rr} - \frac{\nu}{a} z\bar{\sigma}_{rz}\, dz], \tag{3.7}$$

determines the only remaining C' and therefore the solution $F_E(r)$ of the problem of axisymmetric extension in terms of the edge data $\bar{\sigma}_{rr}(z)$ and $\bar{\sigma}_{rz}(z)$. For an annular disc, the condition (3.4') applied to both edges gives two conditions for B' and C'; the interior solution for the axi-symmetric problem is again completely determined.

The condition (3.4') (or its immediate consequence (3.7)) agrees with the approximate condition (correct to order h/a) obtained by Kolos [9] for a plate of general shape (except for ν missing in Kolos' formula, presumably a slip or a typographic error). As a necessary condition for a decaying state of extension (3.4) (or (3.4')) is exact (up to exponentially small terms) for circular plates.

For mixed data in which u_r, σ_{rz}, $\sigma_{r\theta}$ are prescribed on $r = a$, the rigid body displacement (3.1) once again leads to the decaying state condition (3.2). A second admissible suffix (2) state is

$$\begin{aligned} &\sigma_{rr}^{(2)} = (1+\nu) + (1-\nu)\frac{a^2}{r^2}, \quad \sigma_{\theta\theta}^{(2)} = (1+\nu) - (1-\nu)\frac{a^2}{r^2}, \\ &\sigma_{r\theta}^{(2)} = \sigma_{rz}^{(2)} = \sigma_{\theta z}^{(2)} = \sigma_{zz}^{(2)} = 0, \\ &Eu_r^{(2)} = (1-\nu^2)(r - \frac{a^2}{r}), \quad Eu_z^{(2)} = -2\nu(1+\nu)z, \quad Eu_\theta^{(2)} = 0, \end{aligned} \tag{3.8}$$

for which the decaying state condition (2.3) becomes

$$\int_{-h}^{h} \left[E\bar{u}_r^d + \nu(1+\nu)z\bar{\sigma}_{rz}^d\right] dz = 0. \tag{3.9}$$

It follows from (3.2), (3.9) that the conditions which should be applied to the stress function F at $r = a$ when the edge data $\bar{u}_r(z)$, $\bar{\sigma}_{rz}(z)$, $\bar{\sigma}_{r\theta}(z)$ are prescribed there are (3.2') and

$$2h\left[rF_{,rr} - \nu F_{,r}\right]_{r=a} = \int_{-h}^{h} \left[E\bar{u}_r + \nu(1+\nu)z\bar{\sigma}_{rz}\right] dz. \tag{3.10}$$

We have not found elementary suffix (2) states for the case of pure displacement edge data. However, canonical boundary value problems may be formulated for numerical determination of appropriate suffix (2) states which, upon insertion in (2.1) - (2.3) (or their axisymmetric counterparts), yield the necessary conditions for a decaying state to be satisfied by the axisymmetric edge data for problems of extension and torsion. The details of this procedure and the derivation of displacement boundary conditions for plate theories are similar to the developments in [5] for displacement edge data in the case of a semi-infinite plate in a state of plane strain. Further discussion of these details for axisymmetric extension and torsion problem will not be pursued here. We do note however an important consequence of the necessary conditions for decaying residual states such as (3.7) or (3.9). It is evident from these conditions that an interior state of axisymmetric extension may be the consequence of a transverse shear edge stress $\bar{\sigma}_{rz}(z)$ (or a transverse displacement $\bar{u}_z(z)$) which are odd functions of z. Furthermore, depending on the relative magnitude of the edge data, the interior state associated with the transverse shear stress (or transverse displacement) may be dominant compared to that associated with the in-plane radial stress $\bar{\sigma}_{rr}$ (or displacement $\bar{u}_r$). In the next three sections, we apply the stress boundary conditions (3.7) obtained above and the conditions (2.5) and (2.7) obtained in [6] to determine the correct interior solution for a plate under a point load. The solution process will require additional developments involving decaying states.

4. A CIRCULAR PLATE UNDER A CONCENTRATED FORCE AT THE CENTER OF ITS UPPER FACE

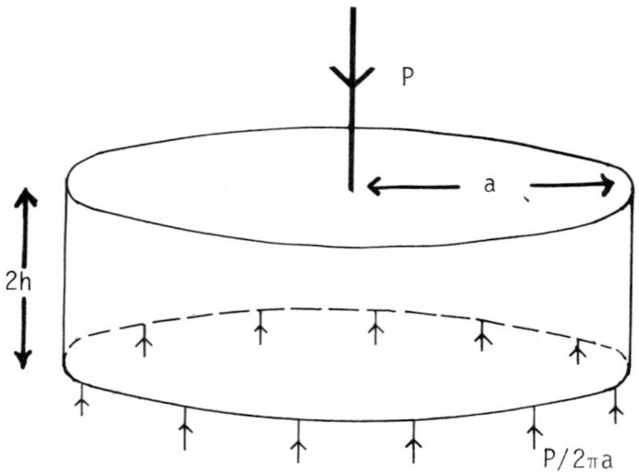

Fig. 1. The plate, the loading and the support

Consider a homogeneous, isotropic and linear elastic circular plate occupying the region $\{r \leqslant a, |z| \leqslant h\}$. As shown in Figure 1, the plate is under a concentrated load P acting at the center of the upper face ($r = 0$, $z = h$), and is "simply supported" around its lower edge ($r = a$, $z = -h$)*. We will

* As we shall see later, it is important to specify the precise nature of this support; other forms of simple support do not lead to the same solution even in the interior of the plate.

regard this support as supplying a uniform vertical line load distribution of magnitude $P/2\pi a$ (with resultant force P) around the lower edge of the cylindrical edge surface so that the plate is in overall equilibrium. Thus, the relevant elastostatic boundary value problem is one of <u>prescribed surface tractions</u>.

The exact (three-dimensional elastostatic) solution $\{u_j, \sigma_{ij}\}$ of this boundary value problem is quite intractable. However, in the spirit of plate theory, we will determine the interior component $\{u_j^I, \sigma_{ij}^I\}$ of the exact solution whose difference from the exact solution is <u>exponentially small</u> as $(h/a) \to 0$ in the region $\eta_0 < r < a - \eta_a, |z| \leq h$ for any fixed positive η_0 and η_a. The fact that this difference is exponentially small as $(h/a) \to 0$ is of great significance; it means that our theory is the best possible "thick plate" theory in the sense that it is accurate to $O([h/a]^N)$ for arbitrarily large N. It will also turn out that earlier "thick plate" solutions of this problem, e.g., [2](see p. 475), which claims to be correct to $O(h^2/a^2)$ are <u>not</u> correct to this order and are even inferior to the Kirchhoff "thin plate" solution.

Layer phenomena associated with the exponentially decaying components of the exact solution are known to occur adjacent to the plate edges. For the present point load problem, there is also (as we shall see) a layer phenomenon around the central axis, $r = 0$, of the circular plate where the point load is applied. Both this interior layer and the boundary layer adjacent to the edge $r = a$ extend radially a distance of the order of the plate thickness. We wish to determine the interior solution which is accurate to within exponentially small error in the annular region of the plate excluding the two layer regions. To do so, it is necessary to determine the form of the singularity at $r = 0$ imparted to the interior solution by the point load P. This is one of the interesting features of the present problem; it does not occur in the other problems treated in [5] and [6] and has not been treated correctly in the existing literature.

With the applied forces all in a direction normal to the plate's mid-plane, it is customary to expect the extensional part of the problem to involve only layer phenomena (and its interior solution component to be identically zero). But this is not so; in fact, it will be seen from our results of sections 6 and 7 that the extensional effects are of the same order as the corrections to the Kirchhoff solution in the bending part of the problem. This second interesting feature of the point load problem is a direct consequence of the elastic reciprocal theorem. In what follows, the solution of the bending portion of the solution for the point load problem will be discussed in section 5. The stretching portion of the solution will be obtained in section 6. The full interior solution and comparisons with previous results will be given in section 7.

5. THE BENDING PORTION OF THE SOLUTION FOR THE POINT FORCE PROBLEM

In this section only, all the fields which appear refer to the bending part of the elastostatic boundary value problem and will be denoted by $\{u_j^B, \sigma_{ij}^B\}$. For plate bending with traction-free faces $(|z| = h)$, the interior solution may be derived (see Appendix 1 of [6]) from the midplane transverse displacement $w(r,\theta) = u_z^B(r,\theta,0)$ which satisfies the two-dimensional biharmonic equation, $\nabla^2\nabla^2 w = 0$. The corresponding interior solution is given by equations (I.8) - (I.16) of [6]. These formulae are similar but not identical to the formulae of the Kirchhoff thin plate theory. The additional terms ensure that $\{u_j^B, \sigma_{ij}^B\}$ is a genuine three-dimensional elastostatic state.

Because of the obvious axisymmetry of the present problem about $r = 0$, it follows that w must have the form

$$W = Ar^2 \ell nr + B \ell nr + Cr^2 + d \ , \tag{5.1}$$

where the constants A, B, C are yet to be determined. The constant d corresponds to rigid body vertical translation and will be chosen so that $w(a) = 0$.

Theorem 1 The constants A and B, in the expression (5.1) for the midplane transverse displacement, are related to the load magnitude P by

$$A = -\frac{P}{8\pi D} \ , \qquad B = \frac{2}{5}\frac{8 - 3\nu}{1 - \nu} h^2 \frac{P}{8\pi D} \tag{5.2,3}$$

where the flexural rigidity D is given in terms of Young's modulus E, Poisson's ratio ν (of the homogeneous and isotropic plate medium) and plate thickness $2h$ by $2Eh^3/3(1-\nu^2)$.
The proof of Theorem 1, given in Appendix II, is also by the reciprocal theorem.

We observe that the value of A as given by (5.2) is identical to that obtained in the Kirchhoff thin plate theory. The coefficient B is taken to be zero in the Kirchhoff theory, on the grounds that "$w(r)$ must be bounded at $r = 0$" and other authors (e.g. Love [2] p. 475) have applied the same reasoning when $w(r)$ is the midplane deflection in the interior of a thick plate. We now see that this is not so, as $B \neq 0$ for thick plates. At the same time with $B = O(h^2 A)$ it is indeed correct to take $B = 0$ in the Kirchhoff thin plate limit. It may seem strange at first that the interior midplane deflection $w = u_z(r,0)$ should be unbounded at $r = 0$. However, it should be remembered that the interior solution only approximates the exact solution (as $h/a \to 0$) in regions bounded away from $r = 0$ and $r = a$; thus the singularity at $r = 0$ in (5.1) need not be present in the exact solution.

The assumption that "$w(r)$ must be bounded at $r = 0$" is not made in a solution by Reissner's plate theory [9] obtained in Appendix IV. This solution gives the same value of B as (5.3) except for a multiplicative factor 8 instead of $8-3\nu$. The numerically small difference may be attributed to the omission of transverse normal stress effects in our use of Reissner's theory. In contrast, an argument given in [8] gives $B = Ph^2/8\pi D(1+\nu)$.

It remains to determine the coefficient C in (5.1); this step has also been performed incorrectly in previous solutions (e.g., Love [2] and Lur'e [7] Chap. 4, pp. 226-230). In these solutions it is assumed that the fields $\{\sigma_{ij}^d, u_j^d\}$, which are the difference between the exact bending solution and the interior bending solution, must satisfy the 'Saint-Venant type' conditions

$$\int_{-h}^{h} [\sigma_{rz}^d]_{r=a}\, dz = 0 , \qquad \int_{-h}^{h} z[\sigma_{rr}^d]_{r=a}\, dz = 0 \tag{5.4,5}$$

on the outer edge $r = a$. They are equivalent to requiring the interior bending solution to have the same transverse shear resultant and bending (stress) couple at $r = a$ as the prescribed data. However we have previously shown in Section 4 of [6] that, while the condition (5.4) is appropriate for the problem, the condition (5.5) is not and should in fact be replaced by the condition

$$\int_{-h}^{h} [z\sigma_{rr}^d - \frac{\nu}{2a} z^2\sigma_{rz}^d]_{r=a}\, dz = 0 . \tag{5.6}$$

The additional term $-\nu z^2\sigma_{rz}^d/2a$ represents a $O(h/a)$ correction to (5.5), which turns out to be the same order as the error in the Kirchhoff solution; thus to improve on the Kirchhoff solution it is essential to use (5.6) and not (5.5). In our problem

$$[\sigma_{rr}^d]_{r=a} = - \frac{Ez}{1-\nu^2} [A\{2(1+\nu)\ell na + (3+\nu) - \frac{4}{a^2} (h^2 - \frac{2-\nu}{6} z^2)\}$$

$$- B(1-\nu) \frac{h^2}{a^2} + 2C(1+\nu)] , \tag{5.7}$$

$$[\sigma_{rz}^d]_{r=a} = - \frac{2E}{a(1-\nu^2)} (h^2-z^2)A - \frac{P}{4\pi a} \{\delta(z-h) + \delta(z+h) , \tag{5.8}$$

where $\delta(z)$ is the Dirac delta function. (The delta functions in (5.8) arise from the bending part of the applied line load around $z = -h$, $r = a$.) With the value of A given by (5.2), we see that (5.4) is satisfied identically. Upon substituting (5.7), (5.8) into (5.6) and using (5.2), (5.3), the value of C is found to be

$$C = \frac{P}{8\pi D} [\ell na + \frac{3+\nu}{2(1+\nu)}] . \tag{5.9}$$

We now substitute (5.2), (5.3), (5.9) into (5.1) to get

$$w = \frac{P}{8\pi D}\left[r^2 \ell n\left(\frac{a}{r}\right) - \frac{2(8-3\nu)}{5(1-\nu)} h^2 \ell n\left(\frac{a}{r}\right) - \frac{3+\nu}{2(1+\nu)}(a^2-r^2)\right] , \tag{5.10}$$

where we have adjusted the constant d so that $w(a) = 0$ to permit easy comparison with earlier solutions.

It should be noted that since the extensional part of the solution (i.e. the part which is symmetrical about $z = 0$) leaves the midplane $z = 0$ undeflected, the expression (5.10) also represents the midplane deflection in the full interior solution $u_z^I(r,0)$.

The interior bending fields $\{u_j^B, \sigma_{ij}^B\}$ can now be calculated from (5.20) by way of the formulae in Appendix 1 of [6]. In particular the important radial tensile stress on the lower face $z = -h$ is given by

$$\sigma_{rr}^B(r,-h) = \frac{3P}{8\pi h^2}\left[(1+\nu)\ell n\left(\frac{a}{r}\right) - \frac{2(2-7\nu)}{15}\frac{h^2}{r^2}\right] \tag{5.11}$$

The leading terms of (5.10) and (5.11) are identical to the customary expressions predicted by the Kirchhoff thin plate theory. The additional terms of $O(h^2)$ in our expressions represent the true correction to the Kirchhoff theory in the interior of the plate. It should be noted that there are no higher order corrections of $O(h^N)$, $N \geqslant 3$; the expressions (5.10), (5.11) are complete as they stand.

6 THE STRETCHING PORTION OF THE SOLUTION FOR THE POINT FORCE PROBLEM

In this section only, all the fields which appear refer to the extensional part of the problem. The extensional part of the interior solution will be denoted by $\{u_j^E, \sigma_{ij}^E\}$. Since the loading corresponding to the extensional part of the problem is symmetrical about the midplane $z = 0$ and perpendicular to that plane, it is often supposed that the extensional part of such a problem makes no contribution to the interior solution. We shall see, however, that the extensional part also makes a correction to the Kirchhoff solution of $O(h^2)$.

For the case of extension (in regions of the plate where the faces $|z| = h$ are traction free) the interior solution may be derived from a scalar function $F(r,\theta)$, which is related to the Airy stress function (see Appendix I). F satisfies the plane bi-harmonic equation

$$\nabla^2(\nabla^2 F) = 0 , \tag{6.1}$$

and the corresponding interior solution is given by equations (I.11)-(I.16) of

Appendix I. These formulae are similar to, but not identical with, the formulae of generalized plane stress. The additional terms which appear ensure that $\{u_j^E, \sigma_{ij}^E\}$ is a genuine three-dimensional elastostatic state. From the axisymmetry of the problem it follows from (6.1) that F must have the form

$$F = A'r^2\ell nr + B'\ell nr + C'r^2 + d' . \tag{6.2}$$

Now $A' = 0$, for otherwise the displacement field in the plate would not be single-valued; also d' makes no contribution to the fields and may be taken to be zero. Thus it remains to determine the constants B', C'. As in the case of bending, B' (the coefficient of the singular term ℓnr) is determined solely by the point loads.

Theorem 2. In (6.2), the constant B' is given by

$$B' = -\frac{\nu P}{4\pi} . \tag{6.3}$$

The proof of Theorem 2 is given in Appendix III. [The interior solution corresponding to $F = -\nu P\ell nr/4\pi$ is given by

$$\begin{aligned} \sigma_{rr}^I &= -\frac{\nu P}{4\pi}\frac{1}{r^2} , \quad \sigma_{rz}^I = \sigma_{zz}^I = 0 , \\ Eu_r^I &= \frac{\nu(1+\nu)P}{4\pi}\frac{1}{r} , \quad Eu_z^I = 0 , \end{aligned} \tag{6.4}$$

from which it is clear that the extensional part of the problem must contribute to the interior solution and must also be unbounded at $r = 0$.]

It now remains to determine the constant C' from the prescribed data on the edge $r = a$. We found in section 4 that for the extensional part of the solution, the condition corresponding to (5.4), (5.6) is (3.7), or

$$\int_{-h}^{h} [\sigma_{rr}^d - \frac{\nu z}{a}\sigma_{rz}^d]_{r=a} dz = 0 \tag{6.5}$$

and not the commonly assumed 'Saint-Venant' condition

$$\int_{-h}^{h} [\sigma_{rr}^d]_{r=a} dz = 0 . \tag{6.6}$$

Upon applying (6.5) to (6.2) and using (6.3) we obtain

$$C' = \frac{\nu P}{4\pi a^2} . \qquad (6.7$$

The corresponding extensional interior solution can now be found from (I.11)-(I.16). In particular, we have

$$\sigma_{rr}^{E}(r,z) = \frac{\nu P}{4\pi a^2}\left[2 - \frac{a^2}{r^2}\right] , \qquad u_z^{E}(r,0) = 0 \tag{6.8,9}$$

7. THE INTERIOR SOLUTION

The full interior solution, denoted by $\{u_j^I, \sigma_{ij}^I\}$, is now just given by

$$\{u_j^I, \sigma_{ij}^I)\} = \{u_j^B, \sigma_{ij}^B\} + \{u_j^E, \sigma_{ij}^E\} . \tag{7.1}$$

In particular it follows from (5.10), (6.9) that the interior midplane deflection is given by

$$u_z^I(r,0) = \frac{P}{8\pi D}\left[r^2 \ell n(\tfrac{a}{r}) - \frac{2(8-3\nu)}{5(1-\nu)} h^2 \ell n(\tfrac{a}{r}) - \frac{3+\nu}{2(1+\nu)}(a^2-r^2)\right] . \tag{7.2}$$

It similarly follows from (5.11), (6.8) that the radial tensile stress on the lower face is given by

$$\sigma_{rr}^I(r,-h) = \frac{P}{8\pi h^2}\left[3(1+\nu)\ell n(\tfrac{a}{r}) - \frac{4}{5}(1-\nu)\frac{h^2}{r^2} + 4\nu\frac{h^2}{a^2}\right] . \tag{7.3}$$

The expressions (7.2), (7.3) differ from the exact values by an <u>exponentially</u> small error (as $h/a \to 0$) except near $r = 0$ and $r = a$, where there are 'boundary layers' of width $O(h)$.

In the limit as $h/a \to 0$, (7.2), (7.3) tend to the values predicted by the Kirchhoff thin plate theory, that is

$$u_z^k(r,0) = \frac{P}{8\pi D}\left[r^2 \ell n(\tfrac{a}{r}) - \frac{3+\nu}{2(1+\nu)}(a^2-r^2)\right] , \tag{7.4}$$

$$\sigma_{rr}^k(r,-h) = \frac{3(1+\nu)P}{8\pi h^2}\,\ell n(\tfrac{a}{r}). \tag{7.5}$$

For thick plates the difference between u_z^I and u_z^k may be quite substantial. For instance

$$\frac{u_z^I(r,0)-u_z^k(r,0)}{u_z^k(0,0)} = \frac{4(1+\nu)(8-3\nu)}{5(1-\nu)(3+\nu)}\frac{h^2}{a^2}\,\ell n(\tfrac{a}{r}) \doteq -4.46\,\frac{h^2}{a^2}\,\ell n(\tfrac{a}{r}) \tag{7.6}$$

when $\nu = {}^1/2$. For $h = 0.2a$, (7.6) has the value 0.29 .... at $r = h$; thus in this case the Kirchhoff theory <u>underestimates</u> the deflection of the plate by about 30%. Here we are assuming that the boundary layer around the point load has a negligible influence at the point $r = h$. See [6], section 5 for a discussion.

The expression (7.2) is quite different from the expression derived by Love [2], p. 475, for this same problem. Both expressions reduce to (7.4) as $h/a \to 0$, but the other terms in Love's expression are incorrect, for the reasons explained earlier. In the work of Lur'e [6] the concentrated load is represented exactly as a series of Bessel functions. While this approach has an advantage in that it allows the fields to be calculated near the load, its complicated form does obscure the essential simplicity of the interior solution, which (as a result) is never discovered. Lur'e did not treat the precise problem of the present paper; however it is clear from his treatment of the corresponding problem with uniform pressure applied to the upper face that the stress boundary conditions at $r = a$ are applied incorrectly, as was also done by Love [2], and Timoshenko and Goodier [8] p. 351. In all these works, the condition (5.5) is applied instead of the true condition (5.6).

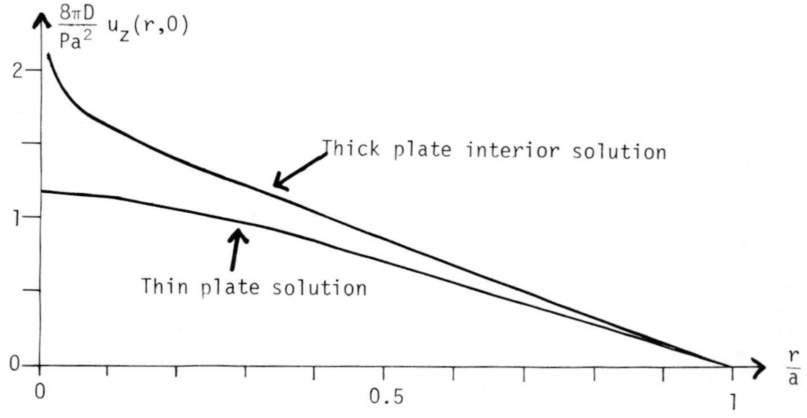

Figure 2. The transverse deflection of the mid-plane of the plate as predicted by (i) the thick plate interior solution (ii) the thin plate Kirchhoff solution, in the case $h/a = 0.2$, $\nu = 0.5$

In Figure 2 we plot $u_z^I(r,0)$, the transverse deflection of the midplane $z = 0$ (as predicted by the interior solution), against r for the case $\nu = 0.5$, $h/a = 0.2$. The difference between u_z^I and the Kirchhoff thin plate solution (shown also) is substantial. u_z^I may be expected to be a good approximation to u_z, the exact deflection, for $0.2 < r/a < 0.8$; in this range we note that the plate profile is quite straight (giving a cone-like shape), unlike the "parabolic" shape predicted by the thin plate theory.

8. DIFFERENT FORMS OF LOADING AND OF SIMPLE SUPPORT

The results (7.2), (7.3) are influenced by the precise method of loading and support employed. If the plate were simply supported by a vertical line load around its upper edge (or around its midplane circle $r = a$, $z = 0$), then this would affect the solution, even in the interior of the plate, by terms of order $O(h^2/a^2)$. The same is true if the concentrated load were applied at $r = 0$, $z = -h$ instead of $r = 0$, $z = h$.

Thus in thick plate theory it is not meaningful to refer to "a concentrated load" or "simple support'; the nature of the loading and support must be specified precisely, for otherwise errors will be introduced which are of the same order as the thick plate corrections being determined. In particular, it is not possible to determine the interior solution when only the stress resultants and couples are prescribed at the edge.

APPENDIX I INTERIOR SOLUTION FOR EXTENSION AND TORSION OF PLATES WITH TRACTION-FREE FACES

For an isotropic and homogeneous plate with traction-free faces at $z = \pm h$, the interior solution for in-plane extension and torsion portion of its elastostatic fields may be expressed in terms of a biharmonic (Airy stress) function F by

$$\sigma^E_{xx} = \frac{\partial^2}{\partial y^2}\left[1 + \frac{\nu(h^2-3z^2)}{6(1+\nu)}\nabla^2\right]F , \qquad \sigma^E_{yy} = \ldots \tag{I.1,2}$$

$$\sigma^E_{xy} = -\frac{\partial^2}{\partial x \partial y}\left[1 + \frac{\nu(h^2-3z^2)}{6(1+\nu)}\nabla^2\right]F , \qquad \sigma^E_{xz} = \sigma^E_{yz} = \sigma^E_{zz} = 0 \tag{I.3,4}$$

$$Eu^E_x = E\tilde{u}_x - \frac{\nu}{6}(h^2-3z^2)\frac{\partial}{\partial x}(\nabla^2 F) \tag{I.5}$$

$$Eu^E_y = \ldots\ldots\ldots , \qquad Eu^E_z = -\nu z^2\nabla^2 F \tag{I.6,7}$$

where $\nabla^2(\nabla^2 F) = 0$ and

$$E\frac{\partial u_x}{\partial x} = \nabla^2 F - (1+\nu)\frac{\partial^2 F}{\partial x^2} , \qquad E\frac{\partial \tilde{u}_y}{\partial y} = \ldots \tag{I.8,9}$$

$$E\left(\frac{\partial \tilde{u}_y}{\partial x} + \frac{\partial \tilde{u}_x}{\partial y}\right) = -2(1+\nu)\frac{\partial^2 F}{\partial x \partial y} , \tag{I.10}$$

∇^2 being the two-dimensional Laplacian in x and y. The formulae for σ^E_{yy}, u^E_y and $\tilde{u}_y$ are obtained from σ^E_{xx}, u^E_x and $\tilde{u}_x$ by interchanging x and y.

In cylindrical coordinates, the corresponding formulae for the interior solution of plate extension and torsion are

$$\{\sigma^E_{\theta\theta}, \sigma^E_{rr}\} = \{\frac{\partial^2}{\partial r^2}, \quad \nabla^2 - \frac{\partial^2}{\partial r^2}\}[1 + \frac{\nu(h^2-3z^2)}{6(1+\nu)} \nabla^2]F \tag{I.11}$$

$$\sigma^E_{r\theta} = -\frac{\partial^2}{\partial r\partial\theta}\{\frac{1}{r}[1 + \frac{\nu(h^2-3z^2)}{6(1+\nu)} \nabla^2]F\}, \quad \sigma^E_{rz} = \sigma^E_{\theta z} = \sigma^E_{zz} = 0 \tag{I.12,13}$$

$$Eu^E_r = E\tilde{u}_r - \frac{\nu}{6}(h^2-3z^2)\frac{\partial}{\partial r}\nabla^2 F \tag{I.14}$$

$$Eu^E_\theta = E\tilde{u}_\theta - \frac{\nu}{6}(h^2-3z^2)\frac{1}{r}\frac{\partial}{\partial\theta}\nabla^2 F, \quad Eu^E_z = -\nu z\nabla^2 F \tag{1.15,16}$$

where $\nabla^2\nabla^2 F = 0$ with ∇^2 being the two-dimensional Laplacian in polar co-ordinates and where

$$E\frac{\partial\tilde{u}_r}{\partial r} = \nabla^2 F - (1+\nu)\frac{\partial^2 F}{\partial r^2} \tag{I.17}$$

$$E(\frac{1}{r}\frac{\partial\tilde{u}_\theta}{\partial\theta} + \frac{1}{r}\tilde{u}_r) = (1+\nu)\frac{\partial^2 F}{\partial r^2} - \nu\nabla^2 F \tag{I.18}$$

$$E(\frac{\partial\tilde{u}_\theta}{\partial r} - \frac{1}{r}u_\theta + \frac{1}{r}\frac{\partial\tilde{u}_r}{\partial\theta}) = -2(1+\nu)\frac{\partial^2}{\partial r\partial\theta}(\frac{1}{r}F) \tag{I.19}$$

APPENDIX II THE BENDING PART OF THE INTERIOR SOLUTION NEAR A POINT LOAD

Consider the bending part of the plate problem depicted in Figure 1. We apply the elastic reciprocal theorem

$$\iint_S \{\sigma^{(1)}_{ij}u^{(2)}_i - \sigma^{(2)}_{ij}u^{(1)}_i\}n_j ds = 0, \tag{II.1}$$

where S is the boundary of the region $r \leqslant b$ $(<a)$ $|z| \leqslant h$, and $b(>0)$ is independent of h. We take as the suffix (1) state the exact solution $\{u_j, \sigma_{ij}\}$ of the plate problem. For the suffix (2) state we take the axi-symmetric state

$$u^{(2)}_r = -2rz, \quad u^{(2)}_z = r^2 - \frac{2\nu}{1-\nu}(h^2-z^2), \quad u^{(2)}_\theta = 0$$

$$\sigma^{(2)}_{rr} = -\frac{2Ez}{1-\nu}, \quad \sigma^{(2)}_{rz} = \sigma^{(2)}_{zz} = \sigma^{(2)}_{r\theta} = \sigma^{(2)}_{\theta\theta} = 0 \tag{II.2}$$

Note that we have arranged that $u^{(2)}_z = 0$ at the points $r = 0$, $z = \pm h$. If we regard the point force P as being the limit of a normal pressure $P/\pi\delta^2$ applied over the area $z = h$, $r \leqslant \delta$ as $\delta \to 0$, then the contribution to the left side of (II.1) from the face $z = h$ is

$$2\pi \int_0^\delta \left(-\frac{P}{\pi\delta^2}\right)(r^2)rdr = -\frac{1}{2}P\delta^2$$

which tends to zero as $\delta \to 0$. Thus there is no contribution from the face $z = h$ in the limit of a point load at $r = 0$, $z = h$. There is certainly no contribution from the lower face $z = -h$ and so, taking into account the axi-symmetry, (II.1) reduces to

$$\int_{-h}^{h} [\sigma_{rr}u_r^{(2)} + \sigma_{rz}u_z^{(2)} - \sigma_{rr}^{(2)}u_r - \sigma_{rz}^{(2)}u_z]_{r=b} \, dz = 0 \, . \qquad \text{(II.3)}$$

Now in (II.3), $u_r^{(2)}$ etc., are given by (II.2) while the exact solution $\{u_j, \sigma_{ij}\}$ may be replaced (with exponentially small error as $h\to 0$) by the bending part of the interior solution $\{u_j^I, \sigma_{ij}^I\}$, derived from (5.1). (The extensional part of $\{u_j^I, \sigma_{ij}^I\}$ gives no contribution, in view of the symmetry of $\{u_j^{(2)}, \sigma_{ij}^{(2)}\}$.) On making these substitutions, we obtain after some simplification that

$$B = -\frac{2(8-3\nu)}{5(1-\nu)} h^2 A \, . \qquad \text{(II.4)}$$

Now it easily follows from the overall equilibrium of the region $r \leqslant b$, $|z| \leqslant h$ that $A = -P/8\pi D$ (where D is the flexural rigidity) as in the Kirchhoff theory. As a consequence, we have from (II.4)

$$B = \frac{2}{5} h^2 \left(\frac{8-3\nu}{1-\nu}\right) \frac{P}{8\pi D} \, . \qquad \text{(II.5)}$$

APPENDIX III THE EXTENSIONAL PART OF THE INTERIOR SOLUTION NEAR A POINT LOAD

Consider the extensional part of the plate problem depicted in Figure 1. From (3.5c) the corresponding interior solution is derivable from the scalar function

$$F = B'\ell n r + C'r^2 \qquad \text{(III.1)}$$

where B', C' are constants. The corresponding stresses and displacements are derived from F by the formulae given in Appendix I.

The method of determining B' follows closely the method of determining B in Appendix II. We apply the elastic reciprocal theorem (II.1) to the same region, with the suffix (1) state being the exact solution $\{u_j, \sigma_{ij}\}$ of the plate problem. However, the suffix (2) state is taken this time to be

$$\sigma_{rr}^{(2)} = 1 \, , \quad \sigma_{rz}^{(2)} = \sigma_{zz}^{(2)} = \sigma_{r\theta}^{(2)} = 0, \qquad \text{(III.2)}$$

$$Eu_r^{(2)} = (1-\nu)r \ , \quad u_\theta^{(2)} = 0 \ , \quad \bar{E}u_z^{(2)} = -2\nu z \ . \tag{III.3}$$

This state is symmetrical about midplane $z = 0$, and so the bending part of the exact solution gives no contribution. The extensional part then gives the result

$$B' = -\frac{\nu P}{4\pi} \ . \tag{III.4}$$

APPENDIX IV SOLUTION FOR THE POINT LOAD PROBLEM BY REISSNER'S PLATE THEORY

In the framework of Reissner's Plate Theory, the relevant boundary value problem for a circular plate with a point force of magnitude P applied at the center of the plate may be formulated in terms of a (meridional slope) angle change variable ϕ which satisfies the second order differential equation

$$\frac{d^2\phi}{dr^2} + \frac{1}{r}\frac{d\phi}{dr} - \frac{1}{r^2}\phi = \frac{P}{2\pi Dr} \qquad (0 < r < a) \tag{IV-1}$$

By symmetry, we expected to have

$$\phi(0) = 0 \tag{IV-2}$$

and for a simply supported edge

$$\left[\frac{d\phi}{dr} + \frac{\nu}{r}\phi\right]_{r=a} = 0 \tag{IV-3}$$

which is simply the condition of vanishing edge moment resultant.

The solution of the above two point boundary value problem for ϕ is

$$\phi(r) = -\frac{Pr}{4\pi D}\left[\ln\left(\frac{a}{r}\right) + \frac{1}{1+\nu}\right]$$

From the stress strain relation $\phi + dw/dr = {}_QA_rQ$ and the overall equilibrium condition $Q_r = P/2\pi r$, we get

$$w = -\int_r^a \left\{\frac{A_QP}{2\pi r} + \frac{Pr}{4\pi D}\left[\ln\left(\frac{a}{r}\right) + \frac{1}{1+\nu}\right]\right\}dr$$

$$= \frac{P}{8\pi D}\left\{r^2\ln\left(\frac{a}{r}\right) - \frac{3+\nu}{2(1+\nu)}(a^2-r^2) - \frac{16}{5(1-\nu)}h^2\ln\left(\frac{a}{r}\right)\right\}$$

This expression is to be compared with (5.10). The only difference is the factor 16 in the last term; the corresponding factor in (5.10) is $2(8-3\nu)$.

REFERENCES

1 C.O. Horgan and J.K. Knowles, "Recent Developments Concerning Saint-Venant's Principle," Adv. in Appl. Mech., Vol. 23, 1983, Academic Press, 179-269.
2 A.E.H. Love, The Mathematical Theory of Elasticity (4th Ed.), Dover Publications, New York, 1944 (p. 475).
3 S. Timoshenko and J.N. Goodier, Theory of Elasticity (2nd Ed.), McGraw-Hill, New York, 1951, (p.33 and pp 351-352).
4 A.V. Kolos, "Methods of refining the classical theory of bending and extension of plates," P.M.M. 29 (4), 1965, 777-781.
5 R.D. Gregory and F.Y.M. Wan, "Decaying states of plane strain in a semi-infinite strip and boundary conditions for plate theory," J. of Elasticity, 14, 1984, 27-64.
6 ________________, "On plate theories and Saint-Venant's principle," Int'l. J. Solids & Structures, 1985, to appear.
7 A.I. Lur'e., Three-Dimensional Problems of the Theory of Elasticity, Interscience Publishers (a division of John Wiley & Sons, Inc.) New York-London-Sydney, 1964 (chapter 4, particularly pp. 226-230).
8 S. Timoshenko and S. Woinowski-Krieger, Theory of Plates and Shells (2nd Ed.), McGraw-Hill, New York, 1959 (p. 77).
9 ________________, "The effect of transverse-shear deformation on the bending of elastic plates," J. Appl. Mech., 12, 1945, A69-A77.
10 E. Reissner, "On the equations for finite symmetric deflections of thin shells of revolution, "Progress in Appl. Mech. (Prager Anniversay Volume), Macmillan Co., 1963, 171-178.

STRESS CONCENTRATION FOR DEFECTS DISTRIBUTED NEAR A SURFACE

N. NGUETSENG and **E. SANCHEZ-PALENCIA**

Laboratoire de Mécanique Théorique, Université Pierre et Marie Curie
4 place Jussieu, 75230 Paris Cedex 05 (France)

ABSTRACT

We consider the mechanical behavior of an elastic body containing small cavities or holes periodically distributed in the vicinity of a surface Σ interior to the body. It constitutes a model of two pieces imperfectly stiked together. The method of analysis is some sort of homogenization with respect to the variables tangent to Σ and matched asymptotic expansions for the normal variable. This furnishes the local problem to be solved numerically in actual situations to obtain the local stress field. Of course the very stress concentration depends on the shape of the hole or crack. The particular problem of two pieces sticked together with a narrow uniform layer of another medium is completely worked out.

1. - INTRODUCTION

In this paper we study the behavior of an elastic body containing small cavities (holes) periodically distributed in the vicinity of a surface Σ interior to the body. As the holes may be in particular cracks, this problem is a model for the study of two pieces sticked together in an imperfect way.

The method of study is some sort of homogenization method with respect to the tangential variables to Σ and a matching asymptotic expansion method for the normal variable (see (ref. 9) for this kind of problems for the Laplace equation).

The concentration of stress is given by the solution of some "local" elasticity problems which only depend on the elastic coefficients and on the form and location of the holes. There local problems are in fact Neumann elasticity problems with periodicity conditions on a part of the boundary ; these periodicity conditions may of course be imposed in the finite element method as in homogenization problems.

The study is performed in sections 2 to 6 for a model problem containing non essential features which may be easily modified : for instance, the body is fixed by its exterior surface, and submitted to body forces, but the case of a body submitted to surface stresses leads to the same local behavior in the vicinity of Σ. The practical results and the explicit problem to be solved numerically in actual computations are given in section 5. Some complements are given in sections 7 and 8. The first one is devoted to the non linear case of cracks or fissures with one-side constraints, and the last one to the problem of two different bodies sticked together with a narrow layer of another material : the local problem is then completely worked out.

We point out that some complements about convergence of the asymptotic expansion towards the unperturbed problem may be found in (ref.8, chap. 4).

Let us give some indications about notation :

The classical convention of summation of repeated indexes is used. Latin (resp. Greek) indexes run from 1 to 3 (resp. to 2) thus

$$u_i v_i = \sum_1^3 u_i v_i \quad , \quad u_\alpha v_\alpha = \sum_1^2 u_\alpha v_\alpha$$

$$\delta_{ij} = 1 \quad \text{if} \quad i = j, \quad = 0 \quad \text{otherwise.}$$

Underlined symbols denote vectors of R^3, for instance

$$\underline{u} = (u_1, u_2, u_3)$$

but as an exception, the current point of R^3 is not underlined :

$$x = (x_1, x_2, x_3) \quad ; \quad y = (y_1, y_2, y_3)$$

When two scale variables are involved, e_{ijx}, e_{ijy} denote the strain tensor in the corresponding variables for instance :

$$e_{ijy}(\underline{u}) = \frac{1}{2}\left(\frac{\partial u_i}{\partial y_j} + \frac{\partial u_j}{\partial y_i}\right)$$

Upper indexes, as in u^1, u^2, are used to denote the terms of a sequence.

If Ω is a domain, $\partial\Omega$ denotes its boundary and $\underline{n}$ is the outer unit normal.

The brackets $[u]$ denote the jump of the function u across a discontinuity.

2. - SETTING OF THE PROBLEM

Let Ω be a boundary domain of R^3, with boundary $\partial\Omega$. We consider it as an elastic body with clamped boundary, submitted to a given force field $f_i(x)$. Let a_{ijlm} be the elastic coefficients, which we shall consider constant all over Ω, satisfying the standard symmetry and positivity conditions :

$$a_{ijlm} = a_{jilm} = a_{lmij} \tag{2.1}$$

$$a_{ijlm}\, e_{lm}\, e_{ij} \geqq \gamma\, e_{ij}\, e_{ij} \qquad \forall\, e_{ij} = e_{ji} \tag{2.2}$$

for some $\gamma > 0$.

We denote by $\underline{u}^o(x)$ the displacement field corresponding to this "unperturbed" (or "without holes") problem. It is the unique solution of

$$- \frac{\partial\, \sigma_{ij}(\underline{u}^o)}{\partial\, x_j} = f_i \quad \text{in } \Omega \tag{2.3}$$

$$\underline{u}^o = 0 \quad \text{on } \partial\Omega \tag{2.4}$$

with

$$\sigma_{ij}(\underline{u}^o) \equiv a_{ijlm}\, e_{lm}(\underline{u}^o) \quad ; \quad e_{lm}(\underline{u}^o) \equiv \frac{1}{2}\left(\frac{\partial\, u_i^o}{\partial\, x_j} + \frac{\partial\, u_j^o}{\partial\, x_i}\right) \tag{2.5}$$

where σ_{ij} e_{ij} denote the stress and strain tensors

We now define the "perturbed" (or "with holes") problem. Let Σ be a section by a plane, $x_3 = 0$, say. Let us consider the auxiliar space of the variable (y_1, y_2, y_3), where the plane $y_3 = 0$ is divided into rectangles ω which will be considered as periods (for instance the squares whose edges have integer coordinates). Let $\mathcal{H}$ (as "hole") be a domain of R^3 intersecting a period ω, as in fig. 2 ; we also consider the holes obtained by ω-periodic displacement, i.e. a hole associated with each period. Here and in the sequel, $\mathcal{H}$ will denote either a hole or the set of all holes.

In order to define the "small" holes in Ω, we consider a small positive parameter ε and the set $\varepsilon\mathcal{H}$ homothetic of $\mathcal{H}$ with ratio ε, and we define the "hollowed" domain $\underline{\Omega_\varepsilon \text{ as } \Omega - \varepsilon\mathcal{H}}$, i.e. the domain Ω less the $\varepsilon\mathcal{H}$ holes.

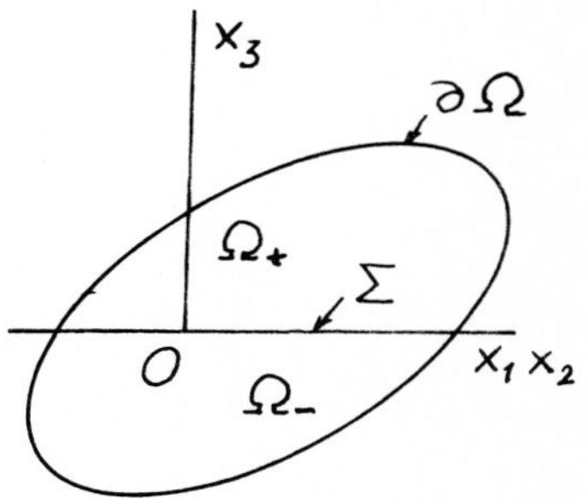

Figure 1

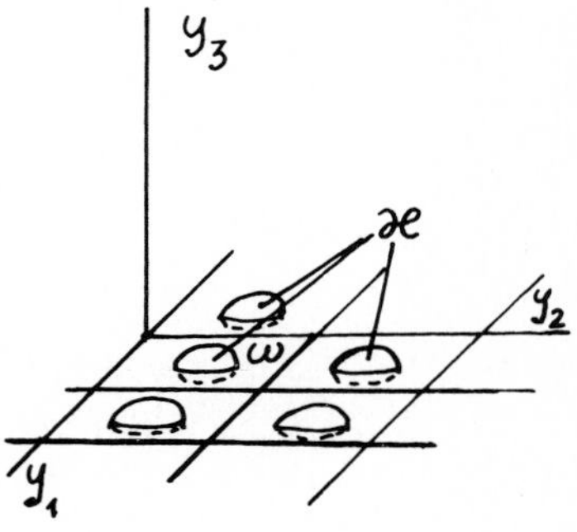

Figure 2

Our aim is to study, for small ε, the solutions of the elasticity problem in Ω_ε, when the $\varepsilon\,\partial\mathcal{H}$ boundaries of the ε-holes are free, i.e. :

$$- \frac{\partial\, \sigma_{ij}(\underline{u}^\varepsilon)}{\partial\, x_j} = f_i \qquad \text{in } \Omega_\varepsilon \tag{2.6}$$

$$\underline{u}^\varepsilon = 0 \qquad \text{on } \partial\Omega \tag{2.7}$$

$$\sigma_{ij}(\underline{u}^\varepsilon)\, n_j = 0 \qquad \text{on } \partial\varepsilon\mathcal{H} \tag{2.8}$$

3. - ASYMPTOTIC EXPANSION

Coming back to figure 1, it is clear that some sort of layer holds in the vicinity of Σ. According to the classical techniques of inner and outer expansions (cf. Van Dyke (ref.13)) we define an inner variable

$$y_3 = x_3/\varepsilon \tag{3.1}$$

in order to dilate the layer, and we shall search for two different asymptotic expansions out of the layer (outer expansion) and in the layer (inner expansion). On the other hand, each of the expansions depend on the variables x_1, x_2 tangential to the layer. Denoting by Ω^+, Ω^- the domains formed by the parts of Ω with $x_3 > 0$, < 0, respectively, we try an <u>outer expansion</u>

$$\underline{u}^\varepsilon(x) = \underline{u}^o(x) + \varepsilon\, \underline{u}^1(x) + \varepsilon^2 \ldots \quad ; \quad x = (x_1, x_2, x_3) \tag{3.2}$$

in Ω^+ and Ω^- (in fact two different expansions of the form (3.2) in Ω^+ and Ω^-). We note that the first term of the outer expansion is denoted by $\underline{u}^o$ as the solution of the unperturbed problem (2.3)-(2.5) ; we shall see that in fact they are the same.

As for the inner expansion, the dependence on the variables x_1, x_2 will be somewhat involved. Indeed, in the layer we must take into account the presence of the small holes. According to asymptotic homogenization techniques (cf. Bensoussan-Lions-Papanicolaou (ref.1) or Sanchez-Palencia (ref.10)) we search for a two-scale asymptotic expansion in the variables x_1, x_2, by considering

$$y_1 = x_1/\varepsilon \quad ; \quad y_2 = x_2/\varepsilon \tag{3.3}$$

and searching for an <u>inner expansion</u>

$$\left.\begin{aligned} &\underline{u}^\varepsilon(x) = \underline{v}^o(x_1,x_2,y_1,y_2,y_3) + \varepsilon\, \underline{v}^1(x_1,x_2,y_1,y_2,y_3) + \varepsilon^2 \ldots \\ &\text{for } y_1 = x_1/\varepsilon\,,\ y_2 = x_2/\varepsilon),\ \text{with } \underline{v}^j\ \omega\text{-periodic in } y_1, y_2. \end{aligned}\right\} \tag{3.4}$$

where it is to be noticed that the dependence on y_1, y_2 is of a very different nature than the dependence on y_3. The functions $\underline{v}^j$ depend on the inner variable y_3 but not on the outer one x_3 ; oppositely, they depend on x_1, x_2 and also on y_1, y_2, (macro- and micro-variables) ; the former ones describing the influence of the macroscopic data of the problem, the later the influence of the local ω-periodic microstructure, the functions v^j are searched ω-periodic in the variables y_1, y_2, and consequently $\underline{v}^i(x_1, x_2, x_1/\varepsilon, x_2/\varepsilon, y_3)$ are "locally periodic", i.e. for small ε, they take almost the same values on contiguous ω-periods but very different values on two distant periods.

It will be useful considering the functions $\underline{v}^j$ of (3.4) as functions of the variables $x_1, x_2 \in \Sigma$ and $y = (y_1, y_2, y_3)$ belonging to the "period" G

$$G = \{\omega \times]-\infty, +\infty[\} - \overline{\mathcal{H}} \qquad (3.5)$$

i.e. the "infinite prism" formed by the points y such that (y_1, y_2) belongs to the period ω and y_3 takes arbitrary values, less the hole $\mathcal{H}$. We shall say indiscriminately ω-periodic or G-periodic.

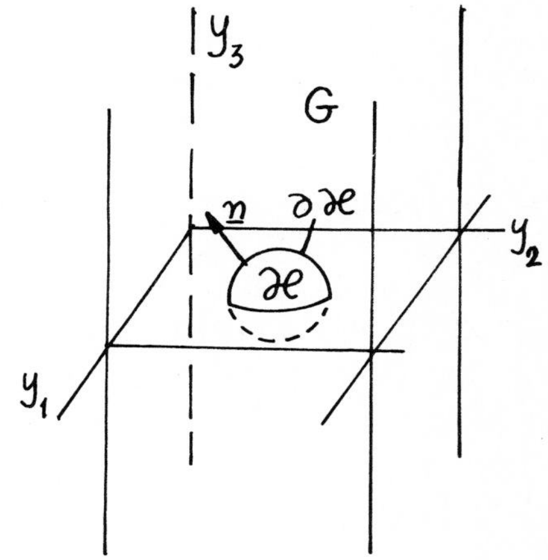

Figure 3

4. - STUDY OF THE FIRST TERMS $\underline{u}^o$, $\underline{v}^o$

From (3.2) and (2.6) we obtain at order 1 :

$$- \frac{\partial}{\partial x_j} \left(a_{ijlm} \, e_{lmx}(\underline{u}^o) \right) = f_i \quad \text{in} \quad \Omega_+ , \Omega_- \qquad (4.1)$$

where e_{lmx} denote the expression of the form (2.5) taking x as variables and y as parameter (the symbol e_{lmy} which will be used later is of course the opposite).

We now write the matching conditions between the boundary layer and the outer expansion at order 1. We recall that the outer (resp. inner) limit is defined as the limit as $\varepsilon \to 0$ with fixed outer variable x_3 (resp. fixed inner variable y_3). The matching condition (ref.13) may be writen :

$$\text{Inner limit of the outer limit} \;=\; \text{Outer limit of the inner limit} \tag{4.2}$$

The outer (resp. inner) limit is $\underline{u}^o$ (resp. $\underline{v}^o$). In order to compute its inner (resp. outer) limit, we write it after replacing $x_3 = \varepsilon\, y_3$ (resp. $y_3 = x_3/\varepsilon$) and (4.2) gives :

$$\underline{u}^o(x_1,x_2,\pm 0) \;=\; \underline{v}^o(x_1,x_2,x_1/\varepsilon,x_2,\pm\infty) \tag{4.3}$$

where ± 0 (resp. $\pm\infty$) denote the limits as x_3 (resp. y_3) tends to 0 (resp. ∞) with positive or negative values. Moreover, we shall see later in the study of $\underline{v}^o$ that it is independent of y, and consequently

$$\underline{u}^o(x_1,x_2,+0) = \underline{u}^o(x_1,x_2,-0) \tag{4.4}$$

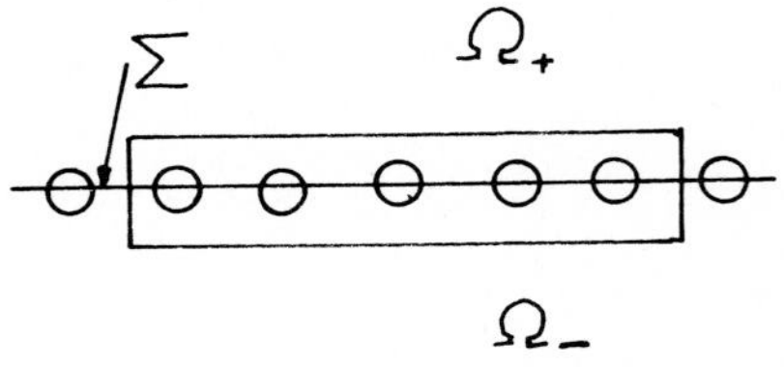

Figure 4

On the other hand, taking a penny-shaped control volume located as in fig. 4 and writing the equilibrium equation at order 1 (neglecting of course the flux across the lateral surface) we easily obtain

$$\sigma_{1i}(\underline{u}^o)\Big|_{x_3=+0} \;=\; \sigma_{1i}(\underline{u}^o)\Big|_{x_3=-0} \quad ; \quad i = 1,2,3. \tag{4.5}$$

But (4.4) and (4.5) show that the jumps of $\underline{u}^o$ and $\sigma_{ij}n_j$ across Σ are zero, and consequently equation (4.1) is satisfied all over Ω and not only in Ω_+ and Ω_- (this is merely an exercise in distribution theory). Of course from (2.7) and (3.2) at order 1 we obtain (2.4) ; <u>the first term $\underline{u}^o$ of the outer expansion (3.2) is the solution of the problem without holes (2.3)-(2.5)</u>.

Now in order to compute $\underline{v}^o$, we search for the equation and boundary conditions for it. We put (3.4) into (2.6) and we obtain at order ε^{-2} :

$$-\frac{\partial}{\partial y_j}\left(a_{ijlm}\, e_{lmy}(\underline{v}^o)\right) \;=\; 0 \qquad \text{in } G \tag{4.6}$$

which is an elliptic equation in the y variables; we then consider x_1,x_2 as parameters. Moreover, from (3.4), (2.8) at order ε^{-1} we have :

$$a_{ijlm}\, e_{lmy}(\underline{v}^o)\, n_j \;=\; 0 \qquad \text{on } \partial\mathcal{H} \tag{4.7}$$

which is the homogeneous Neumann boundary condition on the boundary of the hole. We of course also know that $\underline{v}^o$ is G-periodic in the y variables. This, with the matching conditions (4.3) at infinity suffices to show that $\underline{v}^o$ do not depend on y_1,y_2,y_3. Indeed we multiply (4.6) by v_i^o and we integrate on the part of G with $|y_3| < C$ for some constant C, which will be denoted by G_c. Integrating by parts and taking into account the symmetry properties (2.1) we obtain :

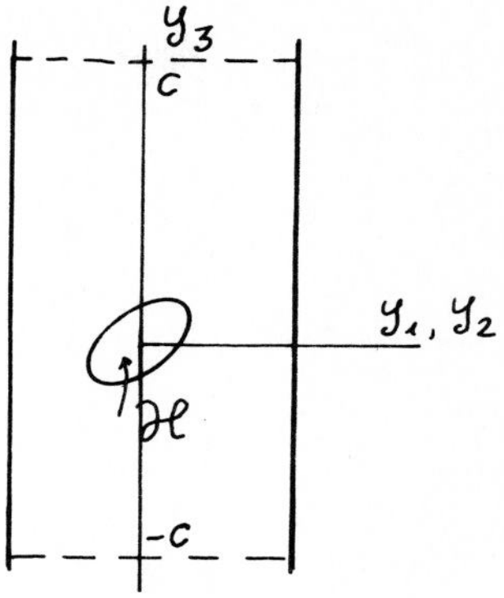

Figure 5

$$0 = - \int_{G_c} \frac{\partial \sigma_{ij}(\underline{v}^o)}{x_j} v_i^o \, dy = - \int_{\partial G_c} \sigma_{ij}(\underline{v}^o) n_j v_i^o \, ds + $$

$$+ \int_{G_c} \sigma_{ij}(\underline{v}^o) e_{ij}(\underline{v}^o) \, dy \qquad (4.8)$$

where we note that the integral over ∂G_c has a zero contribution from $\partial \mathcal{H}$ because of (4.7), and zero also from the lateral boundaries of G by the G-periodicity condition ; then, (4.8) gives :

$$\int_{G_c} a_{ijlm} e_{lm}(\underline{v}^o) e_{ij}(\underline{v}^o) \, dy = \int_{G \cap y_3 = \pm C} a_{ijlm} e_{lm}(\underline{v}^o) n_j v_i^o \, ds \qquad (4.9)$$

but from (4.3) we see that taking the limit as $C \to \infty$, $v_i^o(x_1,x_2,y_1,y_2,\pm C)$ tends to the values $\underline{u}^o(x_1,x_2,\pm 0)$; the values $e_{lm}(\underline{v}^o)$ also tend to zero as $C \to \pm\infty$ (we of course admit that the limits (4.3) hold in a smooth fashion and consequently the gradient tend to zero as $y_3 \to \pm\infty$). As a result, the integral in the left hand side of (4.9) takes the form

$$\int_G a_{ijlm} e_{lm}(\underline{v}^o) e_{ij}(\underline{v}^o) \, dy = 0 \qquad (4.10)$$

and by virtue of (2.2), we have $e_{ij}(\underline{v}^o) = 0$, and consequently $\underline{v}^o$ as a function of y is a solid displacement ; moreover, it is G-periodic, thus $\underline{v}^o = \underline{cte}$ (which of course depend on the parameters x_1,x_2) ; finally (4.3) gives

$$\underline{v}^o(x_1,x_2,y_1,y_2,y_3) \equiv \underline{u}^o(x_1,x_2,0) \qquad (4.11)$$

5. - THE LOCAL PROBLEM IN THE VICINITY OF THE HOLES. STRESS CONCENTRATION

At the present state, the outer and inner expansions are respectively:

$$\left.\begin{aligned} u_i^\varepsilon(x) &\simeq u_i^{\,o}(x_1,x_2,x_3) + \varepsilon\, u_i^{\,1}(x_1,x_2,x_3) + \dots \\ u_i^\varepsilon(x) &\simeq u_i^{\,o}(x_1,x_2,0) + \varepsilon\, v_i^{\,1}(x_1,x_2,y_1,y_2,y_3) + \dots \end{aligned}\right\} \tag{5.1}$$

and it is worthwhile writting the corresponding expansions for the gradients :

$$\left.\begin{aligned} \underline{\text{grad}}\; u_i^\varepsilon(x) &\simeq \underline{\text{grad}}_x\; u_i^{\,o}(x) + \varepsilon\; \underline{\text{grad}}_x\; u_i^{\,1}(x) + \dots \\ \underline{\text{grad}}\; u_i^\varepsilon(x) &\simeq \underline{\text{grad}}_{x_1,x_2}\; u_i^{\,o}(x_1,x_2,0) + \underline{\text{grad}}_y\; v_i^{\,1} + O(\varepsilon) \end{aligned}\right\} \tag{5.2}$$

where the symbol $\underline{\text{grad}}_{x_1x_2}$ expresses the obvious fact that we consider the gradient of the function $u_i^{\,o}(x_1,x_2,0)$ which only depends on the variables x_1,x_2. Then, the matching rule (4.2) applied to the gradients gives (note that the component 3 of the gradient gives an equation different from the others) :

$$\left.\begin{aligned} \frac{\partial u_i^o}{\partial x_\alpha}(x_1,x_2,\pm 0) &= \frac{\partial u_i^o}{\partial x_\alpha}(x_1,x_2,\pm 0) + \lim_{y_3\to\pm\infty} \frac{\partial v_i^1}{\partial y_\alpha}(x_1,x_2,y) \;; \qquad \alpha = 1,2 \\ \frac{\partial u_i^o}{\partial x_3}(x_1,x_2,0) &= \lim_{y_3\to\pm\infty} \frac{\partial v^1}{\partial y_3}(x_1,x_2,y) \end{aligned}\right\} \tag{5.3}$$

and it is evident that some simplifications will appear in the sequel if we consider instead of $\underline{v}^1$ the new unknown :

$$\underline{v}^*(x_1,x_2,y_1,y_2,y_3) \equiv \underline{v}^1(x_1,x_2,y_1,y_2,y_3) - y_3\,\frac{\partial \underline{u}^o}{\partial x_3}(x_1,x_2,0) \tag{5.4}$$

for which the matching (5.3) becomes

$$\lim_{y_3\to\pm\infty} \frac{\partial \underline{v}^*}{\partial y_j}(x_1,x_2,y_1,y_2,y_3) = 0 \tag{5.5}$$

and the inner expansion for the gradient, obtained from the second of (5.2) is

$$\frac{\partial u_i^\varepsilon}{\partial x_j}(x) \simeq \frac{\partial u_i^o}{\partial x_j}(x_1,x_2,0) + \frac{\partial v_i^*}{\partial y_j}(x_1,x_2,y) + O(\varepsilon)\;, \quad j = 1,2,3 \tag{5.6}$$

We now search for the equations and boundary conditions to be satisfied by $\underline{v}^*$. Putting (5.6) into (2.6), we obtain at order ε^{-1} :

$$\frac{\partial}{\partial y_j} \left(a_{ijlm}\, e_{lmy}(\underline{v}^*)\right) = 0 \qquad \text{in } G \tag{5.7}$$

where we note that $\underline{u}^o$ do not appear because the coefficients a_{ijlm} do not depend on the variables y (in fact they are constant). In the same way from (2.8) at order 1 we have

$$n_j\, a_{ijlm} \left(e_{lmx}(\underline{u}^o) + e_{lmy}(\underline{v}^*)\right) = 0 \qquad \text{on } \partial \mathcal{H} \tag{5.8}$$

which may also be written

$$n_j\, \sigma_{ijy}(\underline{v}^*) = - n_j\, \sigma_{ijx}(\underline{u}^o) \qquad \text{on } \partial \mathcal{H} \tag{5.9}$$

that, with the matching condition (5.5) and the G-periodicity condition constitute a boundary value problem for $\underline{v}^*$.

We note that this boundary value problem contains x_1, x_2, as parameters. Moreover, if $\underline{v}^*$ is a solution, $\underline{v}^*$ + any function of x_1, x_2 is also a solution ; consequently, the boundary value problem only gives $\underline{v}^*$ up to an additive constant (which may depend on the "parameters" x_1, x_2). But we note that this do not modify the stress and strain fields at the leading order (i.e. order 1) which are given by the symmetric part of (5.6) and the tensor in the bracket of (5.7), respectively. This is a classical feature of local solutions in homogenization methods, we shall consider $\underline{v}^*$ as a function of y defined up to an additive constant ; we shall see in the next section that (5.5), (5.7), (5.9) effectively define $\underline{v}^*$ up to an additive constant. The non-homogeneous data of the problem are the $\sigma_{ijx}(\underline{u}^o)$ in the right hand side of (5.9) : as the problem is linear we may write

$$\underline{v}^*(x_1, x_2, y) = \sigma_{lmx}(\underline{u}^o)\, \underline{v}^{*lm}(y) \tag{5.10}$$

where the $\underline{v}^{*lm}(y)$ only depend on the variable y ; they are the solutions corresponding to (note that the σ are symmetric tensors) :

$$\sigma_{ijx}(\underline{u}^o) = \frac{1}{2}\left(\delta_{il}\, \delta_{jm} + \delta_{im}\, \delta_{jl}\right) \tag{5.11}$$

Summing up the results of this section (and of the next one), the term $\underline{v}^*(x_1, x_2, y)$ is defined by (5.10) where $\sigma_{lmx}(\underline{u}^o)$ is of course the value of the stress tensor corresponding to the unperturbed solution $\underline{u}^o$, taken at the points $(x_1, x_2, 0)$ of Σ, and the $\underline{v}^{*lm}(y)$ are the solutions of the local G-periodic problem (defined up to an additive constant vector) :

$$- \frac{\partial}{\partial y_j} \left(a_{ijpq} \; e_{pqy} \; (\underline{v}^{*lm})\right) = 0 \qquad \text{in } G \tag{5.12}$$

$$n_j \; a_{ijpq} \; e_{pqy} \; (\underline{v}^{*lm}) = - \frac{1}{2} \left(\delta_{il} \; n_m + \delta_{im} \; n_l\right) \qquad \text{in } \partial \mathcal{H} \tag{5.13}$$

$$\lim_{y_3 \to \pm\infty} \; \text{grad}_y \; \underline{v}^{*lm} \; (y_1, y_2, y_3) = 0 \tag{5.14}$$

Then knowing the six local solutions $\underline{v}^{*lm}$, and the unperturbed solution, we may construct $\underline{v}^{*}(x_1,x_2,y)$. The leading term (of order 1) of the stress tensor in any point of the vicinity of the surface Σ (defined by the coordinates x_1,x_2,y_1,y_2,y_3) is

$$a_{ijpq} \left(e_{pqx} \; (\underline{u}^{o}) + e_{pqy} \; (\underline{v}^{*})\right) \tag{5.15}$$

or equivalently

$$\sigma_{ijx}(\underline{u}^{o}) + \sigma_{lmx}(\underline{u}^{o}) \; a_{ijpq} \; e_{pqy} \; (\underline{v}^{*lm}) \tag{5.16}$$

then the functions $\underline{v}^{*lm}(y)$ furnish the concentration of stress near Σ.

From the form (5.12)-(5.14) we see that $\underline{v}^{*lm}$ is the solution of some elasticity problem in G with G-periodicity condition and some Neumann (i.e. stress-given) non homogeneous condition on $\partial\mathcal{H}$. According to classical theory (see Grisvard (3), (4) for instance), the geometric form of the boundary of the hole $\partial\mathcal{H}$ plays an important role on the smoothness of the solution. If $\partial\mathcal{H}$ is a C^{∞}-class surface, $\underline{v}^{*lm}$ is also of class C^{∞} ; on the other hand, if $\partial\mathcal{H}$ exhibits convex (on the G-side) edges, there is in general a singularity of the stress.

6. - EXISTENCE, UNIQUENESS AND BEHAVIOR AT INFINITY OF THE LOCAL SOLUTIONS

In this section we prove that the local problem in G for the $\underline{v}^{*lm}(y)$ is well posed, i.e. it has a solution which is unique up to an additive constant vector. Moreover, under the additional hypothesis of isotropy of the material, we shall show that the $\underline{v}^{*lm}$ tend very smoothly as $y_3 \to \pm\infty$ to some constant values ; in particular, the corresponding stresses tend smoothly to zero as $y_3 \to \pm\infty$ and on the basis of (5.16) we see that the terms containing $\underline{v}^{*lm}$ are non zero only in a layer of thickness of order $O(\varepsilon)$ in the vicinity of Σ.

Let us defined the space $\dot{V}$ formed by the G-periodic vector fields defined on G (or, by G-periodicity, on R^3 less the holes) taking value zero in in some neighbourhood of $y_3 = \pm\infty$ (the neighbourhood may be different for different vectors). We also consider each vector $\underline{v} \in \dot{V}$ defined up to an arbitrary additive constant vector, and we define on $\dot{V}$:

$$(\underline{v},\underline{w})_V = \int_G e_{ij}(\underline{v})\, e_{ij}(w)\, dy \tag{6.1}$$

which is a scalar product ; indeed, if $\|\underline{v}\|_V^2 \equiv (v,v)_V = 0$, $\underline{v}$ is a solid displacement, and because of the G-periodicity it is a constant vector, i.e. the zero-element of $\dot{V}$. Then, <u>we define the Hilbert space V as the completion of $\dot{V}$ for the norm associated with (6.1)</u>, i.e. the Hilbert space obtained by joining to $\dot{V}$ the ideal elements which are limits of the Cauchy sequences of $\dot{V}$.

It is then easy to obtain a variational formulation of the local problem (5.12)-(5.14). Let $\underline{v}^{*lm}$ be a classical solution of this problem. By multiplying (5.12) by w_i for any given $\underline{w} \in \dot{V}$ and integrating by parts, we have

$$\int_G \sigma_{ij}(\underline{v}^{*lm})\, e_{ij}(\underline{w})\, dy = \int_{\partial G} n_j\, \sigma_{ij}\, (\underline{v}^{*lm})\, w_i\, ds \tag{6.2}$$

with

$$\sigma_{ij}(\underline{v}^{*lm}) \equiv a_{ijlm}\, e_{lm}\, (\underline{v}^{*lm}) \tag{6.3}$$

and we of course interpret the integral on G as the limit of the integral on the part of G bounded by $|y_3| < C$ as $C \to \infty$; the symbol ∂G in (2.6) thus denotes $\partial\mathcal{H}$ and also the "sections of G by $y_3 = \pm\infty$", of course, the integrals on the lateral parts of ∂G cancel by G-periodicity. But, by virtue of (5.14) and the fact that $\underline{w} \in \dot{V}$, the integrals on $y_3 = \pm\infty$ are zero, then with (5.13) we obtain :

$$\int_G a_{ijpq}\, (\underline{v}^{*lm})\, e_{pq}\, (\underline{w})\, dy = \frac{1}{2} \int_{\partial} (w_l\, n_m + w_m\, n_l)\, ds \tag{6.4}$$

where $\underline{n}$ denotes the outer unit normal to $\mathcal{H}$. Conversely, if $\underline{v}^{*lm}$ is an element of V (this is taken as the generalized form of (5.14)) satisfying (6.4) for any $\underline{w} \in \dot{V}$, it is immediately seen that it is a solution of (5.12), (5.13).

Moreover the right hand side of (6.4) is a linear and continuous functional on V. Indeed it takes the same value for any element of the equivalence class of $\underline{w}$ (i.e. for a $\underline{w}$ or $\underline{w} + \underline{c}$), because

$$\int_{\partial\mathcal{H}} n_m \, ds = \int \frac{\partial 1}{\partial y_m} dy = 0.$$

As for the continuity of the functional, it suffices to remark that, taking the restriction of the functions of V to a bounded part of G, the Korn's inequality [2] easily shows that the norm induced by V is equivalent to the norm of $(H^1/R)^3$, and the continuity follows from the trace theorem. As a consequence, we may take in (6.4) $\underline{w} \in \dot{V}$ as well as $\underline{w} \in V$, and we arrive to

Variational formulation of the local problem : find $\underline{v}^{*lm} \in V$ such that (6.4) holds for any test function $\underline{w} \in V$.

On the basis of the Lax-Milgram theorem, we then have :

Proposition 6.1 - The solutions $\underline{v}^{*lm}$ of the local problem exist and are unique as elements of the space V.

Now we emphasize that the matching condition (5.14) was replaced by the weaker condition $\underline{v}^{*lm} \in V$ in order to define the variational formulation. Now we are proving (in the case of an isotropic material) that the solution of the variational problem is in fact a classical solution (unless for the above mentioned regularity of $\partial\mathcal{H}$), and moreover, the matching condition (5.14) is satisfied in the very smooth sense which we explicite in the following proposition :

Propositon 6.2 - Let $\underline{v}$ be the indexes $*pq$ will not be written) the solution of the local problem given by Proposition 6.1. If the material is isotropic, i.e. if the coefficients are

$$a_{ijlm} = 2\,\mu\delta_{il}\,\delta_{jm} + \lambda\,\delta_{ij}\,\delta_{lm} \tag{6.5}$$

Then grad $\underline{v} \to 0$ as $y_3 \to \pm\infty$ in the sense that grad $\underline{v}$ and all its derivatives with respect to y_3 considered as functions of y_3 with values in $L^2(\omega)$ tend to zero faster than any power of $1/y_3$.

In order to prove this proposition we shall use the following lemma, which is proved in [11] or [12] (see also [5,6] for related questions) :

Lemma 6.3 - Let $u(y_1,y_2,y_3)$ a G-periodic function which may be considered, for large y_3 as a temperated distribution with values in $L^2(\omega)$, satisfying :

$$-\Delta u = f \tag{6.6}$$

where $f(y_1,y_2,y_3)$ is a function of y_3 with values in $L^2(\omega)$ tending to zero, with all its derivatives, faster than any power of $1/y_3$. Then, there exist constants α, β such that

$$u = \alpha y_3 + \beta + u^{res} \tag{6.7}$$

where u^{res} tends to zero with all its derivatives, faster than any power of $1/y_3$.

The proof of proposition 6.2 is then easy. For an isotropic material, equation (5.12) is

$$(\lambda + \mu) \frac{\partial}{\partial x_i} (\text{div } \underline{v}) + \mu \Delta v_i = 0 \tag{6.8}$$

and taking the divergence

$$\Delta (\text{div } \underline{v}) = 0 \tag{6.9}$$

But div $\underline{v}$ is G-periodic and from $\underline{v} \in V$ we see that div $\underline{v} \in L^2(G)$; thus it is obviously a temperated distribution of y_3 with values in $L^2(\omega)$ and we may apply lemma 6.3 to it. Consequently, div $\underline{u}$ has an expansion of the form (6.7) where $\alpha = \beta = 0$ as a consequence of div $\underline{u} \in L^2(G)$:

$$\text{div } \underline{u} \equiv \phi \in \mathcal{S}(a, +\infty; L^2(\omega)) \tag{6.10}$$

where the symbol $\mathcal{S}(a,+\infty;L^2(\omega))$ means that for large y_3 it is a function with values in $L^2(\omega)$ which decreases, as well as all its derivatives faster than any power of $1/y_3$.

We shall see that grad (div $\underline{u}$) belongs to $\mathcal{S}$. From (6.10) we obviously have

$$\frac{\partial}{\partial y_3} \text{div } \underline{u} \in \mathcal{S}(a, +\infty ; L^2(\omega)) \tag{6.11}$$

as for the derivatives with respect to y_1, y_2, from (6.9) and (6.10) we have

$$\Delta_\omega (\text{div } \underline{v}) = - \frac{\partial^2}{\partial y_3^2} (\text{div } u) \in \mathcal{S}(a, +\infty ; L^2(\omega)) \tag{6.12}$$

where Δ_ω means the Laplacian with respect to y_1, y_2, with ω-periodicity conditions. We then apply the pseudo-resolvent $(-\Delta_\omega)^{-1}$ to (6.12) noting that 0 is a simple eigenvalue of $-\Delta_\omega$, the pseudo-resolvent sends $L^2(\omega)/R$ (i.e. the subspace of L^2 formed by the functions satisfying the compatibility condition $\int_\omega dy = 0$) into $H^2(\omega)/R$ (i.e. the space of the equivalence classes defined up to an additive constant). Consequently div $\underline{v}$ has first and second derivatives with respect to y_1, y_2 belonging to $\mathcal{S}$; we have expressions analogous to (6.11) for $\partial/\partial y_1$, $\partial/\partial y_2$. From this and (6.8) we see that :

$$\Delta \; v_i = \psi_i \in \mathcal{S}(a, +\infty ; L^2(\omega)) \quad , \quad i = 1,2,3$$

and we may apply lemma 6.3 to each v_i (the v_i are obviously temperated distributions because of $\underline{v} \in V$ and the Korn's inequality in ω) ; it comes :

$$v_i = \alpha_i \, y_3 + \beta_i + v_i^{res} \tag{6.13}$$

and we easily deduce that $\partial u_i/\partial y_j \in \mathcal{S}$. Indeed, from (6.13) for $i = 3$, taking account that $e_{33}(\underline{v}) \in L^2(G)$, we have $\alpha_3 = 0$ and

$$\frac{\partial u_3}{\partial y_3} \in \mathcal{S}(a, +\infty ; L^2(\omega)) \tag{6.14}$$

now, for each u_i :

$$\Delta_\omega \; u_i = - \frac{\partial^2 u_i}{\partial y_3{}^2} + \psi_i \in \mathcal{S}(a, +\infty ; L^2(\omega))$$

and we deduce as before that

$$\frac{\partial u_i}{\partial y_j} \in \mathcal{S}(a, +\infty ; L^2(\omega)) \quad \text{for } i = 1,2,3 \quad , \quad j = 1,2 ; \tag{6.15}$$

we presently consider $\partial u_i/\partial y_3$ for $i = 1,2$; from (6.13), (6.15) and $e_{i3} \in L^2(G)$ we see that $\alpha_1 = \alpha_2 = 0$ and

$$\frac{\partial u_i}{\partial y_3} \in \mathcal{S}(a, +\infty ; L^2(\omega)) \quad , \quad i = 1,2. \tag{6.16}$$

and proposition 6.2 is proven.

7. - COMPLEMENTS

Minor modifications of the preceeding study allow us to consider analogous situations. For instance, if the hole is in fact a fissure F, i.e. it is flatened to become a surface, the only modification to take into account is that the surface $\partial\mathcal{H}$ is in fact twice the surface F, with the corresponding outer unit vectors $\underline{n}$ (Fig. 6).

But, in the cas of a fissure, it is more convenient to consider the non linear problem with the one-side constraint corresponding to the fact that the two lips of the fissure may be open or closed, but they cannot interpenetrate. If there is no friction, the problem may be handled on the basis of the general considerations of (ref.10), p. 106-108. The case with friction is much more difficult, and in fact unsolved in the general case ; see (ref.7) for a particular case. We consider here the case without friction. The consideration of sections 2, 3, 4 hold without modification ; in order to express the one-side cons-

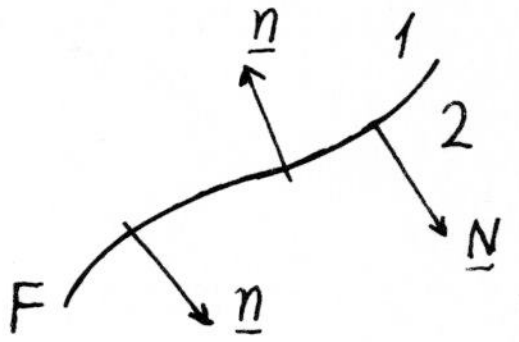

Figure 6

traint, we define (see figure 6) two sides (denoted 1 and 2) of the fissure F and the normal $\underline{N}$ from 1 into 2. As $\underline{u}^o = \underline{u}^o(x)$, the fact that the lips cannot interpenetrate is asymptotically given by the fact that the term $\underline{v}^*$ satisfies

$$v_i N_i|_2 - v_i N_i|_1 \geqq 0 \tag{7.1}$$

Of course the stress conditions on both lips of the fissure are :

$$\sigma_{ij} n_j t_i = 0 \tag{7.2}$$

on both sides, 1 and 2, where $\underline{t}$ is any tangential vector to F, and

$$\sigma_{ij} n_j n_i|_1 = \sigma_{ij} n_j n_i|_2 \leqq 0 \tag{7.3}$$

where the sign $<$ is taken at the points where there is $=$ in (7.1).

We define the closed convex section K of V by

$$K = \{\underline{v} \in V \; ; \; v_i N_i|_2 - v_i N_i|_1 \geqq 0\} \tag{7.4}$$

and the local problem for $\underline{v}^*$ is

Local problem - Find $\underline{v}^* \in K$ such that

$$\int_G \sigma_{ij}(\underline{v}^*)\, e_{ij}(\underline{w} - v^*)\, dy \geqq \int_{\partial\mathcal{H}} n_j\, a_{ijlm}\, e_{lm}(\underline{u}^o)\, e_{ijy}(\underline{w} - \underline{v}^*)\, ds \tag{7.5}$$

for any test function $\underline{w} \in K$.

This defines uniquely $\underline{v}^*$. Out of F, (7.5) is the elasticity equation and Proposition 6.2 still holds. But because of the nonlinearity, decomposition (5.10) is not true.

It is to be noticed that the fact that the surface Σ constitutes a section of the body Ω plays no role in the present problems. We may consider a case as in figure 7, where Σ finish at the interior of Ω. All the preceeding considerations hold true in the situation of figure 7, but of course, special phenomena take place in a cylindrical neighbourhood of the boundary of Σ, i.e. a cylinderlike region as Γ (of course, regions as Γ' are also special and they appear in the case of section 2).

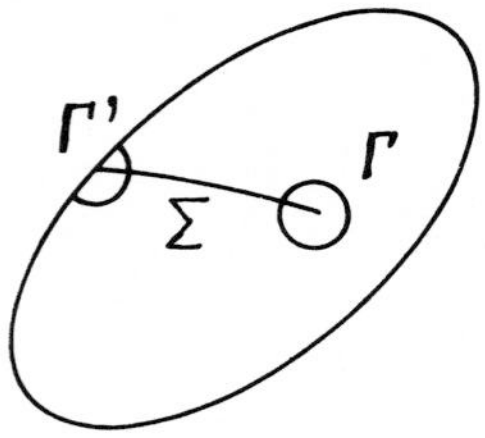

Figure 7

To conclude, we mention that the case where the boundary condition on the holes is of the Dirichlet type :

$$\underline{u}^{\varepsilon} = 0 \qquad \text{on } \partial \varepsilon \mathcal{H} \tag{7.6}$$

instead of (2.8), leads to analogous but different expansions, which may be seen in (ref.8, chap. 4). In particular, the leading term $\underline{u}^{o}(x)$ of the asymptotic expansion take value zero on Σ.

8. - CASE OF A NARROW LAYER BETWEEN TWO DIFFERENT MEDIA

We consider now a problem of practical interest which appears when two bodies of different (or equal) nature are pasted together with a narrow layer of another elastic medium, the "paste". Here the small parameter ε is the thickness of the layer, and the period ω does not exist as the layer is uniform ; in fact we may take any ω as period, and the function v^i must be ω-periodic for any ω ; this amounts to saying that they are constant with respect to y_1, y_2, i.e., they only depend on x_1, x_2, y_3.

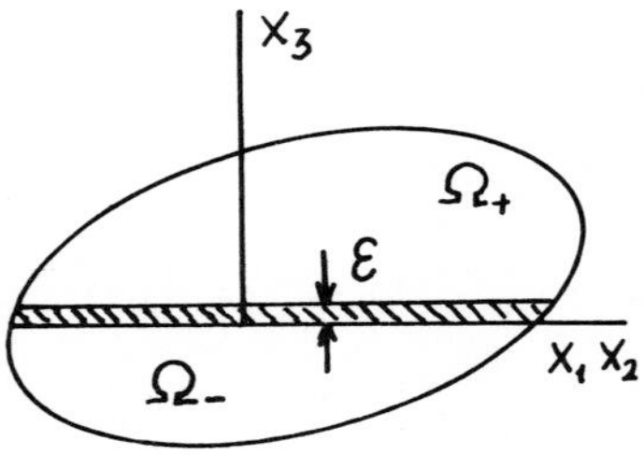

Figure 8

We shall denote by a_{ijlm} with the upper index + - p the elastic coefficients in the three regions Ω^+, Ω^- and in the layer (p as "paste") i.e. for $x_3 \geqq \varepsilon$, $x_3 \leqq 0$, $0 < x_3 < \varepsilon$ or equivalently $y_3 \geqq 1$, $y_3 \leqq 0$, $0 < y_3 < 1$.

The outer expansion is (3.2) as before, and the first term $\underline{u}^o(x)$ is the solution for $\varepsilon = 0$, i.e. the solution of the exact transmission problem (4.1) with the coefficients a_{ijlm} taking the values a^+_{ijlm}, a^-_{ijlm} for $x_3 > 0$ or < 0 respectively. This amounts to solve the elasticity problem with the transmission conditions

$$[\underline{u}^o] = 0 \quad , \quad [\sigma_{i3}] = 0 \quad \text{on} \quad x_3 = 0. \tag{8.1}$$

As for the inner expansion, we have again (3.4) but with the terms independent of y_1, y_2, as we explained above. We obtain as in (4.11) :

$$\underline{v}^o(x_1,x_2) = \underline{u}^o(x_1,x_2,0) \tag{8.2}$$

In order to study the term $\underline{v}^1(x_1,x_2,y_3)$, it will prove useful defining

$$\underline{w}^*(x_1,x_2,y_1,y_2,y_3) \equiv \underline{v}^1(x_1,x_2,y_3) + \frac{\partial \underline{v}^o(x_1,x_2)}{\partial x_\alpha} y_\alpha \tag{8.4}$$

where the Greek index α runs from 1 to 2 (not from 1 to 3), instead of the vector $\underline{v}^*$ of (5.4). Indeed, according to (5.2) we will have

$$\underline{\text{grad}}_x \underline{u}^\varepsilon = \underline{\text{grad}}_y \underline{w}^* + O(\varepsilon) \tag{8.5}$$

and the matching conditions with the outer expansion are :

$$\lim_{y_3 \to \pm\infty} \underline{\text{grad}}_y \underline{w}^* = \underline{\text{grad}}_x \underline{u}^o (x_1,x_2,\pm 0) \tag{8.6}$$

Remark 8.1 - We see that (8.5) and (8.6) are simpler than the corresponding (5.7), (5.3) for $\underline{v}^*$ as we work now with the whole first term of the gradient ; this is now possible because we are obtaining explicit solutions and not using functional spaces as before ; note that (8.6) should imply infinite norms in any suited space.■

The equation satisfied by $\underline{w}^*$, which is obtained as (8.8) is :

$$\frac{\partial \sigma^o_{ij}}{\partial y_j} = 0 \quad ; \quad \sigma^o_{ij} = a_{ijlm}\, e_{lmy}(\underline{w}^*) \tag{8.7}$$

where of course the coefficients a_{ijlm} take the corresponding values with index +-p in the three regions $y_3 \gtrless 1$, $0 < y_3 < 1$. This amounts to the equations (8.7) in each of the three regions with the transmission conditions

$$[\underline{w}^*] = 0 \quad , \quad [\sigma^o_{i3}] = 0 \quad \text{on } y_3 = 0,\ y_3 = 1 \tag{8.8}$$

In these equations and transmission conditions, x_1,x_2 play the role of parameters and the very variables are y_1,y_2,y_3. Moreover, on account of (8.4), equation (8.7) writes

$$\frac{\partial}{\partial y_3}\left(a_{ijlm}\, e_{lm}(\underline{v}^1)\right) = 0$$

Or, on account of the fact that $\underline{v}^1$ is independent of y_1,y_2, we have on each region where the coefficients are constant :

$$0 = a_{i313} \frac{\partial^2 v^1_1}{\partial y_3^2} \Longrightarrow \frac{\partial^2 v^1_1}{\partial y_3^2} = 0 \tag{8.9}$$

as the determinant of the matrix a_{i313} is $\neq 0$ (this is easily seen from the fact that $a_{i313}(\partial u_i/\partial y_3)(\partial u_1/\partial y_3)$ is nothing but the $a_{ijlm}\, e_{lm}\, e_{ij}$ for vectors depending only on y_3). From (8.8), on account of the fact that $\underline{v}^1$ only depends on y_3 (apart from the parameter x_1,x_2) we see that $\underline{v}^1$ is an affine function of y_3 on each region with constant coefficients. We obtain for $\underline{w}^*$ from (8.4) the expression

$$\underline{w}^* = \underline{\beta}(x_1,x_2)y_3 + \underline{\gamma}(x_1,x_2) + \frac{\partial v^o}{\partial x_\alpha} y_\alpha \tag{8.10}$$

where $\underline{\beta}$, $\underline{\gamma}$, are constant (apart from the dependence on the parameters x_1,x_2) in each of the regions $^{+-p}$; we shall write the corresponding indexes if necessary.

We then see that <u>$\text{grad}_y\, \underline{w}^*$ take constant values (apart from x_1,x_2) on each region $^{+-p}$, and the same is true for σ^o_{ij} from (8.7). The problem of computing the stresses in the vicinity of the layer amounts to find the values of $\underline{\beta}^+$, $\underline{\beta}^-$, $\underline{\beta}^p$</u> ; <u>if they are known, from (8.10) we have</u> :

$$\frac{\partial w^*}{\partial y_1} = \frac{\partial v^o}{\partial x_1} \quad ; \quad \frac{\partial w^*}{\partial y_2} = \frac{\partial v^o}{\partial x_2} \quad ; \quad \frac{\partial w^*}{\partial y_3} = \underline{\beta} \tag{8.11}$$

<u>on each of the regions</u> $^{+-p}$ <u>and the corresponding stresses are given by the second of (8.7)</u>.

In order to compute $\underline{\beta}^+$, $\underline{\beta}^-$, $\underline{\beta}^p$, we first use the matching conditions (8.6) which <u>immediately give</u> :

$$\underline{\beta}^{\pm} = \frac{\partial u^o}{\partial x_3}(x_1,x_2,\pm 0) \tag{8.12}$$

Incidentely this shows that the stresses in the regions $y_3 > 1$, $y_3 < 0$ of the layer are exactely the same as the stresses for $\underline{u}^o$ at $(x_1, x_2, \pm 0)$; as the stresses are piecewise constant in the layer, the layer is only relevant for $0 < y_3 < 1$, i.e., $0 < x_3 < \varepsilon$.

To compute $\underline{\beta}^P$ we use the transmission conditions (8.8). We first observe that $[\sigma^o_{ij}] = 0$ involve $\underline{\beta}$ but not $\underline{\gamma}$, if $\underline{\beta}$ is known, the condition $[\underline{w}^*] = 0$ allows us to compute $\underline{\gamma}$, but this is not relevant in order to know the stresses. Consequently, we shall only pay attention to

$$[\sigma^o_{i3}] = 0 \quad \text{on } y_3 = 0, \; y_3 = 1. \tag{8.13}$$

Let us consider for instance $y_3 = 0$; expressing σ^o_{i3} by (8.7), (8.11) we obtain 3 equations for the 3 components of $\underline{\beta}^P$ of the form

$$a_{i3m3} \, \beta^P_m = \text{known terms} \quad , \quad i = 1,2,3 \tag{8.14}$$

and as the determinant of a_{i3m3} is not zero as we pointed out above, this gives in a unique way $\underline{\beta}^P$ and the problem is finished. We remark that, if we take (8.13) for $y_3 = 1$ instead of $y_3 = 0$ we obtain exactely the same equation (8.14), as the outer solution u^o satisfies the second condition (8.1).

As an exercise, we write down (8.14) in the case of isotropic media with Lamé's constants λ^+, μ^+ ; λ^-, μ^- ; λ^P, μ^P :

$$\sigma_{33} = (\lambda + 2\mu)\, e_{33} + \lambda(e_{11} + e_{22})$$

$$\sigma_{23} = 2\,\mu\, e_{23} \quad ; \quad \sigma_{13} = 2\,\mu\, e_{13}$$

and (8.13) becomes :

$$\begin{aligned}
(\lambda^P + 2\,\mu^P)\,\beta^P_3 + \lambda^P\left(\frac{\partial u^o_1}{\partial x_1} + \frac{\partial u^o_2}{\partial x_2}\right) &= (\lambda^- + 2\mu^-)\beta^-_3 + \lambda^-\left(\frac{\partial u^o_1}{\partial x_1} + \frac{\partial u^o_2}{\partial x_2}\right) \\
\mu^P\left(\frac{\partial u^o_3}{\partial x_2} + \beta^P_2\right) &= \beta^-_2\left(\frac{\partial u^o_3}{\partial x_2} + \beta^-_2\right) \\
\mu^P\left(\frac{\partial u^o_3}{\partial x_1} + \beta^P_1\right) &= \mu^-\left(\frac{\partial u^o_3}{\partial x_1} + \beta^-_1\right)
\end{aligned} \tag{8.15}$$

where the derivatives of $\underline{u}^o$ are computed at $(x_1, x_2, -0)$. This system immediately give β^P_1, β^P_2, β^P_3 as β^-_1, β^-_2, β^-_3 are known from (8.12).

REFERENCES :

1 **A. Bensoussan, J.L. Lions, G. Papanicolaou**, "Asymptotic Analysis for Periodic Structures". North-Holland, Amsterdam (1978).

2 **G. Duvaut et J.L. Lions**, "Inéquations variationnelles et applications", Dunod, Paris (1972).

3 **P.Grisvard**, "Boundary value problems in non smooth domains", University of Maryland Lecture Notes (1980).

4 **P. Grisvard**, "Elliptic Boundary Value problems in Non smooth domains", Pitman, London (1984).

5 **E.M. Landis et S.S. Lahturov**, "On the behavior at infinity of the solutions of elliptic equations, periodic in all variables except one", Dokl. Akad. Neuk 250 (1980) pp. 803-806 (= Sov. Math. Dokl. 21 (1980) pp. 211-214).

6 **E.M. Landis et G.P. Panasenko**, "A theorem on the asymptotics of solutions of elliptic equations with coefficients periodic in all variables except one" Dokl. Akad. Nauk 235, n°6 (1977), pp. 1253-1255 (= Sov. Math. Dokl. 18 (1977) n° 4, pp. 1140-1143.

7 **D. Leguillon and E. Sanchez-Palencia**, "On the behavior of a cracked elastic body with (or without) friction" Jour. Méc. Théor. Appl. 1 (1982), pp. 195-209.

8 **G. Nguetseng**, "Sur quelques problèmes de perturbations dans des ouverts périodiques et application à la mécanique des composites". Thèse, Univ. Paris VI (1984).

9 **G. Nguetseng**, "Problèmes d'écrans perforés pour l'équation de Laplace". R.A.I.R.O. Analyse Numérique (to be published, 1985).

10 **E. Sanchez-Palencia**, "Non homogeneous media and vibration theory". Springer, Berlin (1980).

11 **E. Sanchez-Palencia**, "Un problème d'écoulement lent d'un fluide visqueux incompressible à travers une paroi finement perforée" in Ecole d'Eté E.D.F.-C.E.A.-I.N.R.I.A. sur l'Homogénéisation, Eyrolles, Paris (1975).

12 **E. Sanchez-Palencia**, "Problèmes mathématiques liés à l'écoulement d'un fluide visqueux à travers une grille" in Colloque De Giorgi, Pitman, London (1985).

13 **M. Van Dyke**, "Perturbation methods in fluid mechanics". Acad. Press, New-York (1964).

ON THE INFLUENCE OF FREE EDGES IN PLATES AND SHELLS

A.M.A. VAN DER HEIJDEN
Laboratory of Applied Mechanics, Department of Mechanical Engineering, Delft University of Technology, Delft, The Netherlands.

ABSTRACT

The classical theory of shells yields, in general, shear stresses at the nominally free edge of a shell which are of the same order of magnitude as the maximum stresses in the shell. This is due to the contracted boundary condition for the transverse shear force at the edge. As a consequence the classical theory of shells overestimates the actual energy in the shell. It turns out to be possible to give a simple correction for the energy, and once that correction is known, refined solutions can be determined for the stresses and displacements sufficiently far away from the edge of the shell and thence for its overall properties.

Furthermore a simple formula will be derived for the normal-stress in the corner points at the edges, which yields an improved expression for the stress concentration factor at the free edge of a shell.

1. INTRODUCTION

Although the problem of determining the stresses and displacements in a thin elastic shell leads in principle to a boundary value problem in the three-dimensional theory of elasticity, approximate solutions obtained from the (two-dimensional) theory of shells are, fortunately, often adequate for practical purposes. The simplest type of shell theory is based on the Kirchhoff-Love hypothesis which states that: i) points on the normals to the undeformed middle surface remain on the normal to the deformed middle surface, ii) changes in length of these normals may be neglected, and iii) the state of stress is approximately plane and parallel to the middle surface. This theory of shells, which does not take into account transverse shear deformation, will be referred to as the classical theory of shells.

Near places where the geometry and/or the loads vary rapidly the Kirchhoff-Love assumptions are violated, and large errors may occur. This happens for instance near the edges of the shell. In practice we hardly ever know more about the edge stresses than the resulting force or moment per unit edge length on the middle surface so that it is usually not meaningful to try and improve the solutions near the edge. A significant exception is, however, the case of a free edge, where we do know the edge tractions exactly. In this paper we shall analyse the effect of free edges. It will be shown that more accurate solutions for the interior of the shell may be obtained by modifying the boundary conditions. Furthermore we shall derive a formula to determine the normal stress in corner points at the edge with an improved accuracy (compared to the solution of classical shell theory), and a simple formula for improved values of the eigenfrequencies of a shell with free edges.

Apart from the approach to be discussed in this paper various other more powerful (but also more complicated) tools do exist for consistent refinements of shell theory. We mention here the asymptotic integration of the equations of the linear theory of elasticity (c.f. e.g. [13, 28, 29]) and the variational perturbation approach [10].

A completely different approach is Reissner's theory of plates [20], which takes into account transverse shear deformation. This theory, which is not consistent in the asymptotic sense, is good example of an engineering approach [12, 21].

2. CLASSICAL SHELL THEORY

In this chapter we shall briefly review the linear theory of thin shells in the form discussed in e.g. [1, 2]. We shall follow the so-called direct approach, which means that the shell is, ab initio, a two-dimensional continuum subject to certain physical principles.

Let us now first describe the geometry of the shell. Let $\underset{\sim}{r}(x^\alpha)$, $\alpha = 1, 2$, denote the radius vector from a fixed point in space to generic point on the middle surface of the undeformed shell as a vector valued function of the pair of Gaussian coordinates x^α. The tangent plane to the surface in that point is determined by the tangential base vectors $\underset{\sim}{a}_\alpha = \underset{\sim}{r}_{,a}$, where the comma preceding the subscript α denotes partial differentiation with respect to the coordinate x^α. The reciprocal base is defined by $\underset{\sim}{a}_\alpha \cdot \underset{\sim}{a}^\beta = \delta^\beta_\alpha$. The covariant (components of the) metric tensor are $a_{\alpha\beta} = \underset{\sim}{a}_\alpha \cdot \underset{\sim}{a}_\beta$ and the contravariant components are $a^{\alpha\beta} = \underset{\sim}{a}^\alpha \cdot \underset{\sim}{a}^\beta$. In the sequal we shall for the sake of brevity refer to "the components of a tensor" as "a tensor". The tensors $a_{\alpha\beta}$ and $a^{\alpha\beta}$ are employed in raising and lowering the indices of surface tensors. The determinant of the covariant metric tensor is denoted by a, the covariant alternating tensor by $\varepsilon_{\alpha\beta} = \sqrt{a}\, e_{\alpha\beta}$ where $e_{12} = -e_{21} = 1$, $e_{11} = e_{22} = 0$. The normal to the middle surface is defined by $\underset{\sim}{n} = \frac{1}{2} \varepsilon^{\alpha\beta} \underset{\sim}{a}_\alpha \times \underset{\sim}{a}_\beta$. The second fundamental tensor is specified by $b_{\alpha\beta} = \underset{\sim}{n} \cdot \underset{\sim}{r}_{,\alpha\beta}$, and the third one by $c_{\alpha\beta} = b^\kappa_\alpha b_{\kappa\beta}$. Covariant surface differentiation with respect to a coordinate x^α is denoted by an additional subscript α preceeded by a vertical stroke. All the derivatives in the analysis are assumed to be continuous.

A point in shell space is identified by its distance z to the middle surface and by the surface coordinates of its projection on the middle surface i.e. $\underset{\sim}{R}(x^\alpha, z) = \underset{\sim}{r}(x^\alpha) + z\underset{\sim}{n}$. The shell faces $z = \pm \frac{1}{2}h$, where h is the constant shell thickness, are surfaces parallel to the middle surface.

An edge of the shell is assumed to be a ruled surface formed by normals to the middle surface along an edge curve on this surface. Let $\underset{\sim}{\nu}$ be the unit vector in the tangent plane to the middle surface at the edge, normal to the edge curve and directed outwards. The positive sense on the edge curve is defined by the tangential unit vector $\underset{\sim}{t} = \underset{\sim}{n} \times \underset{\sim}{\nu}$. The normal curvature $\kappa_{(\nu)}$, the geodetic torsion $\kappa_{(t)}$ and the geodetic curvature $\kappa_{(n)}$ of the edge curve are introduced by the relations

$$\underset{\sim}{\nu}_{,s} = \kappa_{(n)}\underset{\sim}{t} - \kappa_{(t)}\underset{\sim}{n}, \quad \underset{\sim}{t}_{,s} = \kappa_{(\nu)}\underset{\sim}{n} - \kappa_{(n)}\underset{\sim}{\nu}$$
$$\underset{\sim}{n}_{,s} = \kappa_{(t)}\underset{\sim}{\nu} - \kappa_{(\nu)}\underset{\sim}{t}, \quad \kappa_{(\nu)} = b_{\alpha\beta}t^{\alpha}t^{\beta}, \quad \kappa_{(t)} = -b_{\alpha\beta}t^{\alpha}\nu^{\beta} \tag{2.1}$$

where s preceded by a comma denotes the partial derivative with respect to the arc length along the edge curve. A deformation of the middle surface is described the two-dimensional displacement field

$$\underset{\sim}{u}(x^{\kappa}) = u_{\alpha}(x^{\kappa})\underset{\sim}{a}^{\alpha} + w(x^{\kappa})\underset{\sim}{n}. \tag{2.2}$$

Let $\bar{a}_{\alpha\beta}$ and $\bar{b}_{\alpha\beta}$ be the covariant components of the first and second fundamental tensors in the deformed state, then $\bar{a}_{\alpha\beta} - a_{\alpha\beta}$ and $\bar{b}_{\alpha\beta} - b_{\alpha\beta}$ completely specify the deformation of the middle surface. We shall therefore employ the strain measures

$$\gamma_{\alpha\beta} = \frac{1}{2}(\bar{a}_{\alpha\beta} - a_{\alpha\beta}) = \frac{1}{2}(u_{\alpha|\beta} + u_{\beta|\alpha}) - b_{\alpha\beta}w \tag{2.3}$$

$$\rho_{\alpha\beta} = \bar{b}_{\alpha\beta} - b_{\alpha\beta} = \phi_{\alpha|\beta} - c_{\alpha\beta}w + b^{\kappa}_{\beta}\, u_{\kappa|\alpha}\,, \quad *) \tag{2.4}$$

where

$$\phi_{\alpha} = w_{,\alpha} + b^{\kappa}_{\alpha}\, u_{\kappa} \tag{2.5}$$

is a component of the rotation vector

$$\underset{\sim}{\phi} = \varepsilon^{\alpha\beta}\phi_{\beta}\underset{\sim}{a}_{\alpha} + \Omega\underset{\sim}{n}, \quad \Omega = \frac{1}{2}\varepsilon^{\alpha\beta}u_{\beta|\alpha}\,. \tag{2.6}$$

We shall refer to $\gamma_{\alpha\beta}$ as the strain tensor, and to $\rho_{\alpha\beta}$ as the tensor of changes of curvature.

External loads on the shell faces are reduced to statically equivalent loads

---

*) In [1, 2] $\rho_{\alpha\beta}$ is denoted by $\bar{\rho}_{\alpha\beta}$.

on the middle surface. It is assumed that this reduction does not introduce surface couples. Furthermore we shall omit body forces. The reduced tangential loads per unit area of the middle surface are described by a surface vector p^{α}, and the reduced normal load per unit area is a surface invariant p^3. The loads acting on the edge of the shell are likewise reduced to statically equivalent line loads along the edge curve on the middle surface i.e. a force $\underset{\sim}{N}$ and a couple $\underset{\sim}{M}$, both per unit arc length. These quantities are written as

$$\begin{aligned} \underset{\sim}{N} &= N_{(\nu)}\underset{\sim}{\nu} + N_{(t)}\underset{\sim}{t} + Q\underset{\sim}{n} = N^{\alpha}\underset{\sim}{a}_{\alpha} + Q\underset{\sim}{n} \\ \underset{\sim}{M} &= M_{(\nu)}\,\nu + M_{(t)}\underset{\sim}{t} = \varepsilon_{\alpha\beta}M^{\beta}\underset{\sim}{a}^{\alpha}\,. \end{aligned} \tag{2.7}$$

Notice that $\underset{\sim}{M}$ is a vector in the tangent plane. The physical significance of $N_{(\nu)}$, $N_{(t)}$ and Q is that they are respectively the normal, tangential and transverse shear loads per unit length, whereas $M_{(\nu)}$ and $M_{(t)}$ represent the torsional and the bending moments per unit length.

The principle of virtual work now yields

$$\begin{aligned} \int_S (n^{\alpha\beta}\delta\gamma_{\alpha\beta} + m^{\alpha\beta}\delta\rho_{\alpha\beta})\,dS &= \int_S (p^{\alpha}\delta u_{\alpha} + p^3\delta w)\,dS + \\ &+ \int_{\partial S} (N^{\alpha}\delta u_{\alpha} + Q\delta w + M^{\alpha}\delta\phi_{\alpha})\,ds \end{aligned} \tag{2.8}$$

where $n^{\alpha\beta}$ and $m^{\alpha\beta}$ are symmetric tensors of the stress resultants and stress couples respectively. Using the set $\{\delta u_{\alpha},\ \delta w,\ \delta\phi_n\}$ as independent variables along the edge, where we have written $\phi_n = \phi_{\alpha}\,\nu^{\alpha}$, we obtain by standard methods (c.f. e.g. [5]) the equations of equilibrium

$$\begin{aligned} (n^{\beta\alpha} + b^{\alpha}_{\kappa}m^{\beta\kappa})|_{\beta} + b^{\alpha}_{\kappa}m^{\beta\kappa}|_{\beta} + b^{\alpha}_{\kappa}m^{\beta\kappa}|_{\beta} + p^{\alpha} &= 0 \\ -\,m^{\alpha\beta}|_{\alpha\beta} + c_{\alpha\beta}m^{\alpha\beta} + b_{\alpha\beta}n^{\alpha\beta} + p^3 &= 0 \end{aligned} \tag{2.9}$$

and the dynamic boundary conditions

$$\begin{aligned} [n^{\alpha\beta} + m^{\beta\kappa}\,(b^{\alpha}_{\lambda}t_{\kappa}t^{\lambda} + b^{\alpha}_{\kappa})]\,\nu_{\beta}\,\nu_{\alpha} &= N_{(\nu)} - \kappa_{(t)}\,M_{(\nu)} \\ [n^{\alpha\beta} + m^{\beta\kappa}\,(b^{\alpha}_{\lambda}t_{\kappa}t^{\lambda} + b^{\alpha}_{\kappa})]\,\nu_{\beta}\,t_{\alpha} &= N_{(t)} + \kappa_{(\nu)}\,M_{(\nu)} \\ (m^{\alpha\beta}\nu_{\beta}t_{\alpha})_{,s} + m^{\alpha\beta}|_{\beta}\,\nu_{\alpha} &= (M_{(\nu)})_{,s} - Q \\ m^{\alpha\beta}\,\nu_{\alpha}\,\nu_{\beta} &= -\,M_{(t)}\,. \end{aligned} \tag{2.10}$$

We note here that both the equilibrium equations and the boundary conditions are fully exact in the sense that they ensure the overall equilibrium of any shell element of finite thickness h, bounded by the shell faces and by the ruled surface described by the normals to the middle surface through a closed curve of infinitesimal length on this surface. The approximate nature of shell theory enters the picture at the stage where the constitutive equations are introduced between the stress resultants and stress couples $n^{\alpha\beta}$, $m^{\alpha\beta}$ on the one hand and the middle surface deformations $\gamma_{\alpha\beta}$, $\rho_{\alpha\beta}$ on the other hand. Starting from the assumption of an approximately plane state of stress it can be shown [2] that the approximate expression for the strain energy per unit area of the middle surface is given by

$$V(\gamma, \rho) = \frac{Eh}{2(1-\nu^2)} \left[(1-\nu)\, \gamma^\alpha_\beta\, \gamma^\beta_\alpha + \nu\, \gamma^\alpha_\alpha\, \gamma^\beta_\beta + \frac{h^2}{12} \{ (1-\nu)\, \rho^\alpha_\beta\, \rho^\beta_\alpha + \nu\, \rho^\alpha_\alpha\, \rho^\beta_\beta \} \right] \tag{2.11}$$

where E denotes Young's modulus and ν is Poisson's ratio. The relative error this energy is of order $\varepsilon^2 = h^2/L^2 + h/R$ where L is the minimum "wave length" of the deformation pattern on the middle surface, and R is the minimum principal radius of curvature of the middle surface. Since the internal virtual work equals the increment of the stored elastic energy we readily obtain the constitutive equations

$$\begin{aligned} n^{\alpha\beta} &= \frac{Eh}{1-\nu^2} [(1-\nu)\, \gamma^{\alpha\beta} + \nu\, a^{\alpha\beta}\, \gamma^\kappa_\kappa] \\ m^{\alpha\beta} &= \frac{Eh^3}{12(1-\nu^2)} [(1-\nu)\, \rho^{\alpha\beta} + \nu\, a^{\alpha\beta}\, \rho^\kappa_\kappa] \end{aligned} \tag{2.12}$$

which can be shown to also contain relative errors of order ε^2(c.f. [1]). Furthermore it can be proven that the *solutions* of the shell problem contain relative errors of the same order, provided the edge tractions are in agreement with the requirements of shell theory, and that the surface loads are sufficiently smooth [3, 4]. In general, however, near edges, and near places where the load and/or geometry change rapidly, the assumption of plane stress does not hold, and in the vicinity of those places the constitutive equations (2.12) will not hold, not even approximately. The local errors in and near those places may be of order unity whereas the relative errors sufficiently far away from these places are the of order $\varepsilon^* = h/L + h/R$, c.f. [1]. In practice we hardly ever know the exact distribution of tractions or displacements over the thickness, but they will, in general, not comply with the requirements of shell theory, so that the actual relative error is of order ε^* instead of of order ε^2. The only case where we know the exact distribution of the tractions is in the case of a *free edge*. We note here already that although in that case the right-hand members in (2.10) are specified to be zero, this does not imply that in the third boundary condition both $(m^{\alpha\beta}\, \nu_\beta\, t_\alpha)_{,s}$ and $m^{\alpha\beta}|_\beta\, \nu_\alpha$ are equal to zero. This means that, in general, the internal twisting couple $m^{\alpha\beta}\, \nu_\beta\, t_\alpha$ at the edge is nonzero so that the tangential edge tractions are of the same order of magnitude as in the interior shell domain.

3.1. Analysis of a twisted strip

In order to get a better understanding of the state of stress near a free edge we shall first analyse the elementary problem of twisting of a strip with a rectangular cross-section with thickness h and width b. The exact solution to this problem is well known (c.f. e.g. [6]), and for $h \ll b$ the torsional rigidity is given by

$$S_t = \frac{1}{3} Gbh^3 \left[1 - B\frac{h}{b} + O\left(\frac{h}{b} e^{-\pi b/2h}\right)\right] \tag{3.1}$$

where G is the shear modulus and B is a numerical constant given by

$$B = \frac{192}{\pi^5} \sum_{k=1}^{\infty} (2k-1)^{-5} \approx 0{,}630. \tag{3.2}$$

Introducing a cartesian coordinate system (x, y, z) such that the cross section of the strip is given by $-b/2 \le x \le b/2$, $-h/2 \le z \le h/2$, the nonvanishing stress components are given by

$$\begin{aligned}\tau_{xy} &= G\omega h \left[-\frac{2z}{h} + \frac{8}{\pi^2} \sum_{k=0}^{\infty} \frac{(-1)^k}{(2k+1)} \exp\left\{-(2k+1)\frac{\pi}{h}\left||x| - \frac{b}{2}\right|\right\} \sin(2k+1)\frac{\pi z}{h} \right.\\ &\qquad \left. + O\left(\exp\left(-\frac{\pi b}{2h}\right)\right)\right]\\ \tau_{yz} &= G\omega h \left[\frac{8}{\pi^2} \sum_{k=0}^{\infty} \frac{(-1)^k}{(2k+1)^2} \operatorname{sign}(x) \exp\left\{-(2k+1)\frac{\pi}{h}\left||x| - \frac{b}{2}\right|\right\} \cos(2k+1)\frac{\pi z}{h} \right.\\ &\qquad \left. + O\left(\exp\left(-\frac{\pi b}{2h}\right)\right)\right],\end{aligned} \tag{3.3}$$

where ω is the specific twist in uniform torsion of the strip. The first term in (3.1) corresponds to the torsional rigidity calculated by classical plate theory. The second term may be interpreted as the reduction of the torsional rigidity due to the removal of the shear stresses $\tau_{xy} = -2G\omega z$ along the nominally free edges. This means that the elastic energy in the strip is reduced by $\frac{1}{12} BGh^4\omega^2$ per unit length along each edge. In other words we obtain an improved expression for the elastic energy in the strip by subtracting from the elastic energy corresponding to classical plate theory line integrals

$$\frac{1}{12} \int_{\partial S_f} BGh^4\omega^2 \, ds \tag{3.4}$$

along each free edge. We shall in a later section generalize this result to curved edges and shells.

Let us now remove the shear stress along the free edges by applying tractions of opposite sign. Since the correction terms in (3.3) have only contributions near the edges, it is sufficient to consider a semi-infinite plate loaded by the tractions mentioned above and to look for solutions which decay rapidly with increasing distance from the edge. Let the domain of the plate be $y \in (-\infty, \infty)$ $x \in (-\infty, 0]$, $z \in [-h/2, h/2]$, then the resulting equation is

$$\frac{\partial^2 \phi}{\partial x^2} + \frac{\partial^2 \phi}{\partial z^2} = 0 \qquad (3.5)$$

to be solved under the boundary conditions

$$\tau_{yz} = G\omega h \frac{\partial^2 \phi}{\partial z^2} = 0 \quad \text{for} \quad z = \pm h/2$$
$$\tau_{xy} = G\omega h \frac{\partial^2 \phi}{\partial x^2} = 2G\omega z \quad \text{for} \quad x = 0 \qquad (3.6)$$

and one easily shows that the solution to this "torsion problem" is given by

$$\phi = \frac{8}{\pi 4} \sum_{n=0}^{\infty} \frac{(-1)^n}{(2n+1)^4} \exp\{(2n+1)\,\pi\,\frac{x}{h}\}\ \sin(2n+1)\,\pi\,\frac{z}{h} \qquad (3.7)$$

and one easily verifies that this solution is in full agreement with the additional terms in (3.3). The displacement field is given by

$$v = \omega h^2 \phi_{,x}\,, \qquad u = w = 0\,, \qquad (3.8)$$

which shows that near the edges the displacement distribution over the thickness is *nonlinear*. Notice that the domain of (3.5) and (3.6) is a semi-infinite strip which is perpendicular to both the middle surface of the plate and its edge.

Summarizing our results we find that corrections of order unity for the shear stresses near the edge may be obtained by simply adding the solution of the torsion problem to the results of classical plate theory. These results can easily be generalized to cases with curved edges and slowly varying shear stress distributions at the edge.

In the theory of plates and shells it is assumed that the edges form a ruled surface described by the normals to the middle surface. In practice however it often occurs that the edge makes a small angle with respect to the normals to the middle surface. To investigate the influence of non-square free edges we considered the torsion with an isosceles trapezoidal cross section, such that $|z| \leq h/2$, $|x| < b/2 - z\tan\alpha$, where $\alpha << 1$. In [7] this problem was solved by perturbation methods, and for $h << b$ the result for the torsional rigidy is given by

$$S_t = \frac{1}{3} Gbh^3 \left[1 - \frac{h}{b} (B + 0{,}434 \tan^2\alpha) + O\left(\frac{h}{b} \{ \exp(-\frac{\pi b}{2h}) + \tan^4\alpha\}\right)\right] \tag{3.9}$$

in agreement with the result obtained in [8]. It follows that the contribution to the elastic energy of the correction term with $\tan^2\alpha$ can usually be neglected. The corrections to the shear stresses are of relative order α, and are not affected when it is assumed that α is a sufficiently smooth periodic function of y [9]. In that case the normal stresses are of order $\alpha h/b$, and are beyond the scope of refinement we have in mind. Summarizing the results of this section we see that the most important correction to the energy is given by (3.4). This correction leads to a modification of the overall properties. Local effects near the edges can be taken into account by adding the stress and displacement distributions obtained from (3.7).

4. MODIFIED BOUNDARY CONDITIONS FOR FREE EDGES

Let us now generalize the results obtained in the previous section to shells. We consider a thin shell, i.e. $\varepsilon^* << 1$, with smooth edges having a small curvature ($h\,\kappa_{(n)} << 1$). Furthermore we assume that the loading is such that the edge twisting moment along the free edge (calculated by classical shell theory) varies slowly along the edge. Then the most important effect of the removal of the twisting moment along the nominally free edge will again be a reduction of the torsional strain energy by an amount of $\frac{1}{12} BGh^4\omega^2$ per unit length of the edge, where ω is now the local, slowly varying twist at the edge, and is defined by

$$\omega = \rho_{\alpha\beta} \nu^\alpha t^\beta = (\rho_{\alpha\beta} - c_{\alpha\beta} w + b^\kappa_\beta u_{\kappa|\alpha}) \nu^\alpha t^\beta . \tag{4.1}$$

The virtual work equation (2.8) now reads in its modified form

$$\int_S (n^{\alpha\beta} \delta\gamma_{\alpha\beta} + m^{\alpha\beta} \delta\rho_{\alpha\beta}) \, dS - \int_{\partial S_f} Bhm\delta\omega ds =$$
$$\int_S (p^\alpha \delta u_\alpha + p^3 \delta w) \, dS + \int_{\partial S_L} (N^\alpha \delta u_\alpha + Q\delta w + M^\alpha \delta\phi_\alpha) \, ds \tag{4.2}$$

where m is the twisting couple defined by the classical relation of shell theory

$$m = \frac{1}{6} Gh^3\omega \tag{4.3}$$

and ∂s_f and ∂S_L are the arc lengths along the free and loaded edges respectively. It is assumed that the edge loads are applied in accordance with the requirements of shell theory. The formulation (4.2), however, has a minor drawback, namely by subtracting a line integral of the form (3.2) from (2.11) the resulting functional is not convex, which means that the corresponding solution of the boundary value problem is not unique [10]. We must however keep in mind, that the line integral along ∂s_f is of order h/L times the surface integral. A more appropriate formulation is obtained as follows: Introducing the notation $q = q_c + (h/L)\, q_1 + \ldots$, where q stands for any of the quantities, and collecting terms with the same powers of h/L we obtain

$$\int_S (n_c^{\alpha\beta}\, \delta\gamma_{c\alpha\beta} + m_c^{\alpha\beta}\, \delta\rho_{c\alpha\beta})\, dS =$$

$$\int_S (p^\alpha \delta u_{c\alpha} + p^3\, \delta w_c)\, dS + \int_{\partial s_L} (N^\alpha \delta u_{c\alpha} + Q\delta w_c + M^\alpha \delta\phi_{c\alpha})\, ds \tag{4.4}$$

and

$$\int_S (n_1^{\alpha\beta}\, \delta\gamma_{c\alpha\beta} + m_1^{\alpha\beta}\, \delta\rho_{c\alpha\beta})\, dS = \int_{\partial S_f} B\, h\, m_c\, \delta\, \omega_c\, ds. \tag{4.5}$$

These are two classical shell problems. Solving the boundary value problem corresponding to (4.4) we obtain m_c. The right-hand side of (4.5) may be interpreted as the virtual work of fictitious edge loads. Let us now consider this equation in more detail. Deleting the extra subscripts the right-hand side of (4.5) is in the case of one free edge

$$\oint Bhm\, (\delta\phi_{\alpha|\beta} - c_{\alpha\beta}\delta w + b_\beta^\kappa \delta u_{\kappa|\alpha})\, \nu^\alpha t^\beta\, ds \tag{4.6}$$

evaluated along the closed smooth edge curve of the free edge. In the case of more free edges we simply have to sum the contributions of each free edge. As mentioned in section 2, a set of proper independent variables along the edge is $\{\delta u_\alpha, \delta w, \delta\phi_n\}$. The last term in (4.6), however, contains normal derivatives of the tangential virtual displacement components, which cannot be removed by partial integration, and thence cannot be expressed in that set of independent variables along the edge. This last term may now be rewritten as

$$b_\beta^\kappa\, \delta u_{\kappa|\alpha}\, \nu^\alpha t^\beta = -\,\kappa_{(t)} \delta u_{\kappa|\alpha}\, \nu^\kappa \nu^\alpha + \kappa_{(\nu)}\, \delta u_{\kappa|\alpha}\, t^\kappa \nu^\alpha. \tag{4.7}$$

As shown in [2] the tensor of changes of curvature may be modified by adding terms of the order of magnitude of the product of a middle surface strain and a shell curvature, without affecting the accuracy of the elastic energy in the shell. Guided by the structure of (4.7) we now try to remove these normal derivatives by modifying ω as follows

$$\omega^* = \omega + c_1\, \kappa_{(t)}\, \gamma_{\alpha\beta}\, \nu^\alpha \nu^\beta + c_2\, \kappa_{(\nu)}\, \gamma_{\alpha\beta}\, \nu^\alpha t^\beta \tag{4.8}$$

where c_1 and c_2 are unknown numbers of order unity. This modification would be permissible for the entire shell, and is thus certainly permissible in the line integral. After some simple algebra we find that $c_1 = 1$, $c_2 = -2$, so that the modified edge twist is given by

$$\omega^* = \omega + \kappa_{(t)}\, \gamma_{\alpha\beta}\, \nu^\alpha \nu^\beta - 2\, \kappa_{(\nu)}\, \gamma_{\alpha\beta}\, \nu^\alpha t^\beta . \tag{4.9}$$

This expression may after some lengthy algebra be rewritten as

$$\omega^* = \phi_{\alpha|\beta}\, \nu^\alpha t^\beta - \kappa_{(\nu)}\kappa_{(t)}\, w - \kappa_{(\nu)}(u_\alpha \nu^\alpha),_s + \kappa_{(\nu)}\kappa_{(n)}\, u_\alpha\, t^\alpha . \tag{4.10}$$

Some useful formulae are

$$\begin{aligned} a_{\alpha|\beta} &= b_{\alpha\beta}\, \underset{\sim}{n}\, , \quad b^\kappa_\beta\, t^\beta = \kappa_{(\nu)}\, t^\kappa - \kappa_{(t)}\, \nu^\kappa \\ \nu^\alpha\big|_\beta\, t^\beta &= \kappa_{(n)}\, t^\alpha, \quad a^{\alpha\beta} = \nu^\alpha \nu^\beta + t^\alpha t^\beta . \end{aligned} \tag{4.11}$$

Replacing the line integral (4.6) by

$$\oint Bhm^* \delta\omega^* ds \tag{4.12}$$

where $m^* = \frac{1}{6}\, Gh^3 \omega^*$ we now obtain by standard methods the equilibrium equations

$$\begin{aligned} (n^{\beta\alpha} + b^\alpha_\kappa\, m^{\beta\kappa})\big|_\beta + b^\alpha_\kappa\, m^{\beta\kappa}\big|_\beta &= 0 \\ - m^{\alpha\beta}\big|_{\alpha\beta} + c_{\alpha\beta}\, m^{\alpha\beta} + b_{\alpha\beta}\, n^{\alpha\beta} &= 0 \end{aligned} \tag{4.13}$$

and the modified boundary conditions

$$\begin{aligned} &[n^{\alpha\beta} + m^{\beta\kappa}\, (b^\alpha_\lambda\, t_\kappa t^\lambda + b^\alpha_\kappa)]\, \nu_\beta \nu_\alpha = Bh\, [(m^* \kappa_{(\nu)}),_s + m^* \kappa_{(n)} \kappa_{(t)}] \\ &[n^{\alpha\beta} + m^{\beta\kappa}\, (b^\alpha_\lambda\, t_\kappa t^\lambda + b^\alpha_\kappa)]\, \nu_\beta t_\alpha = 0 \\ &(m^{\alpha\beta}\, \nu_\beta\, t_\alpha),_s + m^{\alpha\beta}\big|_\beta\, \nu_\alpha = -\, Bh\, [(m^* \kappa_{(n)}),_s - m^* \kappa_{(\nu)} \kappa_{(t)}] \\ &m^{\alpha\beta}\, \nu_\alpha\, \nu_\beta = -\, Bhm^*,_s \;{}^{1)} \end{aligned} \tag{4.14}$$

The right-hand sides of these boundary conditions may now be viewed as fictitious edge loads given by

---

1) For a discussion of alternate forms of these boundary conditions we refer to [5].

$$\begin{aligned} N^*_{(\nu)} &= B\,h\,[(m^*\kappa_{(\nu)})_{,s} + m^*\kappa_{(n)}\,\kappa_{(t)}] \\ N^*_{(t)} &= 0 \\ Q^* &= B\,h\,[(m^*\kappa_{(n)})_{,s} - m^*\kappa_{(\nu)}\,\kappa_{(t)}] \\ M^*_{(t)} &= B\,h\,m^*_{,s} \end{aligned} \tag{4.15}$$

and one easily verifies that these loads satisfy the equilibrium conditions

$$\begin{aligned} &\oint (N_{(\nu)}\,\underset{\sim}{\nu} + Q^*\underset{\sim}{n})\,ds = \underset{\sim}{0} \\ &\oint [M_{(t)}\,\underset{\sim}{t} + \underset{\sim}{r} \times (N^*_{(\nu)}\,\underset{\sim}{\nu} + Q^*\underset{\sim}{n})]\,ds = \underset{\sim}{0}. \end{aligned} \tag{4.16}$$

For a graphic derivation of these edge loads we refer to [11, 2].

It can be proven that the solution obtained by adding the solutions of the classical shell problem and the one with the modified boundary conditions has a relative error of order ε^2 in the *interior* of the shell [12]. This result also establishes the validity of the second step of Gol'denveizer's asymptotic theory for plates with a free edge, which leads to modified boundary conditions which are in full agreement with the form of (4.14) for plates [13].

We finally note that for axisymmetrically loaded shells of revolution and for plates loaded in their plane, the twisting couple m* vanishes, so that, as a consequence, for those classes of problems the accury of the classical shell or plate solution has relative errors of order ε^2 in the *entire* shell or plate, provided the edge loads are in agreement with the requirements of shell theory.

5. STRESSES IN THE CORNER POINTS AT FREE EDGES

In section 3 we have shown that near the edge the actual stress and displacement distributions differ considerably from the distributions obtained from the classical theory of shells, and also from the distributions obtained from shell theory with modified boundary conditions. At the edge we still have errors of the order of the maximum shear stress at the edge as predicted by classical shell theory. From section 3 it is known that the main error near the edge zone is due to the edge twist. When the edge is curved, and the twisting moment along the edge is a sufficiently slowly varying function of the arc length along the edge the main correction to the shear stresses is still given by the solution of the torsion problem. Since this solution is a rapidly decreasing function with increasing distance to the edge one may introduce geodetic coordinates near the edge, with the lines x^1 = constant parallel to the edge curve and orthogonal trajectories as lines x^2 = constant. We now select the arc length along the edge curve as the coordinate $x^2 = s$, and the negative distance of the edge curve to the parallel curve as the coordinate $x^1 = x$. The positive direction of x^1 at the edge curve is therefore the unit normal vector $\underset{\sim}{\nu}$ in the tangent plane, and the positive direction of the s-coordinate is the unit tangent vector $\underset{\sim}{t}$. Furthermore we introduce the coordinate x^3 $(-h/2 \le x^3 \le h/2)$ perpendicular to the middle surface.

A detailed analysis of the "torsion problem" corresponding to the removal of the shear stresses along the edge yields [12]

$$u_1 = -\frac{6m,_s}{G}\phi, \quad u_2 = \frac{6m}{G}[\phi,_1 + \frac{1}{2}x\,\kappa_{(n)}\,\phi,_1 + \frac{3}{2}\kappa_{(n)}\,\phi] \tag{5.1}$$

$$\begin{aligned}
s^{11} &= -12\,m,_s\,\phi,_1 + 0\left(\frac{m}{hR} + \frac{m}{L^2}\right)\\
s^{12} &= 6\,m\,\phi,_{11}\,(1 - \frac{3}{2}x\,\kappa_{(n)}) + 0\left(\frac{m}{hR} + \frac{m}{L^2}\right)\\
s^{22} &= 12\,m,_s\,\phi,_1 + 0\left(\frac{m}{hR} + \frac{m}{L^2}\right)\\
s^{13} &= -6\,m,_s\,\phi,_3 + 0\left(\frac{m}{hR}\right)\\
s^{23} &= 6\,m\,[(1 - \frac{3}{2}x\,\kappa_{(n)})\,\phi,_{13} + \frac{3}{2}\kappa_{(n)}\,\phi,_3] + 0\left(\frac{m}{hR} + \frac{m}{L^2}\right)\\
s^{33} &= 0\left(\frac{m}{hR} + \frac{m}{L^2}\right)
\end{aligned} \tag{5.2}$$

where

$$\phi = \frac{8}{\pi^2}\sum_{k=1}^{\infty}\frac{(-1)^k}{(2k+1)^4}\exp\left\{\frac{(2k+1)}{h}\pi\,x^1\right\}\sin\frac{(2k+1)}{h}\pi\,x^3 \tag{5.3}$$

and s^{ij} is the pseudo stress tensor defined by

$$s^{ij} = \sqrt{\frac{g}{a}}\,\sigma^{ij} = \left\{1 + x\,\kappa_n + 0\left(\frac{h}{R} + \frac{h^2}{L^2}\right)\right\}\sigma^{ij}. \tag{5.4}$$

σ^{ij} is the Cauchy stress tensor.

At the edge $x = 0$ we obtain by summation of the Fourier series

$$\begin{aligned}
s^{13} &= -6\left(1 - \frac{4z^2}{h^2}\right)m,_s + 0\left(\frac{m}{hR}\right)\\
s^{12} &= \frac{12z}{h^3}\,m + 0\left(\frac{m}{hR} + \frac{m}{L^2}\right).
\end{aligned} \tag{5.5}$$

Superposition of the expressions for the normal stresses at the edge obtained from (5.2) and the fourth of the boundary conditions (4.14) shows that at $x = 0$ the remaining edge tractions are given by

$$s^{11} = 12\,m,_s\left(B\,\frac{z}{h^2} - \phi,_1\right), \quad s^{13} = 0 \quad s^{12} = 0 \tag{5.6}$$

all with errors of order $\varepsilon^2\bar{s}$, where $\bar{s}$ is the maximum stress in the shell. The tractions s^{11} are now removed by applying loads with the opposite sign. This leads to a plane strain problem for a strip $x \leq 0$, $|x^3| \leq h/2$, with boundary conditions

$$s^{11} = -12\, m,_s \left(B \frac{z}{h^2} - \phi,_1\right), \quad s^{13} = 0 \quad \text{on } x = 0$$
$$s^{13} = s^{33} = 0 \quad \text{on } x^3 = \pm h/2. \tag{5.7}$$

For an explicit solution of this problem we refer to [14]. We shall here set ourselves a more modest goal, namely to obtain the normal stresses s^{22} in the corner points at the edge. In those points the solution to the plane strain problem is given by

$$s^{22} = \mp 12\, \nu\, m,_s \left(\frac{B}{2h} - \phi,_1 \,(x = 0,\ z = \pm h/2)\right) \tag{5.8}$$

Adding the various contributions we are now in a position to write down the formula for the stresses in the corner points at the edge

$$\sigma_{tt}(\text{edge},\ z = \pm h/2) = \frac{n_{tt}}{h} \mp 6 \frac{m_{tt}}{h^2} \pm$$
$$\pm \left[(1+\nu) \frac{84}{\pi^3} \zeta(3) - \nu \frac{1116}{\pi^5} \zeta(5)\right] \frac{m,_s}{h} \tag{5.9}$$

where the first two terms represent the stresses obtained from classical shell theory plus the ones from shell theory with modified boundary conditions, and the last terms represent uni-axial stress arizing from constraint warping in non-uniform torsion, and the stress due to the plane strain problem. Further we have used Riemann's zêta function defined by

$$\zeta(n) = \frac{2n}{2^n - 1} \sum_{k=0}^{\infty} \frac{1}{(2k+1)^5} \,. \tag{5.10}$$

We notice here that although the error in (5.9) is expected to be of order ε^2, we are still in need of a rigorous proof. In section 7 it will be shown that for the case of an infinite plate with a circular hole our result is in full agreement with the asymptotic expansion of the exact solution.

6. THE INFLUENCE OF FREE EDGES ON EIGENFREQUENCIES

In this section we shall discuss a simple correction to the eigenfrequencies of shells with free edges, based on properties of the Rayleigh quotient. In the classical theory of vibrations of shells the numeration and denominator of the Rayleigh quotient are given by

$$P\,[u_\alpha,\ w] = \frac{Eh}{2(1-\nu^2)} \int_S \left[(1-\nu)\,\gamma^\alpha_\beta \gamma^\beta_\alpha + \nu\,\gamma^\alpha_\alpha \gamma^\beta_\beta + \frac{h^2}{12} \left\{ (1-\nu)\,\rho^\alpha_\beta \rho^\beta_\alpha + \nu\,\rho^\alpha_\alpha \rho^\beta_\beta \right\} \right] dS$$

$$T\,[u_\alpha,\ w] = \frac{1}{2}\,\rho h \int_S (u_\alpha u^\alpha + w^2)\ dS \tag{6.1}$$

where ρ is the mass density of the material. The squares of the frequencies of free vibrations are characterized by stationary values of the Rayleigh quotient. Let $\bar{u}^k_\alpha$, $\bar{w}^k$, $k = 1,2,..,$ denote the associated modes in the Kirchhoff-Love approximation, then the actual natural frequencies ω_k, $k = 1,2,\ ...$ are approximated by ω^*_κ, given by

$$\omega^2_k \cong \omega^{*2}_k = \frac{P[\bar{u}^k_\alpha,\ \bar{w}^k]}{T[\bar{u}^k_\alpha,\ \bar{w}^k]}\ ,\quad k = 1,2,... \tag{6.2}$$

As we have seen in the previous sections, the Kirchhoff-Love approximation overestimates the elastic energy in the shell by an amount that is approximately given by

$$Q[u_\alpha,\ w] = \frac{1}{12}\,BGh^4 \int_{\partial S_f} (\rho_{\alpha\beta}\ \nu^\alpha t^\beta)^2\ ds. \tag{6.3}$$

Improved natural frequencies ω^{**}_k, $k = 1,2,...,$ are now obtained by subtracting from the numerator in (6.2) the line integral (6.3), evaluated for the appropriate Kirchhoff-Love mode $\bar{u}^k_\alpha$, $\bar{w}^k$ so that

$$\omega^2 \cong \omega^{**2} = \omega^{*2} \left\{ 1 - \frac{Q[\bar{u}^k_\alpha,\ \bar{w}^k]}{P[\bar{u}^k_\alpha,\ \bar{w}^k]} \right\}. \tag{6.4}$$

Notice that this correction is of order h/L compared to unity, and that to obtain this factor all we have to do is to evaluate Q/P i.e. we do not have to solve another eigenvalue problem. The mathematical proof of the validity of (6.4) is given in [15], and although not proven, the conjecture is that for "shell-like" vibration modes the following estimate holds

$$\omega^2_k = \omega^{**2}_k\ [1 + 0(\varepsilon^2)]. \tag{6.5}$$

Since the Rayleigh quotient evaluated for an approximation of the fundamental mode is always an overestimate for the square of the actual *fundamental* frequency we have the rigorous inequality

$$\omega^2_1 \leq \omega^{*2}_1 \left[1 - \frac{Q\ \bar{u}^1_\alpha,\ \bar{w}^1]}{P[\bar{u}^1_\alpha,\ \bar{w}^1]} \right] (1 + 0(\varepsilon^2)). \tag{6.6}$$

From (6.4) it follows that significant corrections to the classical frequencies can only be expected in vibration modes which yield sufficiently large shear stresses along the nominally free edges.

7. APPLICATIONS

For the application of the theory to plate bending problems it is often convenient to make use of complex variable methods as developed by Muskhelishvili [16], and first applied to plate bending by Lechnizky (c.f. e.g. [17]).

In [12] the theory for the bending of plates with modified boundary conditions was formulated with the aid of complex functions, and applied to the plate bending problems to be discussed. For details we refer to [12].

7.1. Bending of a plate with an elliptical hole

We consider an infinite plate with an elliptical hole, which is loaded by bending couples $M_x = \frac{1}{6} Th^2$ at infinity. The origin of the x, y, z coordinate system is situated in the middle plane of the plate in the centre of the ellips, which is the projection of the edge surface on the middle plane, and is given by $\underset{\sim}{r} = R(1+m)\cos\theta\, \underset{\sim}{e}_x + R(1-m)\sin\theta\, \underset{\sim}{e}_y$ where $R = (a+b)/2$ and $m = (a-b)/(a+b)$.

The stresses in the corner points along the edge are given by [12]

$$\frac{\sigma_\theta}{T} = 1 + \frac{2(1+\nu)(1-m)}{(3+\nu)} \frac{m-\cos 2\theta}{1-2m\cos 2\theta+m^2} +$$

$$\frac{(1-m)(1+\nu)}{3+\nu} \frac{h}{R} \left\{\left(\frac{744}{\pi^5}\zeta(5) - \frac{56}{\pi^3}\zeta(3)\right) \frac{(1+m^2)\cos 2\theta - 2m}{(1-2m\cos 2\theta+m^2)^{5/2}} + \right.$$

$$+ \frac{1488}{\pi^6} \frac{\zeta(5)}{(3+\nu)(1-2m\cos 2\theta+m^2)^2} \left[\left\{(3+m^2)\cos 2\theta - \frac{1+3m^2}{m}\right\} K(m) + \right.$$

$$\frac{1-m^2}{m} E(m) - 4\{(1+m^2)\cos 2\theta - 2m\}\, \mathrm{Re}\{\Pi(m\sigma^2, m)\} +$$

$$\left.\left. + 2(1-m^2)\sin 2\theta\, \mathrm{Im}\{\Pi(m\sigma^2, m)\}\right]\right\}, \text{ (for } z = h/2). \tag{7.1;1}$$

Here $K(m)$, $E(m)$ and $\Pi(m\sigma^2, m)$ are complete elliptic integrals of respectively the first, second and third kind, defined by

$$K(m) = \int_0^1 \frac{dx}{\sqrt{(1-x^2)(1-m^2x^2)}}, \quad E(m) = \int_0^1 \sqrt{\frac{1-m^2x^2}{1-x^2}}\, dx,$$

$$\Pi(m\sigma^2, m) = \int_0^1 \frac{dx}{(1-m\sigma^2x^2)\sqrt{(1-x^2)(1-m^2x^2)}} \tag{7.1;2}$$

where $\sigma = \exp(-i\theta)$.

In the figures 1 and 2 the behaviour of σ_θ/T is shown as a function of θ, with the excentricity m as a parameter, for a thickness parameter $\beta = 2R/h = 6$, and Poisson's ratio $\nu = \frac{1}{4}$.

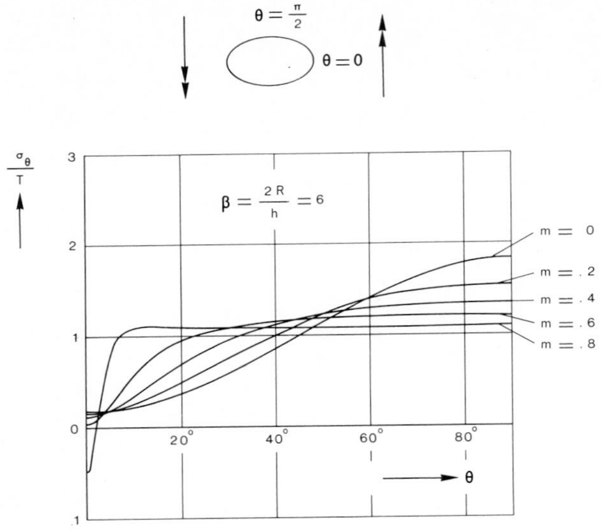

Fig. 1. Edge stresses as a function of $\theta(\nu = \frac{1}{4})$.

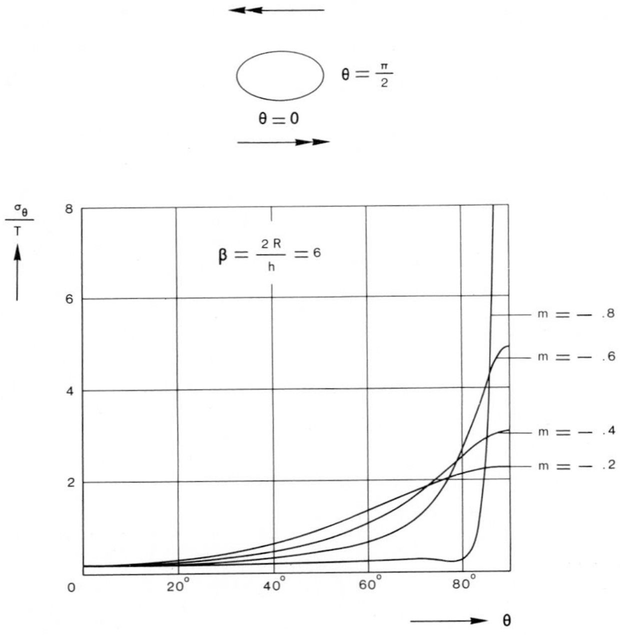

Fig. 2. Edge stresses as a function of $\theta(\nu = \frac{1}{4})$.

From these figures follows that for values of $|m|$ close to unity there is a tendency towards oscillating behaviour of the stresses near the ends of the longest axis of the ellips. However, in those cases our approximate solution ceases to hold since then the plate is locally very thick, as follows from

$$\left(\frac{h}{R}\right)_{Loc} = \frac{1-m}{(1+m)^2}\frac{h}{R} \quad \text{at } \theta = \frac{\pi}{2} \tag{7.1;3}$$

e.g. for $m = -.8$, $h/R = \frac{1}{3}$ we have $(h/R)_{Loc} \doteq 15$.

The factor of stress concentration k is obtained from (7.1;1) for $\theta = \pi/2$, and our result is

$$k = 1 + \frac{(1+\nu)(1-m)}{(3+\nu)(1+m)}\left[2 + \frac{h}{R(1+m)^2}\left\{\frac{56}{\pi^3}\zeta(3) - \right.\right.$$
$$\left.\left. - \frac{744}{3+\nu}\frac{\zeta(5)}{\pi^5}\left(1 + \nu + \frac{2(1-m)}{\pi m}\{(1+m)K(m) - E(m)\}\right)\right\}\right]. \tag{7.1;4}$$

In [18] the results of our analysis for the elliptic hole were used to show that thickness effects are minor in the energy release rate integral for bent plates containing elliptic holes or cracks.

Let us now consider the special case of a circular hole. For $m = 0$, (7.1;1) reduces to

$$\frac{\sigma_\theta}{T} = 1 + \frac{1+\nu}{3+\nu}\left\{-2 + \frac{h}{T}\left(\frac{744}{\pi^5}\zeta(5)\frac{(2+\nu)}{3+\nu} - \frac{56}{\pi^3}\zeta(3)\right)\right\}\cos 2\theta. \tag{7.1;5}$$

This result is in complete agreement with the result obtained by Alblas [19] from an asymptotic series expansion of the exact solution of the three-dimensional problem, and thus confirms our conjecture that (5.9) has errors not exceeding the order of ε^2. In figure 3 we have shown the result for the component of σ_θ/T which varies with $\cos 2\theta$ as a function of $\beta = 2R/h$, for $\nu = \frac{1}{4}$. The result is compared with the exact three dimensional solution, the solution obtained from classical plate theory (Goodier) and Reissner's modified theory of plates [20]. It is kind of surprizing that our approximation still is accurate up to $\beta = 1$.

Further we see that for $\nu = \frac{1}{4}$ Reissner's and our results more or less coincide for sufficiently large value of β. For a further discussion of Reissner's theory we refer to [12, 21].

7.2. Bending and torsion of perforated plates

The problem of calculating stresses and displacements in perforated plates under various kinds of loading has been investigated extensively in the literature (c.f. e.g. [23, 24]). In [23] analytic solutions were obtained for bending and torsion of thin perforated plates with a regular pattern of equal and equally spaced circular holes, and also for plane strain and generalized plane

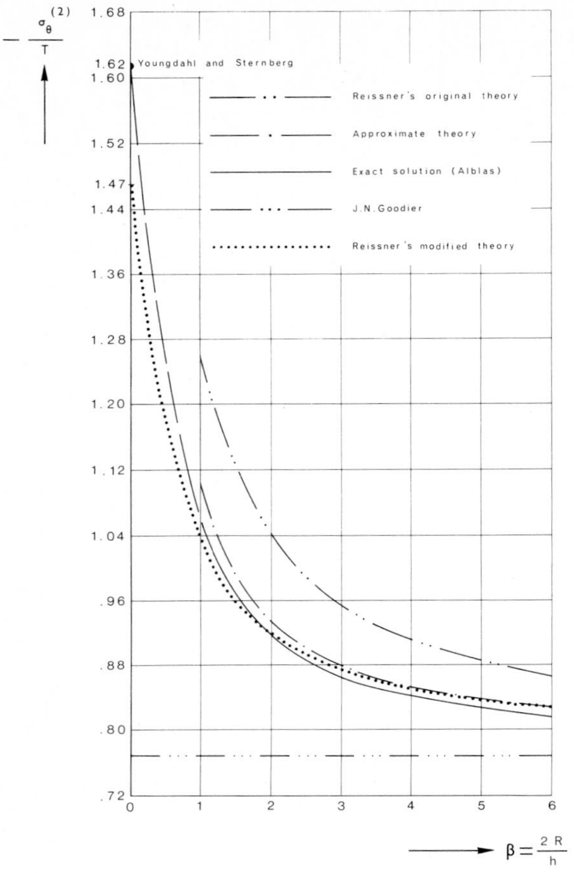

Fig. 3. Edge stresses as a function of β ($\nu = \frac{1}{4}$).

stress conditions. These are approximate solutions for the bending and torsion of very thick plates. Since there are considerable differences in stress concentration factors and effective elastic constants for bending (c.q. torsion) of thin and thick plates it is of interest to know the transition between these limiting cases. In [22] based on the theory presented a first order correction has been determined for the corner stresses and the effective elastic constants for perforated plates with free edges. In addition, results of finite element calculations were used to be able to cover the whole transition from thin to thick plates. The analysis has been restricted to plates with a square or equilateral triangle as a basic pattern.

The interpolation for the stress concentration can only be carried out in an approximate way, since for very thick plates the maximum stress does not occur in the corner points of the hole, but at some distance from the faces of the plate [25].

In fig. 4 the corner stresses have been plotted for a plate with a square basic pattern of circular holes, loaded in bending. Notice that for $h/2\omega = 1$ the curves are near $\phi = 0$ much flatter than would be expected on the basis of a state of plane stress. This is entirely due to the boundary layer effect mentioned above (c.f. [25]). In figure 5 the results are given for a plate loaded in torsion. The dashed lines in figure 4 represent the result of our asymptotic analysis.

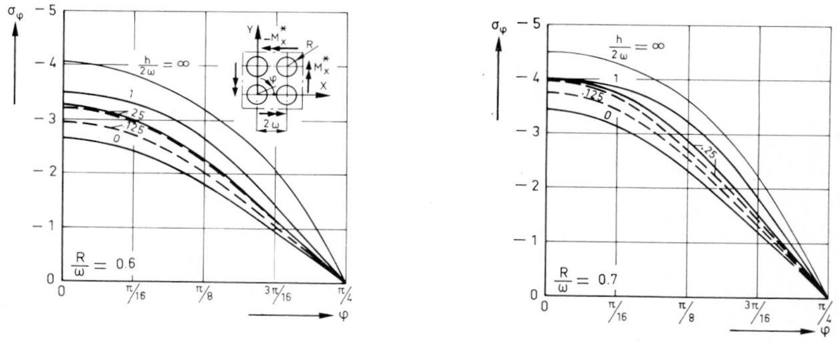

Fig. 4. σ_ϕ for $M_x^* = - M_y^* = \frac{1}{6} h^2$ ($\nu = 0.3$).

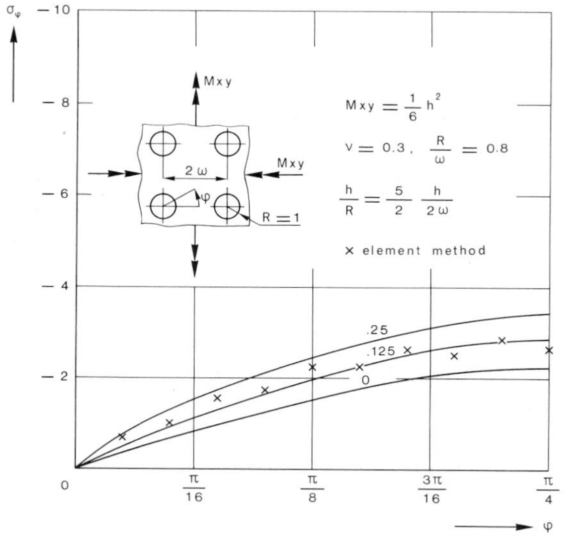

Fig. 5. σ_ϕ for a plate loaded in torsion.

7.3. Shells with circular holes

Two types of shells will be discussed; the circular cilindrical shell and the spherical shell.

A theory for the treatment of cilindrical shells with circular holes, under arbitrary boundary conditions has been presented in [26]. The theory is valid under the restriction that the radius of the hole is much smaller than the radius of the cilinder. For the case of a free edge the modifications for the subsequent refinement were introduced in [27]. It turned out, however, that for the cases considered the corrections were minor, and therefore insignificant from the practical point of view. This does not mean, however, that this holds true for all kinds of loading.

The spherical shell with free edges is still under investigation by the author together with Dr. A. Pisanco [1], but some preliminary results will be reported on.

For the spherical shell the field equations may be reduced to a single complex equation for the curvature-stress function

$$\psi = W + \frac{2ic}{Eh^2} F \qquad (7.3;1)$$

where $c = \sqrt{3(1-\nu^2)}$ and the surface invariants W and F are the curvature function and the stress function respectively. In the absense of surface loads the governing equation is given by [31]

$$\left(\Delta + \frac{2}{R^2}\right)\left(\Delta + \frac{1}{R^2} + \frac{2ic}{hR}\right)\psi = 0. \qquad (7.3;2)$$

Notice that in this equation Poisson's ratio appears only in the form of the number c. Lack of space prevents us to discuss the formulae for the stress-strain relations and the boundary conditions in more detail. We mention here only that in [31] a modified expression for the tensor of curvature has been used, which yields different expressions for the boundary conditions [5]. In the figures 6 and 7 we have shown the results for the corner stresses at the free edge for a half sphere with a circular hole, as a function of the thickness parameter h/R, for the case that the shell is loaded by a twisting couple $M_{(\nu)} = 37.8 \cos 2\phi$ at its equator.

---

[1] Dr. A. Pisanco from Dnepropetrovsk State University visited Delft University on a sabatical leave.

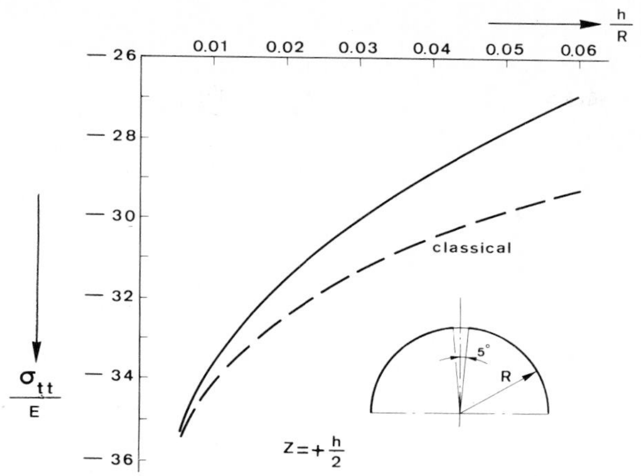

Fig. 6. Corner stresses at the outer corner points.

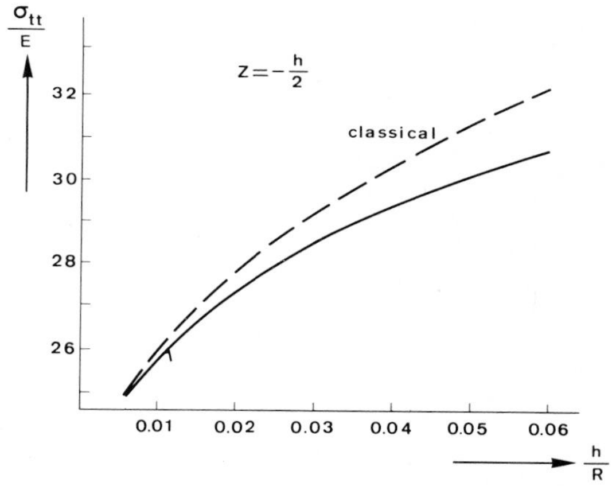

Fig. 7. Corner stresses at the inner corner points.

In these figures we have taken E equal to unity, and Poisson's ratio $\nu = 0.25$. The differences between the classical solution and the refined solution are due to a redistribution of the membrane stresses and the bending stresses caused by the modified boundary conditions, and the contributions of the torsion and the plane strain problem. We see that in this case the classical theory overestimates the stresses, whereas for the plate with the circular hole it underestimated the stresses in the corner points.

7.4. Vibrations of plates

Expression (6.4) for the square of the improved eigenfrequencies may be rewritten in the more transparent form

$$\omega_n^{**2} = \omega_n^{*2}\left(1 - \frac{h}{L} C_n^*\right), \qquad (7.4;1)$$

where C_n^* depends both on n and on Poisson's ratio ν. For shell or plate theory to be applicable it is necessary that $h/L << 1$, so that practically significant corrections will only occur if $C_n^* = O(1)$. For a plate strip $-b/2 \le x \le b/2$, $-\infty \le y \le \infty$, $-h/2 \le z \le h/2$ hinged on the y-axis in a scew-symmetric mode of vibration such that $w(-x, y) = -w(x, y)$ the correction for *large wave-length* torsional vibrations is given by [15]

$$\omega^{**2} = \omega^{*2}\ (1-Bh/b) \qquad (7.4;2)$$

where the constant B is given by (3.2).

For *short wave-lengths* the motion is confined to extremely narrow zones near the free edges, and the result is the same as for a semi infinite plate with a straight free edge, and is for sufficiently large values of $\pi b/2L$ given by

$$\omega_k^{**2} = \omega_k^{*2}\left[1 - 2\,\frac{B\pi h}{L}\,\frac{\{2-4\nu+3\nu^2-2(1-\nu)(1-2\nu+2\nu^2)^{\frac{1}{2}}\}^{5/4}}{3(1-2\nu+2\nu^2) - (3-4\nu)(1-2\nu+2\nu^2)^{\frac{1}{2}}}\right],\quad k >> 1. \qquad (7.4;3)$$

For $\nu = 0$ the correction factor vanishes, and for small values of Poisson's ratio we have

$$\omega_k^{**2} \approx \omega_k^{*2}\left[1 - \frac{\pi}{4}\sqrt{2}\,B\,\nu^4\,\frac{h}{L}\right] \qquad (7.4;4)$$

which shows that for short wave-lengths the corrections are insignificant.

In [15] it was shown that for the fundamental frequencies of free circular plates in modes with nodal diameters the constant C_n^* has a maximum value of approximately 0.05 for $n = 4$ and $\nu = \frac{1}{2}$, so that the correction is always quite small for circular plates. For the limiting case of a large number of nodal diameters the result (7.5;3) is retrieved.

We finally note that some preliminary results for vibrations of cilindrical shells and a spherical cap indicate that also in those cases the corrections are small. The correction for torsional vibrations of a long plate strip in the long-wave mode appears the largest one for the cases we have considered.

BIBLIOGRAPHY

1. Koiter, W.T. and Simmonds, J.G., Foundations of shell theory. Proc. 13th Int. Congr. of Theor. and Appl. Mech., Springer-Verlag, 150-176 (1972).
2. Koiter, W.T., A consistent first approximation in the general theory of thin elastic shells.

Proc. IUTAM Symp. on the Theory of Thin Elastic Shells, Delft 1959. North-Holland Publ. Co., Amsterdam, 12-33 (1960).

3. Koiter, W.T., On the mathematical foundation of shell theory. Proc. Int. Congr. Math. Nice 1970. Gautier-Villars, Paris, vol. 3, 123-130 (1971).
4. Danielson, D.A., Improved error estimates in the linear theory of thin elastic shells. Proc. Kon. Ned. Ak. Wet., B74, 294-300 (1971).
5. Van der Heijden, A.M.A., On the dynamic boundary conditions in the linear theory of thin shells. Rep. 800, Lab. of Eng. Mech., Dept. of Mech. Eng., Univ. of Techn., Delft (1985).
6. Love, A.E.H., A treatise on the mathematical theory of elasticity (fourth edition). Dover Publ., New York.
7. Van der Heijden, A.M.A., Lecture notes on applications of perturbation methods (in Dutch, 1974).
8. Booij, J., The torsional stiffness of a plate with a non-square free edge. Rep. 492, Lab. of Eng. Mech., Dept. of Mech. Eng., Univ. of Techn., Delft (1972).
9. Van der Giessen, E., The influence of scew edges on the stress distribution in plates (in Dutch). Lab. rep. 756, Lab. of Eng. Mech., Dept. of Mech. Eng., Univ. of Techn., Delft (1983).
10. Berdichevskii, V.L., Variational asymptotic method in nonlinear shell theory. In "Theory of shells", Proc. of the Third IUTAM Symp. on Shell Theory. Tbilisi, USSR 1978. North Holland Publ. Cy., 137-161 (1980).
11. Van der Heijden, A.M.A., Benaderingstheorie voor het bepalen van spanningen langs de vrije gatrand. Rep. 456, Lab. of Eng. Mech., Dept. of Mech. Eng., Univ. of Techn. Delft (1971).
12. Van der Heijden, A.M.A., On modified boundary conditions for the free edge of a shell. Thesis, Delft. Delft University Press (1976).
13. Gol'denveizer, A.L. and Kolos, A.V., On the derivation of two-dimensional equations in the theory of thin elastic plates. P.M.M. Vol. 29 no. 1, 151-166 (1965).
14. Van der Heijden, A.M.A., A method for the solution of a class of semi-infinite and finite strip problems in stress analysis. Rep. 551, Lab. of Eng.Mech., Dept. of Mech.Eng., Univ. of Techn., Delft (1975).

15. Koiter, W.T. and Van der Heijden, A.M.A., Improved natural frequencies for plates with a free edge.
Appl. Sc. Res. 37: 91-102 (1981).
16. Muskhelishvili, N.I., Some basic problems of the mathematical theory of elasticity.
P. Noordhoff, Groningen (1953).
17. Savin, G.N., Spannungserhöhung am Rande von Löchern.
V.E.B. Verlag Technik, Berlin (1956).
18. Simmonds, J.G. and Duva, J., Thickness effects are minor in the energy release rate integral for bent plates containing elliptic holes or cracks DAMACS.
Rep. no. 80-12, Dept. of Appl. Math. and Comp. Science, School of Engg. and Appl. Science, University of Virginia (1980).
19. Alblas, J.B., Theorie van de driedimensionale spanningstoestand in een doorboorde plaat.
Thesis, Delft (1957).
20. Reissner, E., On the transverse bending of plates including the effect of transverse shear deformations.
Int. J. Solids Structures, vol. 11, 569-573 (1975).
21. Van der Heijden, A.M.A., Thick plates theories: fiction and fact.
Rep. 609, Lab. of Eng. Mech., Dept. of Mech. Eng., Univ. of Techn., Delft (1977).
22. Meijers, P. and Van der Heijden, A.M.A., Refined theory for bending of perforated plates.
WTHD 125, Delft University of Technology (1980).
23. Meijers, P., Doubly-periodic stress distributions in perforated plates.
Thesis Delft University of Technology (1967).
24. Slot, T., Stress analysis of thick perforated plates.
Thesis, Delft University of Technology (1972).
25. Youngdahl, C.K. and Sternberg, E., Three dimensional stress concentration around a cylindrical hole in a semi-infinite elastic body.
J. Appl. Mech., 33, 855-865 (1966).
26. Lekkerkerker, J.G., On the stress distribution in cilindrical shells weakened by a circular hole.
Thesis, Delft (1965).
27. Dijksman, J.F., Over correcties van spanningen langs de vrije rand van een cirkelvormig gat in een cilinderschaal.
Rep. 508, Lab. of Eng. Mech., Dept. of Mech. Eng., Delft University of Technology, Delft (1973).
28. Van der Heijden, A.M.A., An asymptotic analysis for a plate with a circular hole.

WTHD rep. 78, Dept. of Mech. Eng., University of Technology, Delft (1976).

29. Van der Heijden, A.M.A., An asymptotic theory for flexural vibrations of circular plates.
WTHD rep. 130, Dept. of Mech. Eng., University of Technology (1980).

30. Van der Heijden, A.M.A. and Pisanco, A., On spherical shells with free edges.
To be published.

31. Koiter, W.T., A unified reduction of the intrinsic equations for spherical and circular cylindrical shells.
Rep. 744, Lab. of Eng. Mech., Delft University of Technology (1983).

ON A METHOD TO EVALUATE EDGE EFFECTS IN ELASTIC PLATES

F. PECASTAINGS

Laboratoire de Mécanique Théorique, Université Paris 6, C.N.R.S,

4 place Jussieu, 75230 Paris Cedex 05

It is very important to take into account the so-called "edge effects" in the plate problems and more especially for the composite plates. These effects, unlike "interior effects", are localized in the vicinity of the edge of the plate. The method we apply derives from the form of the solution. It is shown in (ref.19), in the case of isotropic plates, that, for any boundary conditions prescribed on the edge of the plate, the displacement field U may be written in the form:

$$U = U_I + U_L$$

- U_I corresponds to the interior effect,
- U_L is the part of the displacement which is strongly localized. It decays with the distance from the edge, the characteristic decay length being about the thickness of the plate.

Classically, the interior effect is computed by solving a Kirchhoff-Love's or Reissner's plate theory; accurate error estimates were given by Koiter (Ref.1), Danielson (Ref.2), Simmonds (Ref.3), and Ladevèze (Ref.4-6).

In this paper, we intend to give a method to evaluate the edge effects. Many studies already exist: the edge effects are obtained by solving problems in a strip orthogonal to the edge; these problems are computed either by finite element methods, Dong and Goetchel (Ref.7) - or by the Papkovich functions, in a classical or analogous form for the composites -Horgan (Ref.8), Choi (Ref.9), Zwiers, Ting and Spilker (Ref.10). These techniques are very interesting for straight edges and for the edges which have a weak curvature, but they require regular data and geometry. Our method avoids those difficulties, simply because the edge zone is taken into account "in the average".

First it is proved that the localized solution U_L is of the form:

$$U_L = U_L^1 + U_L^2$$

where U_L^1 and U_L^2 are respectively solutions of so-called "generalized plane strain and torsion problems", by analogy to the problems introduced by Friedrichs (Ref.11) and usually used for the plate studies. U_L^1 is expressed in terms of two scalar potentials, U_L^2 only depends on one.

The basic idea consists in using a series representation of U_L^1 and U_L^2, whose coordinate functions are solutions of explicit eigenvalue problems. These functions which are written by means of Papkovich functions were already introduced by Lur'e (Ref.12), then used by Cheng (Ref.13). From the results established by Gregory (Ref.14-16), Joseph, Sturges and Warner (Ref.17) in the case of a semi-infinite strip, we show that the set of the coordinate functions is complete. What distinguishes our approach is the fact that any boundary conditions may be taken into account by the way of a global formulation on the edge.

A code has been realized and numerical results are given.

1. STATEMENT OF THE PROBLEM

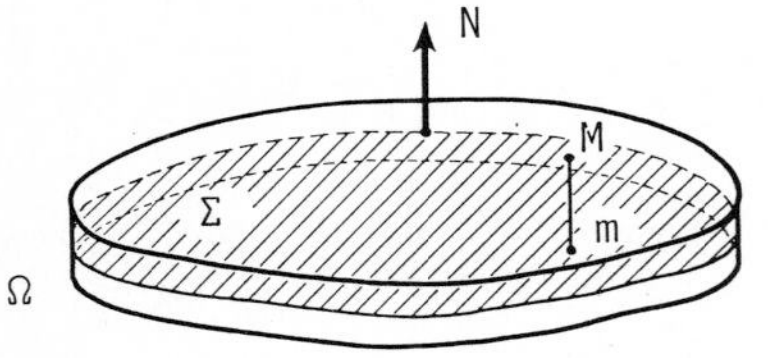

(a)

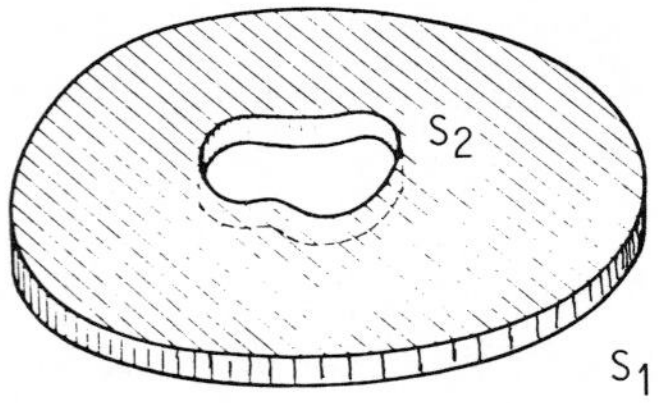

(b)

Fig. 1

Let Ω denote a plate which is elastic, homogeneous and isotropic, Σ its middle surface, 2h its thickness, M a point of , m its projection upon Σ, and z its thickness coordinate (Fig. 1.a).

We consider a Saint-Venant problem, that is, we assume the faces of the plate are stress free and there are no body-forces.

The displacement is specified over the part S_D of the edge, while the load is prescribed over the remaining part S_L.

So, we have to seek a displacement field U and a stress field σ which satisfy:

1. $$\int_\Omega \mathrm{Tr}\, \varepsilon(U)\, K\, \varepsilon(U)\, d\Omega < \infty \qquad (1.1)$$

where . $\varepsilon(U)$ is the strain corresponding to U
. K is the Hooke tensor

2. $$U|_{S_D} = U_d \text{ (kinematic conditions)} \qquad (1.2)$$

3. $$\forall\, U^* \in \mathcal{U},\ \int_\Omega \mathrm{Tr}\,[\sigma\, \varepsilon(U^*)]\, d\Omega = \int_{S_L} F_d \cdot U^*\, dS \text{ (principle of virtual work)} \qquad (1.3)$$

where $\mathcal{U}$ is the set of the displacement fields which have a finite strain energy and such that:

$$U|_{S_D} = 0$$

4. $$\sigma = K\, \varepsilon(U) \quad \text{(Hooke's law)} \qquad (1.4)$$

It is well-known that the solution of such a problem, is split into two parts:

- the Saint-Venant solution which is related to the "interior effect", that is, the effect which diffuses into the interior of the plate,
- the localized solution which describes the edge effect, that is, the effect which is localized in the neighbourhood of the edge.

  It is to be pointed out that, for the plate Fig. 1.b, two edge effects have to be distinguished.
- one is related to the interior edge S_1
- the other is related to the exterior edge S_2.

Consequently, two standard problems will be considered:
- the interior problem

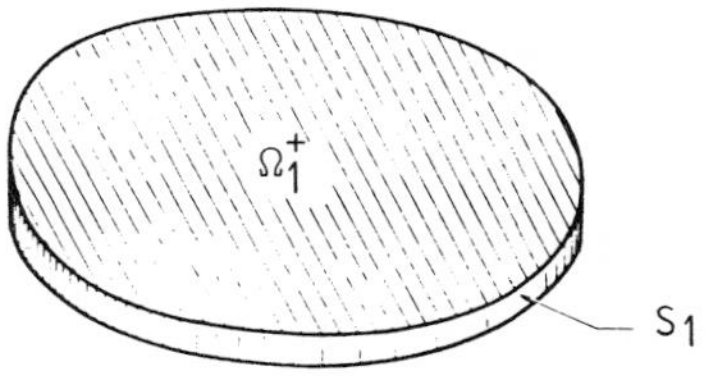

Fig. 2

It is the Saint-Venant problem for the plate Ω_1^+, whose edge is S_1 (Fig. 2)
- the exterior problem

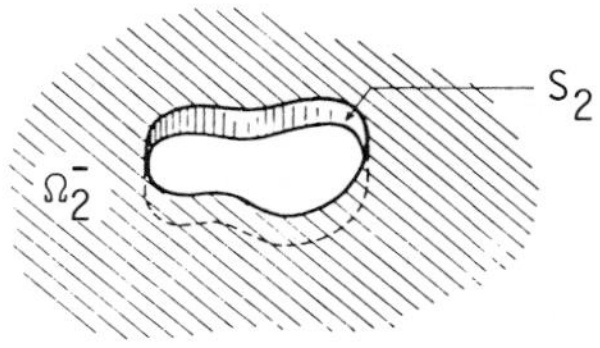

Fig. 3

It is the Saint-Venant's problem for the plate Ω_2^-, whose edge is S_2 (Fig. 3).

2. FUNDAMENTAL RESULT

We consider a cross-section S_0, the interior and exterior problems related to S_0, and we denote by Ω^+ and Ω^- the domain related to each of these problems. From now, we shall assume that the conditions of the Saint-Venant principle for the plates hold (see paper by P. Ladevèze). Consequently, the Saint-Venant solution vanishes and the result established in (Ref. 19) in the general case, leads to:

<u>Property 1</u>

The set of the displacement fields solution of interior (resp. exterior) problems is the space:

$\overline{W_1^+} \oplus \overline{W_2^+}$ (resp. $\overline{W_1^-} \oplus \overline{W_2^-}$)

where W_1^+, W_2^+ (resp W_1^-, W_2^-) are the subspaces of the spaces W_1 et W_2 determined by:

Space W_1

. W_1 is the space of the displacement fields U:

$$U = N \wedge \mathrm{grad}_m f \tag{2.1}$$

where f is a scalar function, solution of the equation

$$\mathscr{H} f = - \begin{bmatrix} -\Delta_m f \\ 0 \\ 0 \end{bmatrix} \tag{2.2}$$

$\Delta_m f$ being the laplacian of f and $\mathscr{H}$ being the differential operator defined on the thickness $]-h, h[$ by:

$$\mathscr{H} f = \begin{bmatrix} \dfrac{\partial^2 f}{\partial z^2} \\ \left.\dfrac{\partial f}{\partial z}\right|_{z=\pm h} \\ \displaystyle\int_{-h}^{h} f \, dz \end{bmatrix} \tag{2.3}$$

Space W_2

. W_2 is the space of the displacement fields U:

$$U = \mathrm{grad}_m \alpha + u N \tag{2.4}$$

where α and u are scalar functions such that the vectorial function:

$$X = (m, z) \longrightarrow \begin{bmatrix} \alpha(m,z) \\ \\ u(m,z) \end{bmatrix} \tag{2.5}$$

satisfies the equation:

$$\mathcal{K} X = - \begin{bmatrix} \Delta_m X \\ 0 \\ 0 \end{bmatrix} \tag{2.6}$$

where $\mathcal{K}$ is the differential operator defined on the thickness $]-h, h[$ by:

$$\mathcal{K} X = \begin{bmatrix} AX \\ BX|_{z=\pm h} \\ \int_{-h}^{h} CX \, dz \end{bmatrix} \tag{2.7}$$

with:

$$A = \frac{1}{\lambda + 2\mu} \begin{bmatrix} \mu \dfrac{\partial^2 .}{\partial z^2} & (\lambda+\mu) \dfrac{\partial .}{\partial z} \\ -(\lambda+\mu) \dfrac{\partial^3 .}{\partial z^3} & (2\lambda+3\mu) \dfrac{\partial^2 .}{\partial z^2} \end{bmatrix} \tag{2.8}$$

$$B = \begin{bmatrix} \dfrac{\partial .}{\partial z} & 1. \\ -\lambda \dfrac{\partial^2 .}{\partial z^2} & (3\lambda+4\mu) \dfrac{\partial .}{\partial z} \end{bmatrix} \tag{2.9}$$

$$C = \begin{bmatrix} \dfrac{\partial .}{\partial z} & 1. \\ 4(\lambda+\mu). + \lambda z \dfrac{\partial .}{\partial z} & \lambda z. \\ 8(\lambda+\mu)z. + \lambda z^2 \dfrac{\partial .}{\partial z} & \lambda z^2. \end{bmatrix} \tag{2.10}$$

λ, μ are Lame's moduli.

W_i^+ (resp. W_i^-) are the subspaces of bound strain energy on Ω^+ (resp. Ω^-) displacement fields which belong to W_i, and $\overline{W_i^+}$ (resp. $\overline{W_i^-}$) is the closure of W_i^+ (resp. W_i^-) with respect to the norm in energy.

3. DIFFERENTIAL EQUATIONS

Any solution of an interior problem, for instance, can be expressed by means of three scalar functions f, α and u. But, on the one hand the function f, on the other hand the function $X = (\alpha, u)$, are solutions of an equation of the form:

$$D_m Y + AY = 0 \tag{3.1}$$

where:

. D_m is the operator : $Y \longrightarrow \begin{pmatrix} \Delta_m Y \\ 0 \\ 0 \end{pmatrix}$, so D_m depends on m alone.

. A is the operator $\mathcal{H}$ or $\mathcal{K}$, therefore A depends on the thickness coordinate z alone.

As a consequence, this equation has such a structure that the variables m and z are "separated". This quality will be the basic point of our method to evaluate the edge effects.

Moreover, it is possible to characterize the fonction f_1 which is solution of the equation:

$$D_m f + \mathcal{H} f = 0$$

and which is related to the "projection" U_1 of the solution U onto the space $\overline{W}_1^+$. Indeed, using the fact that the spaces $\overline{W}_1^+$ and $\overline{W}_2^+$ are "orthogonal", a characterization of f_1 can be provided and in the case of a bending problem, for instance, it is proved that:

Property 2

U_1 is the field: $U_1 = N \wedge \text{grad}_m f_1$
where f_1 is the solution of the equation:

$$\begin{aligned} & D_m f + \mathcal{H} f = 0 \\ & f|_{\partial\Sigma} = \int_0^z dt \int_{-h}^t \omega(m,u)\, du \end{aligned} \tag{3.2}$$

ω being the curl around N: $\omega = \text{div}_m N \wedge \pi U$ and π the orthogonal projection onto Σ.

Consequently, a characterization of U_2, projection of U onto $\overline{W}_2^+$, follows.

4. SPLITTING UP OF THE SPACES $\overline{W}_i$

A classical way to solve a differential equation is used. Since the variables m and z are "separated" in the equations satisfied respectively by f and by (α, u), one looks for the eigenvalues and the eigenfunctions of the operator A.

Then a set of "coordinate functions" is derived, and the solution is approximated by a sum of these functions.

First, the spectra of $\mathscr{H}$ and $\mathscr{K}$ are defined. From the results we have obtained, it follows that, in each case: $A = \mathscr{H}$ or $A = \mathscr{K}$, the set of eigenvalues is countable, each one being of order one.

If α_k is an eigenvalue of A and if y_k is an eigenfunction related to α_k, the function Y_k:

$$Y_k(m,z) = \varphi_k(m)\ y_k(z) \tag{3.3}$$

where φ_k is a function depending on m alone, is obviously an eigenfunction of A since A is independant of m:

$$A\ Y_k = \alpha_k \varphi_k\ y_k \tag{3.4}$$

It follows that Y_k satisfies the equation 3.1 if φ_k is such that:

$$\Delta_m \varphi_k + \alpha_k \varphi_k = 0 \tag{3.5}$$

Among all the functions φ_k which satisfy 3.5, only the functions such that the corresponding displacement field has a bounded strain energy on the domain are kept out. Two sets of functions are obtained for each eigenfunction, therefore a sequence ($\overline{W}_i^{k+}$) of the space $\overline{W}_i^+$ and a sequence ($\overline{W}_i^{k-}$) of the space $\overline{W}_i^-$ are derived.

4.1. Spaces W_1^{k+} and W_1^{k-}

Bending problems (B.P.) and extension problems (E.P.) are investigated separately.

Spectrum of $\mathcal{H}$

	eigenvalue $\alpha_k = -\beta_k^2$	eigenfunction t_k
B.P.	$\beta_k = (2k+1)\frac{\pi}{2h}$, $k \in N$	$t_k(z) = \sin \beta_k z$
E.P.	$\beta_k = k\frac{\pi}{h}$, $k \in N^*$	$t_k(z) = \cos \beta_k z$

(4.1)

Spaces W_1^k

W_1^k is the space of the displacement fields U_k.

$$U_k = t_k(z)\, N \wedge \text{grad}_m f_k \tag{4.2}$$

with
- t_k is an eigenfunction corresponding to $\alpha_k = -\beta_k^2$
- f_k is a scalar function of the variable m which satisfies:

$$\Delta f_k - \beta_k^2 f_k = 0 \tag{4.3}$$

W_1^{k+} (resp. W_1^{k-}) is the set of bounded strain energy on Ω^+ (rep Ω^-) displacement fields which belong to W_1^k.

<u>Remark</u> :

The shear stress $\pi\sigma_k N$ is given by :

$$\pi\sigma_k N = \dot{t}_k(z)\, N \wedge \text{grad}_m f_k$$

Then, it follows that any function f_k related to a field U_k of W_1^k satisfies:

$$\int_\Sigma \text{grad } f_k \cdot \text{grad } f_k \, d\Sigma < \infty$$

4.2. Spaces W_2^{k+} and W_2^{k-}

Spectrum of $\mathcal{K}$

	eigenvalues $\alpha_k = -\beta_k^2$	eigenfunctions $t_k = \begin{pmatrix} u_k \\ \beta_k \quad v_k \end{pmatrix}$
B.P.	β_k root of the equation: $\sin 2hx = 2hx$, $Re(\beta_k) > 0$	$u_k(z) = \beta_k z \cos(\beta_k z) + [\sin^2(h\beta_k) + \frac{\mu}{\lambda+\mu}] \sin(\beta_k z)$ $v_k(z) = -\left\{\beta_k z \sin(\beta_k z) + [\cos^2(h\beta_k) + \frac{\mu}{\lambda+\mu}] \cos(\beta_k z)\right\}$
E.P.	β_k root of the equation: $\sin 2hx + 2hx = 0$, $Re(\beta_k) > 0$	$u_k(z) = \beta_k z \sin(\beta_k z) - [\cos^2(h\beta_k) + \frac{\mu}{\lambda+\mu}] \cos(\beta_k z)$ $v_k(z) = \beta_k z \cos(\beta_k z) - [\sin^2(h\beta_k) + \frac{\mu}{\lambda+\mu}] \sin(\beta_k z)$

(4.4)

Remarks

1. All the roots of the transcendental equations $\sin Z = Z$ and $\sin Z = -Z$, other than $x=0$, are complex. If Z_0 is a root, so is the conjugate complex $\overline{Z_0}$.
2. In each case, the roots Z_k lying in the first quadrant are ordered by increasing real part.
3. The first ten roots Z_k are:

	equation: $\sin Z = Z$, $Z_k = x_k + i\, y_k$		equation: $\sin Z = -Z$, $Z_k = x_k + i\, y_k$	
k	x_k	y_k	x_k	y_k
1	7.497676	2.768678	4.212392	2.250729
2	13.899960	3.352210	10.712537	3.103149
3	20.238518	3.716768	17.073365	3.551087
4	26.554547	3.983142	23.398355	3.858809
5	32.859741	4.193251	29.708120	4.093705
6	39.158817	4.366795	36.009866	4.283782
7	45.454071	4.514640	42.306827	4.443446
8	51.746768	4.643428	48.600684	4.581105
9	58.037662	4.757515	54.892406	4.702096
10	64.327234	4.859917	61.182590	4.810025

(4.5)

and for k sufficiently large:

$x_k \sim (4k+1)\frac{\pi}{2}$, $y_k \sim \ln(4k+1)\pi$	$x_k \sim (4k-1)\frac{\pi}{2}$, $y_k \sim \ln(4k-1)\pi$

4. u_k and v_k are the classical Papkovich-Fadle functions.

Spaces W_2^k

W_2^k is the space of the displacement fields U_k:

$$\begin{cases} \pi U_k = u_k(z)\ \text{grad}_m\ \varphi_k \\ N.U_k = \beta_k\ v_k(z)\ \varphi_k(m) \end{cases} \tag{4.6}$$

with : . u_k, v_k, β_k defined by (4.4)

. φ_k is a scalar function of the variable m which satisfies:

$$\Delta\varphi_k - \beta_k^2\ \varphi_k = 0 \tag{4.7}$$

W_2^{k+} (resp. W_2^{k-}) is the set of bounded strain energy on Ω^+ (resp. Ω^-) displacement fields which belong to W_2^k.

Remark

The shear stress $\pi\sigma_k N$ is given by:

$$\pi\sigma_k\ N = [\dot{u}_k + \beta_k\ v_k]\ \text{grad}_m\ \varphi_k \tag{4.5}$$

Then it follows that any function φ_k related to a field U_k of W_2^{k+} satisfies:

$$\int_{\Sigma^+} \text{grad}\ \bar{\varphi}_k\ .\ \text{grad}\ \varphi_k\ d\Sigma < \infty,$$

where $\bar{\varphi}_k$ is the conjugate function.

4.3. "Orthogonality" of the spaces $\overline{W}_i^{k+}$ et $\tilde{W}_j^{l}$

Let $\tilde{W}_j^{l}$ (j = 1,2) be the set of the displacement fields U wich satisfy:

- U belong to W_j^k
- U have a strain energy bounded on an "interior neighbourhood" V of the end-cross section S_0.

For any fields $U_i \in \overline{W}_i^{k+}$ and $U_j \in \tilde{W}_j^{l}$, it follows from the Stokes formula that the quantity:

$$a(U_i, U_j) = \int_S U_j \cdot \sigma(U_i)\nu \,|_S - U_i \cdot \sigma(U_j)\nu \,|_S \, dS$$

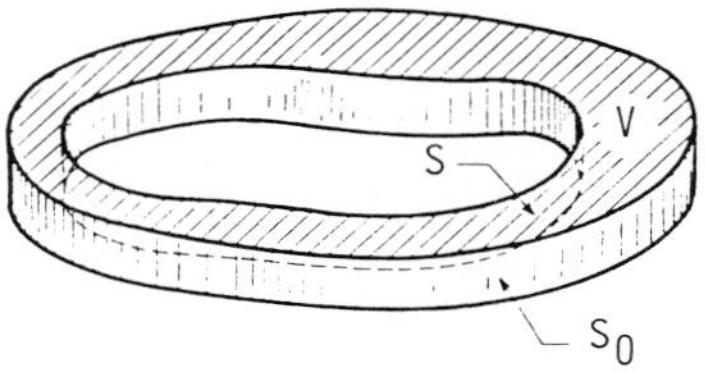

Fig. 4

where ν is a unit vector, orthogonal to S and exterior to Ω^+.

is a constant independant of the cross section S included in V.

When $a(U_i,U_j) = 0$, the fields U_i and U_j are so-called "orthogonal". Then, it can be proved:

Property 3

The spaces $\overline{W}_i^{k+}$ and $\tilde{W}_j^{l}$ ($i \neq j$) are orthogonal.

The spaces $\overline{W}_i^{k+}$ and $\tilde{W}_i^{l}$ ($k \neq l$) are orthogonal.

5. FORMAL SERIES CONNECTED TO A SOLUTION

Any localized solution of an interior problem, for instance, can be written in the unique form:

$$U = U_1 + U_2 \quad \text{where } U_i \in \overline{W}_i^{+}$$

Therefore, series representation of U_1 and U_2 have to be sought. To this end, a method is applied, similar to the one which allows to relate a Fourier series representation to a periodic function. Starting from the "orthogonality" property 3, the "projections" U_k of U onto the spaces $\overline{W}_i^{k+}$ are expressed in terms of U .

Then the formal series related to U is the series:

$$\sum_k U_k$$

5.1. Representation of U_1

Any field $U_k \in \overline{W}_1^{k+}$ is "orthogonal" to any subspaces $\tilde{W}_2^i$ and $\tilde{W}_1^j$ $(j \neq k)$.

Consequently, it seems reasonable to define the projection U_k of U onto $\overline{W}_1^{k+}$ by:

$$\forall\, i,\ \forall\, j,\ (j \neq k),\ \forall\, U^* \in \tilde{W}_2^i + \tilde{W}_1^j,\ a(U^*, U_k) = 0 \tag{5.1}$$

that is,

$$\int_S U_k \cdot \sigma(U^*)\nu - U^*\, \sigma(U_k)\nu \; dS = 0 \tag{5.2}$$

It is proved that this relation really defines a unique field U_k:

Property 4

The formal series representation of U_1 is the series $\Sigma\, U_k$ where:

- $U_k\ (m,z) = t_k(z)\ N \wedge \text{grad}_m\ f_k$ is a field of $\overline{W}_1^{k+}$,
- f_k is the function defined on the middle surface Σ, solution of one of the following problems Pi or Pii:

Problem Pi

$$\Delta_m f_k - \beta_k^2 f_k = 0$$

$$f_k\big|_{\partial\Sigma} = \frac{1}{h\, \beta_k^2} \int_{-h}^{h} t_k\ (z)\ \omega(U)\ dz \qquad 5.3)$$

$$\int_\Sigma \text{grad}_m\ f_k \cdot \text{grad}_m\ f_k\ d\Sigma < \infty$$

Problem Pii

$$\Delta_m f_k - \beta_k^2 f_k = 0$$

$$\frac{\partial f_k}{\partial \nu}\Big|_{\partial\Sigma} = - \frac{1}{\mu h\, \beta_k^2} \int_{-h}^{h} t_k(z)\ [\frac{\partial a}{\partial s}(U) - \mu\beta_k^2\, \tau\, U] - \mu \dot{t}_k(z)\ \frac{\partial \overline{N}U}{\partial s}\ dz \tag{5.4}$$

$$\int_\Sigma \text{grad}\ f_k \cdot \text{grad}\ f_k\ d\Sigma < \infty$$

where . $\frac{\partial \cdot}{\partial s}$ and $\frac{\partial \cdot}{\partial \nu}$ are respectively tangential and normal derivatives,

. τ is the unit tangential vector ($\tau = N \Lambda \nu$)

. $\omega(U) = \mathrm{div}_m \, N \;\Lambda\; \pi U$

$$a(U) = (\lambda + 2\mu)\, \mathrm{div}_m \; \pi U + \lambda \overline{N}\, U_{,z} \tag{5.5}$$

5.2. Representation of U_2

In the same way, the "projections" U_k of U onto the spaces $\overline{W}_2^{k+}$ are introduced:

Property 5

The formal series representation of U_2 is the series

$$\sum_k W_k$$

where

- $\begin{cases} \pi\, U_k = u\,(z)\ \mathrm{grad}_m \varphi_k \\ N.U_k = \beta_k\ v_k(z)\ \varphi_k(m) \end{cases}$

- φ_k is the function defined on the middle surface Σ, solution of one of the following problems Pi or Pii:

Problem Pi

$$\left|\begin{array}{l} \Delta_m \varphi_k - \beta_k^2 \varphi_k = 0 \\ \varphi_k|_{\partial\Sigma} = \dfrac{1}{\int_{-h}^{h} \dot{c}_k\, u_k + \beta_k\, v_k\, c_k\, dz} \cdot \displaystyle\int_{-h}^{h} c_k(z)\ N.U - u_k(z)\ a(U)\ dz \\ \displaystyle\int_\Sigma \mathrm{grad}\, \overline{\varphi_k} \,.\, \mathrm{grad}\, \varphi_k\ d\Sigma < \infty \end{array}\right. \tag{5.6}$$

Problem Pii

$$\left|\begin{array}{l} \Delta_m \varphi_k - \beta_k^2 \varphi_k = 0 \\ \dfrac{\partial\varphi_k}{\partial\nu}\Big|_{\partial\Sigma} = \dfrac{1}{\int_{-h}^{h} \dot{c}_k\, u_k + \beta_k\, v_k\, c_k\, dz} \cdot \displaystyle\int_{-h}^{h} c_k \nu\ .U + \beta_k\ v_k\ N.\sigma(U)\,\nu - u_k \frac{\partial\omega(U)}{\partial s}\, dz \\ \displaystyle\int_\Sigma \mathrm{grad}\, \overline{\varphi_k}\ \mathrm{grad}\, \varphi_k\ d\Sigma < \infty \end{array}\right. \tag{5.7}$$

with :

$$c_k = \dot{u}_k + \beta_k v_k \tag{5.8}$$

5.3. Fundamental properties

Provided that some regularity holds, the formal series related to U_1 converges to U_1, and the formal series related to U_2 converges to U_2. It follows:

$$\begin{cases} \overline{W}_1^+ = \oplus \overline{W}_1^{k+} \\ \overline{W}_2^+ = \oplus \overline{W}_2^{k+} \end{cases} \tag{5.9}$$

To establish these properties, we have used the following statements:

- when U belongs to $\overline{W}_1^+$, the property is derived from classical results about the Fourier series.
- when U belongs to $\overline{W}_2^+$, this property is derived from the results established by Gusein-Zade (Ref.18) and by Gregory (Ref.14-16) in the case of the semi infinite strip.

Remark

The regularity conditions prescribed by Gregory implies the convergence for any point of the cross-section. Obviously, a convergence with respect to the norm in energy needs less data regularity.

6. EVALUATION OF THE EDGE EFFECTS

6.1. Coordinate functions

Any solution of an interior problem can be written as the sum of a series:

$$U = \sum_k U_k$$

For any k, U_k is explicit in the variable z and expressed in terms of a scalar function x_k ($x_k = \varphi_k$ or $x_k = f_k$) defined on the middle surface Σ. In each case, x_k satisfies the two conditions:

$$\left| \begin{array}{l} \Delta x_k - \beta_k^2 x_k = 0 \\ \int_\Sigma \text{grad } \overline{x_k} \cdot \text{grad } x_k \, d\Sigma < \infty \end{array} \right. \tag{6.1}$$

Generally, these functions x_k cannot be evaluated even numerically since the boundary conditions 5.3, 5.4 and 5.6, 5.7 cannot be expressed, in any way, in terms of the data. Consequently, it is necessary to use methods of approximation, directly deduced from the form of the equations 6.1. It is possible to build a set of coordinate functions as follows:

1. For each eigenvalue $-\beta_k^2$, one brings out a basis (x_k^n) of the space E_k of the functions satisfying (6.1).
2. A displacement field belonging to W_i^{k+} is related to each function x_k^n, thus, a basis of the spaces $\overline{W}_i^{k+}$ is obtained.
3. These basis, gathered, form a basis of $\overline{W}_i^+$.

Therefore, we have to characterize a basis of the space E_k. Such a basis depends on the data regularity and on the boundary regularity.

First case : regular geometry and data

Let 0 be an interior point of Σ and (r,Θ) a system of coordinates (Fig. 5)

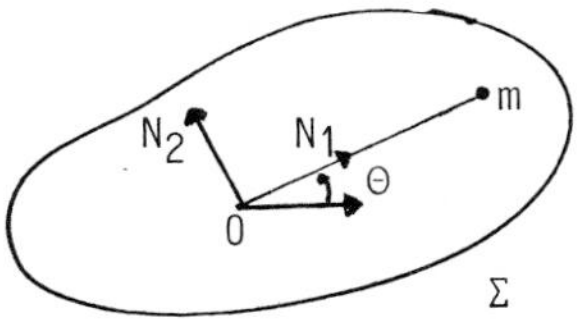

$0m = r\,N_1$

Fig. 5

The functions are 2π Θ-periodic. Then, any function of E_k may be expanded as a Fourier series on the basis:

$$\Theta \longrightarrow e^{in\Theta} \quad , n \in Z$$

It follows:

1. a basis of E_k, for an interior problem, is (x_k^n), where,
   - $x_k^n\,(r,\Theta) = I_n(\beta_k r)\, e^{in\Theta}$ (6.2)
   - I_n is the Bessel's modified function bounded on the vicinity of 0.
2. a basis of E_k, for an exterior problem, is (x_k^n), where,
   - $x_k^n(r,\Theta) = K_n(\beta_k r)\, e^{in\Theta}$ (6.3)
   - K_n is the other Bessel's modified function.

Thence, this set of coordinate functions U_k is obtained:

a. Interior problem

$$\overline{W}_1^+ \left| \begin{aligned} \overline{N}_1 U_{k1}^n &= -\frac{in}{r} I_n(\beta_k r)\, e^{in\Theta}\, t_k(z) \\ \overline{N}_2 U_{k1}^n &= \beta_k \dot{I}_n(\beta_k r)\, e^{in\Theta}\, t_k(z) \\ \overline{N} U_{k1}^n &= 0 \end{aligned} \right. \tag{6.4}$$

$$\overline{W}_2^+ \left| \begin{aligned} \overline{N}_1 U_{k2}^n &= \beta_k \dot{I}_n(\beta_k r)\, e^{in\Theta}\, u_k(z) \\ \overline{N}_2 U_{k2}^n &= \frac{in}{r} I_n(\beta_k r)\, e^{in\Theta}\, u_k(z) \\ \overline{N} U_{k2}^n &= \beta_k I_n(\beta_k r)\, e^{in\Theta}\, v_k(z) \end{aligned} \right. \tag{6.5}$$

b. Exterior problem

The coordinate functions are deduced from the previous one by substituting K_n in place of I_n.

Second case : geometry or (and) data are not regular

Then, one needs to add to the set of previously obtained coordinate functions, a set of functions which take into account the boundary singularities or (and) data singularities.

These functions can be characterized in two ways:

- the first one is analytic. Their expression depends on the kind of singularities. Consequently, their general form cannot be given.
- the second is numerical. Boundary function value, which is non-zero in the vicinity of the singularity point, is prescribed and the problem determining the corresponding coordinate function is solved by a finite element method. It is to be pointed out that such a problem is inexpensive since it is sufficient to solve it in a neighbourhood of the singularity point.

Remark

The previous coordinate functions do not take into account all the kinds of singularities (see. paper P. Ladevèze), and additional coordinate functions have to be introduced according to the specific case.

6.2. Computation method

Let us consider an interior or exterior problem and let us call (u_n) a system of coordinate functions.

The exact solution is approximated by U_n^*:

$$U_n^* = \sum_{k=1}^{n} a_k u_k \tag{6.6}$$

and obviously the problem is to calculate the constants a_k. Since the boundary values characterizing the projections U_k of U onto the subspaces $\overline{W}^k$, are not explicit in terms of the data, it is impossible to look for approximations U_{kn}^* of U_k directly. So, it is first necessary to seek a "global" approximation of the solution.

Let us assume that the displacement U_d is prescribed over the part S_D of the edge, while the load F_d is specified over the remaining part S_L.

Then, the approximation U_n^* is defined by the variational problem:

$$\forall\, U^* \in \mathcal{U}_n, \quad \int_{S_D} (U_n^* - U_d)\cdot\sigma(U^*)\nu \, dS + \int_{S_L} (\sigma(U_n^*)\nu - F_d)\cdot U^* \, dS = 0 \tag{6.7}$$

where $\mathcal{U}_n$ is the space spanned by the n first coordinate functions u_n.

Therefore, a linear system has to be solved to compute the constants a_k.

Thus a sequence of fields U_n^* can be built and a field U_n^* is considered as satisfactory when the values of U_n^* on S_D, and the values of $\sigma(U_n^*)\nu$ on S_L are close to the data. When such an approximate solution U_n^* is known, it is possible to build a better approximation: Approximate boundary values characterizing the functions f_k and φ_k may be evaluated; therefore, using a finite element method, these functions may be computed and a new approximate solution V_n^* is obtained.

6.3. Numerical results

We have built a code which allows to take into account arbitrary boundary conditions.

Some tests have been performed with the following data:

- The plate is circular. Its thickness 2h and its diameter 2R are such that: $\frac{R}{h} = 10$
- The Young modulus E and the Poisson ratio ν are:
  E = 200000 MPa $\qquad \nu = 0.3$
- Bending loads are prescribed over all the edge, and we consider successively:

case i : normal load to the edge: $F_d = F_1 N_1$

case ii : tangential load to the edge: $F_d = F_2 N_2$

case iii: normal load to the plate: $F_d = F_3 N$

In every case, F_i is a Θ-periodic function of the form:

$$F_i(z,\Theta) = f_i(z)\, e^{2i\Theta}$$

the variation in the thickness being defined by:

$$f_1(z) = 5z^3 - 3z$$
$$f_2(z) = 5z^3 - 3z$$
$$f_3(z) = 5z^4 - 6z^2 + 1$$

One will note that all these loads satisfy:

. the resultant and the torque in the thickness are zero

. $f_3(\pm 1) = 0$

The exact solution U is split into two terms:

$$U = U_I + U_L$$

. U_I corresponds to the interior effect. In every case, U_I can be written (Ref.19):

$$U_I = e^{2i\Theta}\,[a_1\, u_I^1(r,z) + a_2\, u_I^2(r,z)]$$

where: . a_1 and a_2 are constants

. u_I^1 and u_I^2 are the fields defined by:

$$\left|\begin{array}{l} N_1 \cdot u_I^1 = -2rz \\ N_2 \cdot u_I^2 = -2rz \\ N \cdot u_I^3 = r^2 \end{array}\right.$$

$$\left|\begin{array}{l} N_1 \cdot u_I^2 = -4zr^3 + \dfrac{4(3\lambda+4\mu)}{\lambda+2\mu} z^3 r \\[2ex] N_2 \cdot u_I^2 = -2zr^3 + \dfrac{4(3\lambda+4\mu)}{\lambda+2\mu} z^3 r \\[2ex] N \cdot {}^3u_I = r^4 + \dfrac{6r^2}{\lambda+2\mu}\,[\lambda z^2 - 4(\lambda+\mu)] \end{array}\right.$$

. U_L corresponds to the edge effects. In every case, U_L may be expanded as a series whose coordinate functions are U_{k1}^2 and U_{k2}^2 defined respectively by 6.4 and 6.5.

An approximate solution U* is sought in the form:

$$U^* = e^{2i\Theta}[a_1^* u_I^1 + a_2^* u_I^2] + \sum_{k=1}^{K} b_k^* U_{k1}^2 + \sum_{l=1}^{N} [c_l^* U_{l2}^2 + \overline{c_l^* U_{l2}^2}]$$

and the constants a_i^*, b_k^*, c_l^* are defined from 6.7.

Obviously, U* and the associated stress σ(U*) can be written:

$$U^* = u^*(r,z)e^{2i\Theta}$$

$$\sigma(U^*) = \sigma^*(r,z)e^{2i\Theta}$$

The values of the data and the values on the edge of the normal stress related to the approximate solution U^* are compared on Fig.6-7-8. The full lines represent the functions $z \longrightarrow f_i(z)$, while the dotted lines are linked to the component $N_i.\sigma(U^*)N_1|_{r=R}$ of the normal stress and represent the functions: $z \longrightarrow N_i.\sigma^*(R,z)N_1$.

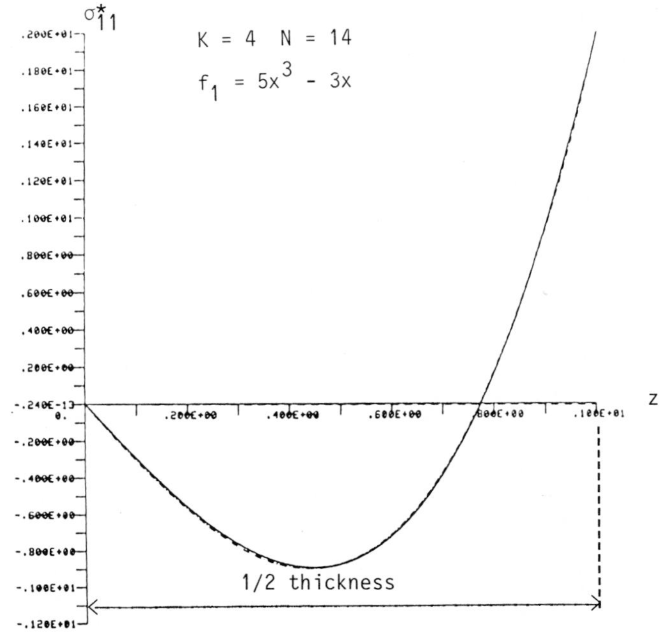

Fig. 6

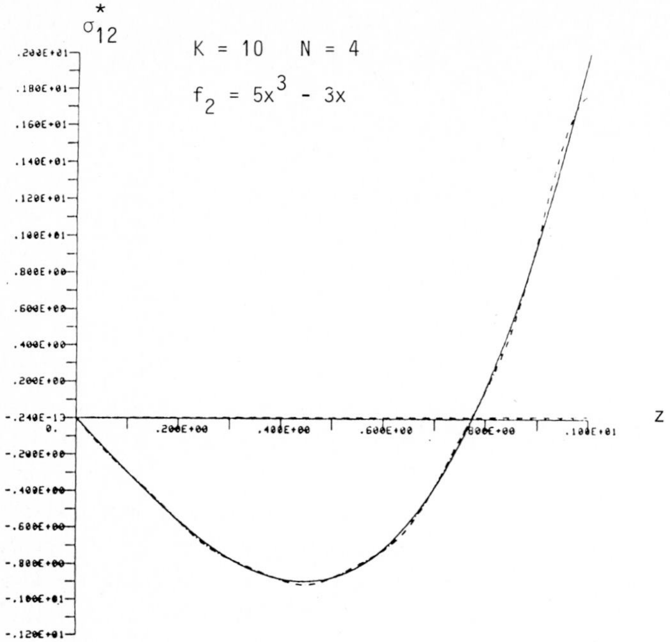

σ^*_{12}

K = 10   N = 4

$f_2 = 5x^3 - 3x$

Fig. 7

σ^*_{13}

K = 4   N = 14

$f_3 = 5x^4 - 6x^2 + 1$

Fig. 8

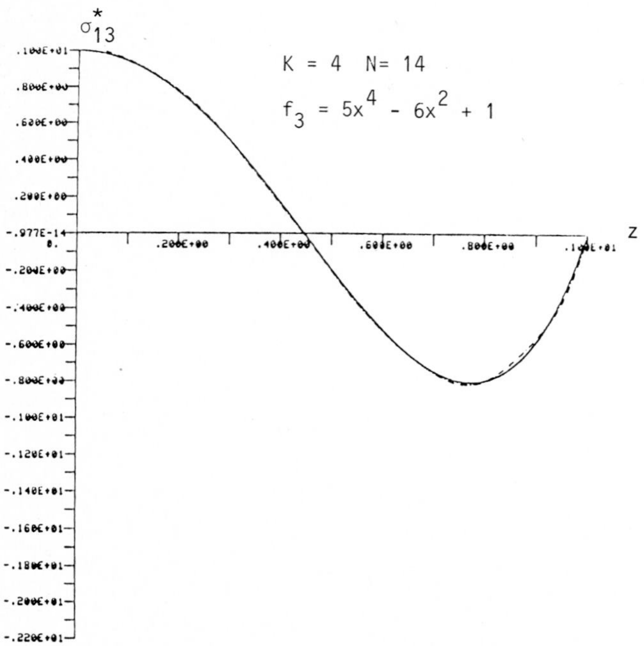

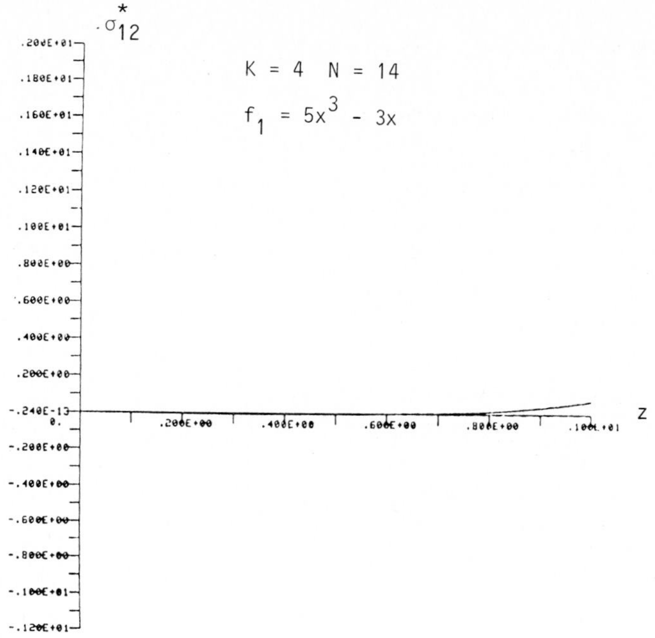

σ*12
K = 4 N = 14
f1 = 5x3 - 3x
.200E+01
.180E+01
.160E+01
.140E+01
.120E+01
.100E+01
.800E+00
.600E+00
.400E+00
.200E+00
-.240E-13
0.
-.200E+00
-.400E+00
-.600E+00
-.800E+00
-.100E+01
-.120E+01
.200E+00
.400E+00
.600E+00
.800E+00
.100E+01
Z

Fig. 9

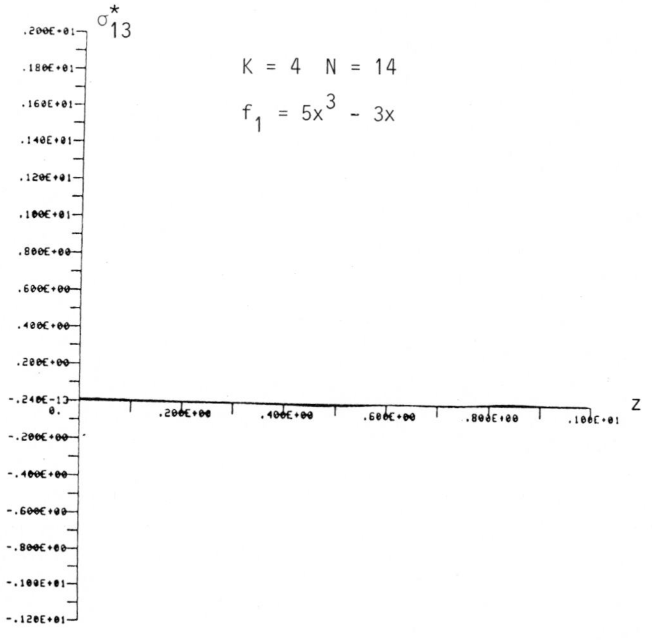

σ*13
K = 4 N = 14
f1 = 5x3 - 3x
.200E+01
.180E+01
.160E+01
.140E+01
.120E+01
.100E+01
.800E+00
.600E+00
.400E+00
.200E+00
-.240E-13
0.
-.200E+00
-.400E+00
-.600E+00
-.800E+00
-.100E+01
-.120E+01
.200E+00
.400E+00
.600E+00
.800E+00
.100E+01
Z

Fig. 10

It is to be pointed out that:

- there is a very good agreement between the calculated values and the prescribed ones.
- a greater number of coordinate functions belonging to $\overline{W}_2^+$ is required when the load is in a plane normal to the edge, whereas one has to take more functions belonging to $\overline{W}_1^+$ when the load is a twisting load. By the way, this result might be expected: it is shown in (Ref. 19) that the exact localized solution is practically the solution of a strip problem when the Saint-Venant Principle holds.

On the Fig.9-10, it is verified, in the case i, that the other components of the normal stress $N_1.\sigma(U^*)N_1$ and $N_2.\sigma(U^*)N_1$ are nearly zero on the edge. Similar results are noticed in the cases i and ii.

The values on the edge of the component $N_1.\sigma(U^*)N_1$ where U^* is the approximate solution corresponding to five eigenvalues only, are shown on the Fig.11. The rapid rate of convergence is obvious.

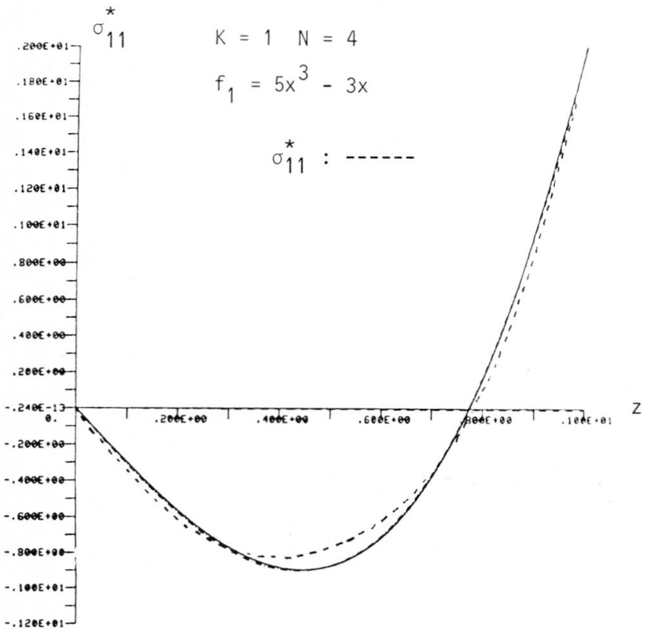

Fig.11

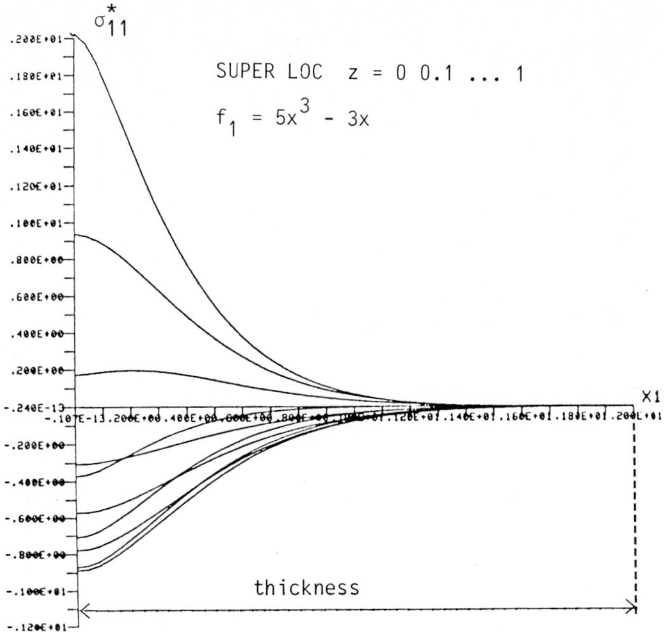

σ*11
SUPER LOC z = 0 0.1 ... 1
f1 = 5x3 - 3x
X1
thickness

Fig. 12

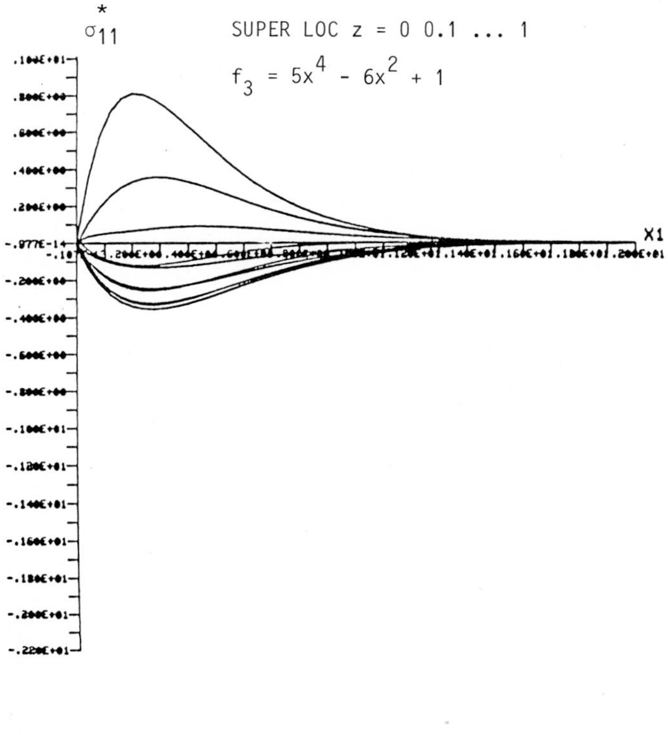

σ*11
SUPER LOC z = 0 0.1 ... 1
f3 = 5x4 - 6x2 + 1
X1

Fig. 13

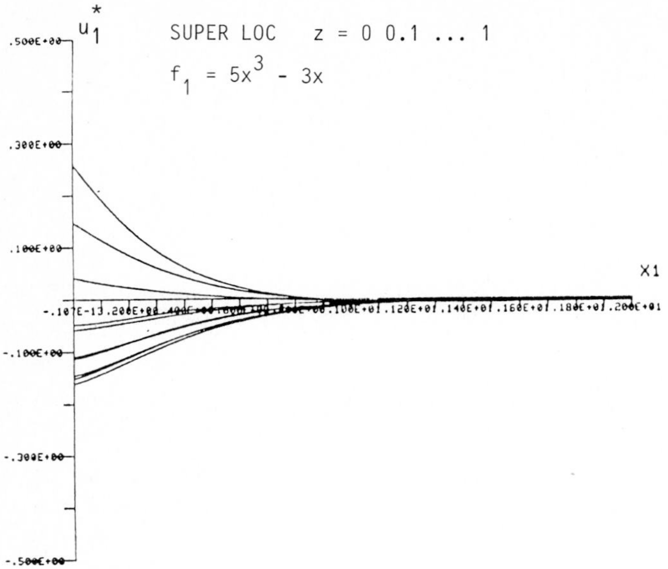
u_1^*
SUPER LOC z = 0 0.1 ... 1
$f_1 = 5x^3 - 3x$
.500E+00
.300E+00
.100E+00
-.100E+00
-.300E+00
-.500E+00
X1
-.107E-13 .200E+00
.100E+01 .120E+01 .140E+01 .160E+01 .180E+01 .200E+01

Fig. 14

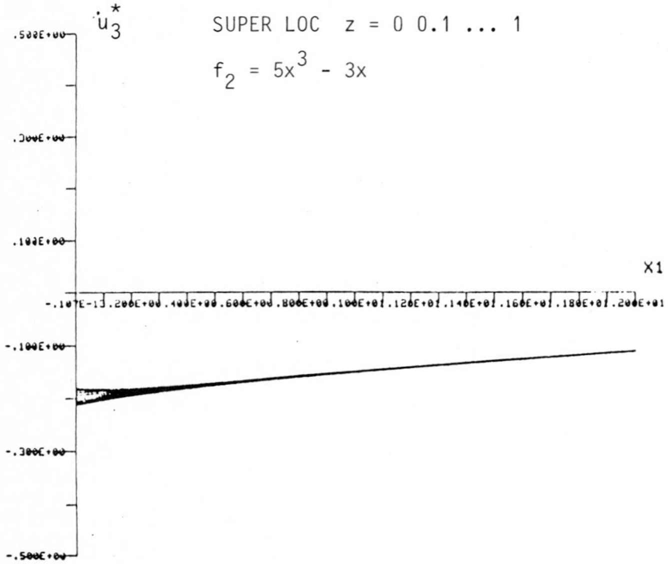
u_3^*
SUPER LOC z = 0 0.1 ... 1
$f_2 = 5x^3 - 3x$
.500E+00
.300E+00
.100E+00
-.100E+00
-.300E+00
-.500E+00
X1
-.107E-13 .200E+00 .400E+00 .600E+00 .800E+00 .100E+01 .120E+01 .140E+01 .160E+01 .180E+01 .200E+01

Fig. 15

On the Fig.12-13, one has drawn the lines corresponding to the variations of the normal stress component $N_1.\sigma(U^*)N_1$ (case i) and $N_1.\sigma(U^*)N_1$ (case iii) on the straight lines $z = 0$, $z = 0.1$, $z = 0.2$, ..., $z = 1$ included in a plane normal to the edge, versus the distance x_1 from the edge of the plate. The very fast decay of the solution is shown clearly, these components are practically zero at a distance from the edge about 1.2. It is to be pointed out that the interior effect is quite negligible.

The graphs of the Fig.14-15 represent the variations of the components of the displacement $N_1.U^*$ (case i) and $N_3.U^*$ (case ii) versus the distance x_1 from the edge. In the case i, the displacement decays very quickly and is practically zero at about $x_1 = 0.8$; the displacement $N_1.U_L$ corresponding to the interior effect is negligible. On the contrary, the component $N_3.U^*$ (case ii) decays very slowly. The edge effect is zero at about $x_1 = 0.2$ and at greater distances from the edge, only the interior effect remains. In this case, it comes from other results that the localized displacement U_L and the "interior" displacement U_I are of the same order of magnitude. These results were foreseen. This load F_dN_2 satisfies the conditions of the first order approximation of the Saint Venant principle but not the conditions of the second order (see paper P. Ladevèze). Consequently, it can be proved that the interior stress σ_I is $O(\frac{h}{r}\sigma_L)$ but the interior displacement U_I is $O(U_L)$.

REFERENCES

1 W.T. Koiter, Error estimates for certain approximate solutions of problems in the linear theory of elasticity, Z.A.M.P., Vol. 21, 1970, p. 534-538.
2 D.A. Danielson, Improved error estimates in the linear theory of thin elastic shells, Proc. Kon. Ned. Ak. Wet. B 73, 1971, p. 294-300.
3 J.G. Simmonds, An improved estimate for the error in the classical linear theory of plate bending, Quart. Appl. Math. Vol. 29, 1972, p. 439-447.
4 P. Ladevèze, Comparaison de modèles de milieux continus, Thèse, Paris 1975.
5 P. Ladevèze, Justification de la théorie linéaire des coques élastiques, J. de Mécanique, Vol. 15, 1976, p. 813-856.
6 P. Ladevèze, On the validity of linear shell theories, Theory of shells - Edit. W.T. Koiter, G.K. Mikhailov, IUTAM, 1980.
7 S.B. Dong and D.B. Goetschel, Edge effects in laminated composite plates, J. of Appl. Mech., Vol. 49, 1982, p. 129-135.
8 C.O. Horgan, Saint-Venant End effects in composites, J. of composite Materials, Vol. 12, 1982, p. 411-422.
9 I. Choi and C.O. Horgan, Saint-Venant's principle and end effects in anisotropic elasticity, J. of Appl. Mech., Vol. 44, 1977, p. 424-430.
10 R.I. Zwiers, T.C.T. Ting and R.L. Spilker, On the logarithmic singularity of free edges stress in laminated composites under uniform extension, J. of Appl. Mech., Vol. 49, 1982, p. 561-569.
11 K.O. Friedrichs and R.F. Dressler, A boundary layer theory for elastic plates, Comm. on Pure and Appl. Math., Vol. 14, 1961, p. 1-33.
12 A.I. Lur'e, Three dimensional problems of the theory of elasticity, Interscience, 1964.
13 S. Cheng, Elasticity theory of plates and a refined theory, J. of Appl. Mech., Vol. 46, 1979, p. 644-650.
14 R.D. Gregory, Green's functions, bi-linear forms, and completeness of the eigenfunctions for the elastostatic strip and wedge, J. of Elasticity, Vol. 9, 1979, p. 283-309.
15 R.D. Gregory, The semi-infinite strip $x > 0$, $-1 < y < 1$, completeness of the Papkovich-Fadle eigenfunctions when $\phi_{xx}(0,y)$, $\phi_{yy}(0,y)$ are prescribed, J. of Elasticity, Vol. 10, 1980, p. 57-80.
16 R.D. Gregory, The traction boundary problem for the elastostatic semi-infinite strip, existence of solution and completeness of the Papkovich-Fadle eigenfunctions, J. of elasticity, Vol. 10, 1980, p. 295-327.
17 D.D. Joseph, L.D. Sturges and W.H. Warner, Convergence of biorthogonal series of biharmonic eigenfunctions by the method of Titchmarch, A.R.M.A., Vol. 78, 1982, p. 223-274.
18. M.I. Gusein-Zade, On necessary and sufficient conditions for the existence of decaying solutions of the plane problem of the theory of elasticity for a semi strip, P.M.M., Vol. 29, n°4, 1965, p. 752-760.
19. F. Pécastaings, Sur le principe de Saint-Venant pour les plaques, Thèse, Paris 1985.

ENERGY STRUCTURE OF LOCALIZATION

V. BERDICHEVSKII[1] and L.TRUSKINOVSKII[2]
[1]The Faculty of Mathematics and Mechanics, Moscow State University, 119899 Moscow, GSP-3 (USSR)
[2]Vernadskii Institute of Geochemistry, the Academy of Sciences, Moscow (USSR)

ABSTRACT

It is shown that a great number of physical objects with localized energy are described by theories with the same mathematical structure: the energy is the sum of two terms - one is a nonconvex functional of field variables and other is a quadratic functional in their gradients. The following examples are considered: bending of rods, dimples in shells, vapor bubbles in a liquid, necking, solitary waves, elementary particles, dislocations.

INTRODUCTION

There is a great number of physical objects in nature with which it is possible to associate energy, localized in space - solitary waves, elementary particles, dislocations in crystal lattice, vapor bubbles in liquid, domen walls in magnetics etc. There arises a question: what is the mathematical structure of the theory which makes it possible to describe such localized congigurations in a natural way as regions of high field gradients. It is clear that such a theory should be nonlinear. This question is deeply connected with the problem of nonlinear generalization of Maxwell's electrodynamics, where charges perform as localized states with a finite energy (cf.[1]). We know a number of theories of localized states (for solitons [2], particle-like solutions in field theory [3], different interfacial boundaries [4-6], dimples on shells [7], dislocations [8]). The aim of the present paper is to draw attention to the fact that all theories suggested have common energy structure: energy is the sum of a nonconvex functional of field variables and a quadratic functional in their gradients. Usually, the density of the energy contains two terms: nonconvex function of field variable and a quadratic form in their derivatives. Some examples are considered here: rod bending, shell buckling, vapor bubbles, necking, solitons, dislocations, elementary particles.

Although almost all the above (excluding rod bending) have already been considered before such a unification and common outlook throw additional light upon the subject. In particular, we discover deep analogy between the dimple edge energy and the energy of interfacial tension.

ENERGETICAL STRUCTURE OF LOCALIZATION

In this paragraph we deal with the main conception in a simplified one-dimensional situation.

Let us consider a scalar field $u(x)$, x changes in the segment $[-1, 1]$. The energy of the field $u(x)$ has the form

$$E = \int_{-\ell}^{\ell} F(u)dx \tag{1}$$

where $F(u)$ is the energy density. If $F(u)$ is strictly convex, the energy E achieves its minimum on the singular minimizing element.

If $F(u)$ is nonconvex, the number of minimizing elements can be infinite.

For example function $F(u) = Au^2(u - 1)^2$ has two local minimums $u = 0$ and $u = 1$ with the same values of the energy density $F = 0$. Then, all the minimizing elements have the next structure: the segment $[-1,1]$ is devided on several intervals and, at each of them, the function $u(x)$ has a constant value: 0 or 1.

Let us add the term $\frac{1}{2}\varepsilon^2(du/dx)^2$ to the energy density. Then the jumps become energetically unprofitable because the energy connected with a jump is infinite. Therefore all the jumps are smoothed and so changed by continuous transition.

In this connection it is interesting to investigate the structure of the stationary points of the energy functional considered.

In order to exclude the influence of boundary points we put $\ell = \infty$

$$E = \int_{-\infty}^{\infty} \left(F(u) + \frac{1}{2}\varepsilon^2 \left(\frac{du}{dx}\right)^2\right)dx . \tag{2}$$

We'll denote the limit values of function $u(x)$ for $x = +\infty$ and $x = -\infty$ by u_- and u_+ respectively. For definiteness, we now suppose that $u_- \leq u_+$.

Stationary points of the functional E (2) are the solutions of the equation

$$\frac{\partial F}{\partial u} - \varepsilon^2 \frac{d^2u}{dx^2} = 0 \qquad (3)$$

As $d^2u/dx^2 \to 0$ for $|x| \to \infty$ from the equation (3) it follows that the limit values u_- and u_+ of the function $u(x)$ are stationary points of the function $F(u)$:

$$\frac{\partial F}{\partial u} = 0, \text{for } u = u_+, \quad u = u_- \qquad (4)$$

Further we shall use the terminology of the theory of phase transitions, so we shall say that the system "is in phase state" u_* if $u(x) \equiv u_*$ and u_* - is the stationary point of $F(u)$. So, one can say with some degree of liberty, that in $x = -\infty$ the system is in u_--phase state and in $x = +\infty$ the system is in u_+ phase state, so the function $u(x)$ describes two-phase state - the continuous transition from u_- phase state to u_+ state (interfacial region).

Let us lower the order of the equation (3), multiplying it by $\frac{du}{dx}$ and integrating on x. We obtain

$$F(u) - \frac{1}{2} \varepsilon^2 \left(\frac{du}{dx}\right)^2 = h \qquad (5)$$

where h is a constant.

As $\frac{du}{dx} \to 0$ when $|x| \to \infty$ from the equation (5) it is seen that the values of F in the points u_- and u_+ are equal to h; so, they are identical

$$F(u_-) = F(u_+) \qquad (6)$$

The equality (6) is highly notable. It shows that "coexsistance" is possible only between phases with equal values of energy. As it will be shown below, the equality (6) appears to be an analogy of the Gibbs condition [9] of chemical potentials equality in thermodynamic equilibrium (in the theory of phase transitions the role of energy density $F(u)$ is played by a chemical potential).

Let us consider the structure of solutions for the functions

F(u), represented in Fig. 1 a,b,c

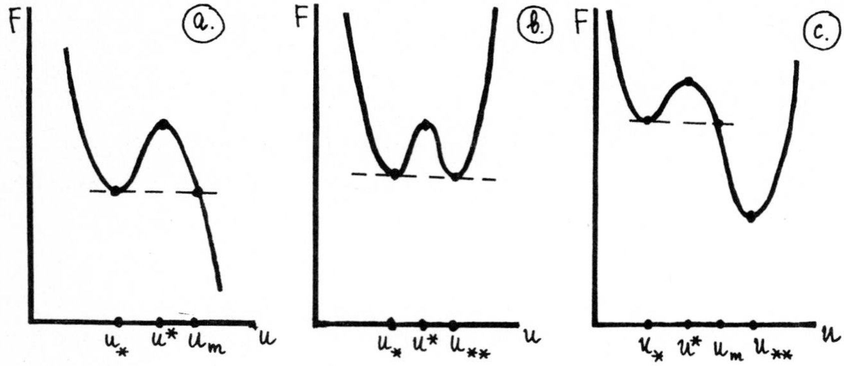

Fig. 1

Functions F(u) 1a, and 1b are particular cases and F(u) 1c is a common one. Function F(u) 1a is a result of drawing the second local minimum of F(u) 1c down in $-\infty$. Function F(u) 1b corresponds to the case of equal values of function F(u) 1c in points of local minimum.

Let us consider only nontrivial solutions, i.e. solutions with $u(x) \neq \text{Const}$.

We shall begin with case 1a, when function F(u) has a local minimum in the point $u = u_*$ and a local maximum in the point $u = u^*$. From the equality (4) and the accepted condition $u_- \leqslant u_+$, it follows that alternatively $u_- = u_*$ and $u_+ = u_*$ or $u_- = u^*$ and $u_+ = u^*$, or $u_- = u^*$ and $u_+ = u^*$. The second variant is impossible because of (6). The third one is also impossible: according to (5) $h = F(u_+)$, so we have the inequality $F(u) - h = F(u) - F(u_+) \leqslant 0$, which is in contradition with the inequality $\frac{1}{2}\,\varepsilon^2(\frac{du}{dx})^2 = F(u) - h \geqslant 0$. The first variant can be realized and the type of the solution is presented in Fig. 2a.

The solution u(x) increases from the value $u_- = u_*$ up to the value u_m, which is defined by the equation $F(u_*) = F(u_m)$, and then descreases to the value $u_+ = u_*$. This solution has the following physical meaning - the system as a whole is in u_* - phase state which corresponds to the local minimum of the energy. The existance of the minimum lying nearby which is deeper (in our case it is equal to $-\infty$) "gains over" part of the system into the phase state with $u \neq u_*$; the last one is localized in space.

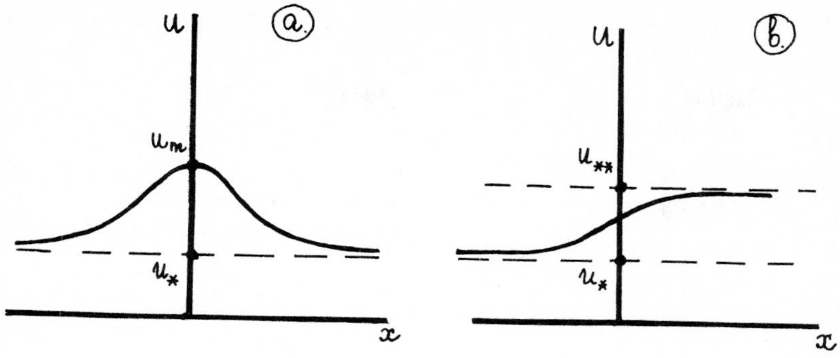

Fig. 2

Function $u(x)$ can be found from (5) by quadrature. In particular, for the function $F(u) = \frac{u^2}{2} - \frac{u^3}{6}$ the solution $u(x)$ is expressed through elementary functions

$$u = 3\ \mathrm{sech}^2 \frac{x}{2\varepsilon} \tag{7}$$

We see from (7) that localized state has the width of order ε. This also follows from (2), because the solution does not depend on ε after the change of veriables $x \to x/\varepsilon$

Let us turn now to case 1b. For the same reason as in case 1a there are no solutions joining the state u_* with u^* and u_{**} with u^*; there is a nontrivial solution with $u_- = u_*$, $u_+ = u_{**}$, and its qualitative structure is presented in Fig 2b. This solution describes the following physical situation: at $-\infty$ the system is in the phase state u_*, at $+\infty$ the system is in another phase state u_{**} and the function $u(x)$ describes gradual transition from one phase to another - the structure of interfacial boundary. The width of the boundary is of order ε. The investigated solution $u(x)$ can be expressed in terms of elementary functions for $F(u) = \frac{u^4}{4} - \frac{u^2}{2}$.

$$u = \mathrm{th} \frac{x}{\sqrt{2}\,\varepsilon}$$

In case 1c there is only one nontrivial solution: $u_- = u_*$, $u_+ = u_*$, $\max u(x) = u_m$. It looks just the same as in case 1a. It is worth mentioning that the solution analysed does not depend on the depth of the minimim in point u_{**}: the only important thing is that the value of F in point u_{**} should be less than in point u_*.

The solutions considered will have finite energy if we regularize the energy functional (2) by subtracting the constant h from the energy density F ($F(u_-) = F(u_+) = h$).

Thus, we have obtained two solutions: one of them is monotone and the other is not. The first is usually called kink and the second is known as a solitary wave. It can be demostrated [3] that the first is stable in the sense that it corresponds to the minimum of the functional, but the second is unstable and corresponds to the saddle point. Nevertheless the solitary wave solutions are of considerable interest, because they make it possible to calculate the minimal energy of the fluctuation, transforming the system from one phase state to another.

Until now we investigated one-dimensional case which of couse is a model one. In situations of physical interest the energy usually depends on a number of functions and the solution is seldom expressed in quadratures. Nevertheless the main conception may be used for constructing a qualitative solution in multidimensional case as well.

ROD BENDING

The energy structure which we have described - nonconvex function + higher derivatives can be observed in the theory of rods and shells. First we shall analize the theory of rods.

Let us consider the rod which is straight in the nondeformed state, and let us load it axially. We assume that the rod is deforming in one plane and that the symmetry of the cross-section is such that no torsion occurs. Hence, the kinematics of the rod can be described with two functions - the components of the displacement vector. We denote the longitudinal displacement by $u(x)$, the transverse displacement by $w(x)$; x is the length of the axis line in nondeformed state (Fig. 3a).

The measure of rod extension γ and the measure of rod bending ϱ are given by the formulas

$$\gamma = \frac{du}{dx} + \frac{1}{2}\left(\frac{dw}{dx}\right)^2, \qquad \varrho = \frac{d^2w}{dx^2} \tag{8}$$

We do not mention restrictions here which make the use of (8) possible.

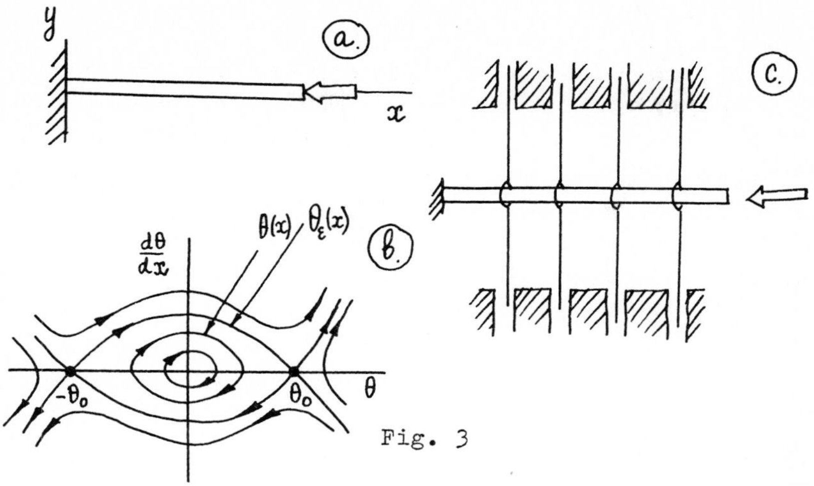

Fig. 3

The density of the rod elastic energy Φ can be expressed as a sum of the strain and bending energy:

$$2\Phi = ES\gamma^2 + EI\rho^2 \tag{9}$$

Here E is Young's modulus, S and I are the area and the inertia moment of a cross-section.

Let us introduce parameter ε through the equality

$$\varepsilon^2 = I/S$$

Parameter ε has the dimension of length. Now we can rewrite the expression for the energy density in the next form:

$$2\Phi/ES = \left(\frac{du}{dx} + \frac{1}{2}\left(\frac{dw}{dx}\right)^2\right)^2 + \varepsilon^2\left(\frac{d^2w}{dx^2}\right)^2 \tag{10}$$

Let us assume that the boundary points of the rod are not fixed. Then the finding of energy stationary points relative to $w(x)$ is equivalent to the same problem relative to $\theta(x) = dW/dx$; the energy depends on W through θ and $d\theta/dx$ only and we can construct admissible function $w(x)$, if we know $\theta(x)$.

So, we get

$$2\Phi/ES = \left(\frac{du}{dx} + \frac{1}{2}\theta^2\right)^2 + \varepsilon^2\left(\frac{d\theta}{dx}\right)^2$$

Now consider the rod under axial load with a fixed linear longitudinal displacement $u = ax$, a is a fixed negative constant (the rod is compressed). Let us mention, that transverse displacement is free. We introduce new constant θ_o by the formula $a = -\frac{1}{2}\theta_o^2$ and write down the energy density of the system + constraint in the form

$$2\Phi/ES = F(\theta) + \varepsilon^2(\frac{d\theta}{dx})^2, \quad F(\theta) = \frac{1}{4}(\theta^2-\theta_o^2)^2 \qquad (11)$$

We see that the energy density contains two terms: nonconvex function $F(\theta)$ *), possesing two local minimums with equal values of F and the square of the first derivative of $\theta(x)$.

The structure of solution on infinite interval has been investigated above. We consider now the case of finite interval.

The equilibrium rod configurations are the stationary points of the functional

$$\int_0^\ell \Phi \, dx + M_o\theta(0) - M_1\theta(\ell) \qquad (12)$$

where M_o and M_1 are the moments applied to the rod ends. Note that M_o and M_1 should not be equal. From (11) and (12) we find

$$F(\theta) = \varepsilon^2(\frac{d\theta}{dx})^2 + h, \quad ES\,\varepsilon^2\frac{d\theta}{dx}\Big|_{x=o,1} = M_{o,1}$$

It is obvious, that for zero moments the solution can be written down in the form (Fig. 2b)

$$x(\theta) = 2\varepsilon\int_{-\sqrt{\theta_o^2-h_o^2}}^{\theta}\left[(\theta'^2-\theta_o^2)^2-h_o^4\right]^{-\frac{1}{2}}d\theta'$$

The constant $h_o = \sqrt[4]{4h}$ can be obtained from the condition $x(\sqrt{\theta_o^2-h_o^2}) = 1$. If the rod is thin enough, that means $\varepsilon/\ell \ll 1$, the integration limits in (12) can be referred to infinity. Then the solution has the form

$$\theta_\varepsilon = \theta_o \, th\frac{\theta_o x}{2\varepsilon}$$

*) It should be noted that for $a > 0$ the function $F(\theta) = (a + \frac{1}{2}\theta^2)^2$ is convex and the effect of localization is absent.

This solution describes the structure of the fin. When ε tends to zero, the continuous solutions θ_ε a.e. uniformly converge to the discontinuous solution with the jump

$$\theta = \begin{cases} -\theta_0, & x < 0 \\ \theta_0, & x > 0 \end{cases}$$

If we calculate the energy we obtain $\frac{2}{3} ES\varepsilon\theta_0^2$.
It is obvious that all the energy is concentrated in the fin, because in the region with zero curvature the strain energy is also equal to zero.

The kinematic constraint for the function u(x), introduced before, can be realized with the help of a simple physical model. Let us fasten to the points of the rod (with the help of hinges) a great number of rigid fibres and, after the rod is compressed, we introduce fibres inside cylinders as is shown in Fig. 3c. The fibres slide inside cylinders without friction. If the number of fibres is big enough, we can describe such a situation using the theory considered.

The kinematic constraint, of course, is significant for the appearance of localized states. It is well known that the Euler's rod has no such states. The reason is that function u(x) , which was fixed in considered model, varies and has to be found in the case of Euler's rod. It tunes in order to minimize the energy and "softens" the nonconvexity of the strain energy, making localized states impossible. The classical problem without constraints can be solved in exact way and the energy minimum is realized at the solution

$$w = 0, \quad u = -\frac{1}{2}\theta_0^2 x \qquad \text{for} \quad \theta_0 < \sqrt{2}\,\frac{\pi\varepsilon}{l}$$

with the energy $1/8\, ES\theta_0^4 l$ and $w = 4\left(\frac{\theta_0^2 l^2}{2\pi^2} - \varepsilon^2\right)\sin\frac{\pi x}{l}$,

$$u = -\frac{1}{2}\theta_0^2 x - \frac{\pi}{2l}\left(\frac{\theta_0^2 l^2}{2\pi^2} - \varepsilon^2\right)\sin\frac{2\pi}{l}x \qquad \text{for} \quad \theta_0 > \sqrt{2}\,\frac{\pi\varepsilon}{l}$$

with the energy $\frac{1}{2} ES\left(\theta_0^2 - \frac{\pi^2\varepsilon^2}{l^2}\right)\frac{\pi^2\varepsilon^2}{l}$ and the last one bifurcates from the trivial first one at $\theta_0 = \sqrt{2}\,\frac{\pi\varepsilon}{l}$. We see that in both cases the localization is absent. That is also true for the constrained rods with $\varepsilon/l \sim 1$.

In the following example, kinematic constraint has not been put from the "outside", but appears in a natural way from the geometrical features of the construction.

DIMPLES ON SHELLS

The energy structure, considered above, can be found in shells theory. We show that the strain energy and the bending energy correspond to nonconvex and differential parts of energy, respectively. The suggested analogy introduces the problem of finding the structure of the dimple boundary.

Let us write down the expression for the elastic energy density of a shell. To simplify, we consider a shell which is a sphere with radius R in nondeformed state.

It's section accross a meridian is presented in Fig. 4a. The zero value of the angular meridianal coordinate α corresponds to the vertical direction, α being increased in the clockwise direction.

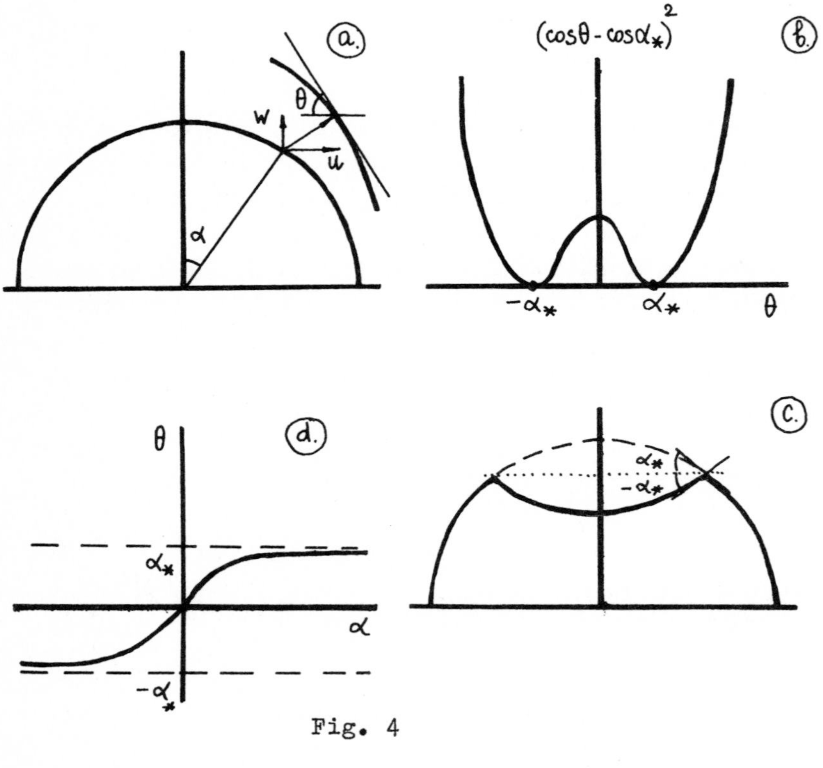

Fig. 4

Let us suppose that the deformation is rotationally symmetric. As a function, characterizing deformed state of the shell, it is convenient to take $\theta(\alpha)$, where θ - is the angle between the horizontal direction and the tangential direction to the deformed middle surface, α can be used as a Lagrange coordinate of the meridian points (see Fig. 4a). By our condition, the angle θ increases when the tangential line rotates clockwise.

If the meridian deformation is nonextensional, the function $\theta(\alpha)$ will be the only characteristic of deformation that we need (up to rigid rotation). If the meridian is subject to extensions we need another function to characterize the deformation; for such function we choose the horisontal component of displacement divided by R , $u(\alpha)$. It is a simple task to calculate the strains γ and γ_1 and the curvature variations Δk and Δk_1 of the meridian and of the parallel, respectively, in terms of $\theta(\alpha)$ and $u(\alpha)$:

$$\gamma = \frac{u' - \cos\theta + \cos\alpha}{\cos\theta} ; \quad \gamma_1 = \frac{u}{\sin\alpha}$$

$$\Delta k = \frac{1}{R}(\theta' - 1); \qquad k_1 = \frac{1}{R}\,\frac{\sin\theta - \sin\alpha}{\sin\alpha} \tag{13}$$

Prime indicates differentiation with respect to α .

The elastic energy of the shell $\mathcal{E}$ is given by the formula

$$\mathcal{E} = \int \Phi \, d\alpha$$

$$\Phi = \frac{\pi E h R^2}{1-\nu^2}\left[\gamma^2 + \gamma_1^2 + 2\nu\gamma\gamma_1 + \frac{h^2}{12}\left(\Delta k_1^2 + \Delta k^2 + 2\nu\,\Delta k\,\Delta k_1\right)\right]\sin\alpha \tag{14}$$

where h is the shell thickness, ν - is Poisson's ratio, E is Young's modulus.

The limits of integration depend on the shell geometry and their values are not important for us. We assume for simplicity that $-\alpha_0 \le \alpha \le \alpha_0$, $(\alpha_0 \sim 1)$. One can think that the deformation is caused by a consentrated force, applied at the shell's pole.

The expression (14) is far from being simple. Let us try to simplify it. We shall show that for a deformation state localized in the neighbourhood of an angle α_* , the expression (14) can be changed by:

$$\Phi = \pi E h R^2 \left[\frac{u^2}{\sin^2\alpha_*} + \varepsilon^2\left(\frac{d\theta}{d\alpha}\right)^2\right]\sin\alpha_* \tag{15}$$

The dimensionless parameter ε is defined by the equality $\varepsilon^2 = h^2/12R^2(1-\nu^2)$. The functions $u(\alpha)$ and $\theta(\alpha)$ satisfy the nonholonomic constraint

$$u' = \cos\theta - \cos\alpha_* \qquad (16)$$

If we "remove" the prime in (16), i.e. if we put $u = \cos\theta - \cos\alpha_*$, we obtain the familiar energy with the function $F(\theta)$, shown in Fig. 4b. The derivative in (16) changes to some extent the character of the nonconvexity; however, as it will be seen below, localized states still appear to be possible.

We give now a proof of formulae (15), (16). In the localized state, the functions to be found $u(\alpha)$ and $\theta(\alpha)$ vary at the distance of order ε, hence after the differentiation the large parameter ε^{-1} appears, so we can investigate the functional (15) by the variational-assymptotic method [10].

Let us fix the function $\theta(\alpha)$ and look for the function $u(\alpha)$. In accordance with the general scheme of the variational-assymptotic method, we keep in the energy the main terms, containing $u(\alpha)$, and the main interacting terms between $u(\alpha)$ and $\theta(\alpha)$. As $u' \gg u$, we have

$$= \frac{EhR^2}{1-\nu^2} \quad \frac{1}{\cos^2\theta} (u'^2 - 2u'(\cos\theta - \cos\alpha))$$

Varying u', we obtain the equality (16).

Denoting the solution of the equation (16) by u_0, we present $u(\alpha)$ in the form

$$u = u_0 + u_1 \qquad (17)$$

where $u_1 \ll u_0$. Substituting the expression (17) into (14) and keeping the main terms, containing u_1, and the main interacting terms, we get

$$\Phi = \frac{EhR^2}{1-\nu^2}\left[\frac{1}{\cos^2\theta} u_1'^2 + \frac{2u_0u_1}{\sin^2\alpha_*} + 2\nu\frac{u_1'}{\cos\theta} - \frac{u_0}{\sin\alpha_*}\right] \qquad (18)$$

The second term in (18) may be omitted in comparison with the third term, because, $u_1' \gg u_1$. Varying u_1, we find

$$\frac{u'}{\cos\theta} = -\nu\frac{u_0}{\sin\alpha_*} \qquad (19)$$

Substituting again (17) into (14), we see that the main term in the strain energy coincides with the first term in (15). The proof is finished by omitting 1 in the expression for Δk (because of $\theta' \gg 1$) as well as Δk_1 in comparison with θ' (because of $\theta' \gg \Delta k_1$).

Now we turn to the analysis of formulae (15) and (16). We suppose that the width of the localized state is far less than α_*, so the localized state is placed far from the shell's pole. In this case we can consider α as a formal variable, changing at the whole axis: $-\infty < \alpha < \infty$. So we can find the localized states by looking for stationary points of the functional

$$I = \int_{-\infty}^{\infty} \left[\frac{u^2}{\sin^2\alpha_*} + \varepsilon^2 \left(\frac{d\theta}{d\alpha} \right)^2 \right] \sin \alpha_* \, d\alpha \qquad (20)$$

with the functions $u(\alpha)$ and $\theta(\alpha)$ subject to the constraint (16).

The finitness of the functional I involve the conditions $\theta' \to 0$, $u \to 0$ when $|\alpha| \to \infty$. So, according to (16), $\cos\theta|_{\alpha=\pm\infty} = \cos\alpha_*$ and we have $\theta|_{\alpha=\pm\infty} = \pm\alpha_*$. This means, that the localized state has the form of a circular edge (Fig. 4c) and can be considered in the first approximation as an isometric deformation of the sphere: the upper spherical segment is reflected with regards to the horizontal plane. This fully coincides with the theory of shell dimples due to A.Pogorelov[7]

A qualitative form of the dependance of θ upon α is presented in Fig. 4d.

In order to obtain the edge energy as function of the parameters ε and α_*, it is convenient to make a change of variables $u \to \bar{u}$, $\theta \to \bar{\theta}$, $\alpha \to x$ using the formulae $u = a\bar{u}$, $\theta = \alpha_* \bar{\theta}$, $\alpha = bx$. The values of the constants a and b are choosen in a special way to make the integrand of (20) independent on ε and α_* and the constraint (16) independent on ε and α_* in the limit of small α_*. It is easy to find that we have to set

$$a = b\alpha_*^2 \cos\alpha_* , \quad b = \sqrt{\frac{\varepsilon \, tg\,\alpha_*}{\alpha_*}}$$

Under this choice of the values of the constants the minimal value $\underline{I}$ of the functional I will be equal to

$$\underline{I} = \varepsilon^{3/2} \alpha_*^{5/2} \sqrt{tg\,\alpha_*} \cos \alpha_* \; \underline{I}_1(\alpha_*) \tag{21}$$

where $\underline{I}_1(\alpha_*)$ is the minimal value of the functional

$$I_1 = \int_{-\infty}^{+\infty} (\bar{u}^2 + (\frac{d\theta}{dx})^2)\,dx \tag{22}$$

subject to the constraint

$$\frac{d\bar{u}}{dx} = \frac{1}{\alpha_*^2}(\frac{\cos \alpha_* \bar{\theta}}{\cos \alpha_*} - 1) \tag{23}$$

The constraint (23) in the limit $\alpha_* \to 0$, corresponding to shallow shells, takes the form, independent of α_*

$$\frac{d\bar{u}}{dx} = \frac{1}{2}(1 - \bar{\theta}^2) \tag{24}$$

Tending also α_* to zero in (21) and taking into consideration (14), (15), we obtain the final expression for the edge energy:

$$\mathcal{E} = \pi\, EhR^2\, \varepsilon^{3/2} \alpha_*^3\, \underline{I}_1(0)$$

The equivalent expression was obtained in other terms in the pioneer work by A.Pogorelov [7].

The width of the localization region has in terms of the angle variables the order of $\sqrt{\varepsilon} \sim \frac{h}{R}$. In the terms of arc lenth of the meridian the corresponding order is $R\sqrt{\varepsilon} \sim \sqrt{hR}$.

This analysis makes it possible to get the next corrections for the energy with respect to α_* . To perform this we have to calculate the lower limit of the functional (22), when the constraint contains the first correction term with respect to α_*

$$\frac{d\bar{u}}{dx} = \frac{1}{2}(1 - \bar{\theta}^2) + \frac{5}{24}\alpha_*^2 \; (1 - \bar{\theta}^2) \tag{25}$$

Let us find the order of magnitude of the energy correction term. Presenting the solution in the form $u = u_0 + u_1$, where u_0, u_1 - the solution of our problem for $\alpha_* = 0$, and $u_1 \ll u_0$, $\theta_1 \ll \theta_0$ we get

$$I_1 = \underline{I}_1(0) + \int_{-\infty}^{+\infty} 2(u_0 u_1 + \frac{d\theta_0}{dx}\frac{d\theta_1}{dx})\,dx + \int_{-\infty}^{+\infty} (u_1^2 + (\frac{d\theta_1}{dx})^2)\,dx \to \inf_{u_1, \theta_1} \tag{26}$$

where u_1 and θ_1 satisfy the linearized restrictions (25)

$$\frac{du_1}{dx} = - \theta_0\theta_1 + \frac{5}{24} \alpha_*^2 (1 - \theta_0^2) \tag{27}$$

The Euler's equations for the variational problem (22), (24) have the form

$$2u_0 = \frac{d\lambda}{dx}, \quad \lambda\theta_0 - 2 \frac{d^2\theta}{dx^2} = 0 \tag{28}$$

Here λ is a Lagrangean multiplier for the constraint (24). Because of (28), the first integral in (26) does not depend upon u_1 and θ_1:

$$\int_{-\infty}^{\infty} 2(u_0 u_1 + \frac{d\theta_0}{dx}\frac{d\theta_1}{dx})dx = -\int_{-\infty}^{\infty} \lambda (u_{1x} + \theta_0\theta_1)dx =$$
$$= -\int_{-\infty}^{\infty} \lambda \frac{5}{24}\alpha_*^2 (1 - \theta_0^2)dx = c\,\alpha_*^2 \tag{29}$$

Therefore the problem of finding u_1 and θ_1 comes to the minimization of the second integral in (26) subject to the constraint (27). It is convenient to make the change of variables: $u_1 = \alpha_*^2 u_2$, $\theta_1 = \alpha_*^2\theta_2$. Then there appears the multiplier α_*^4 in front of the second integral in (26), though in the constraint (27) parameter α_* disappears. Hence the problem of determining u_2 and θ_2 does not contain the small parameter and the second integral in (26) has the order α_*^4. Thus, when α_* is sufficiently small the function $I_1(\alpha_*)$ has the asymptotic form

$$I_1(\alpha_*) = I_1(0) + c\,\alpha_*^2$$

According to (29) the constant c can be found after we solve the problem (22), (24), corresponding to case $\alpha_* = 0$.

Let us estimate the error of the expression $\varepsilon^2\theta'^2$ for bending energy. Because of $\theta' = \frac{\alpha_*}{b} d\bar{\theta}/dx$, $b \sim \sqrt{\varepsilon}$, the replacement of $(\theta' - 1)^2$ by θ' brings us to error of order $\varepsilon\alpha_*^{-2}$ in comparison with unity (we note that the integral of the term, linear on θ', vanishes). The same error is due to omitting the bending energy of the parallel $(\Delta k_1)^2$. The quantity $\varepsilon\alpha_*^{-2}$ has the order of hR/r^2, where r is the distance between the fin and the shell axis. So, our expression is admissible if $r \gg hR$.

The position of the localized state (of the fin) can be determined by the procedure, suggested by Pogorelov [7]: we have to equalize the variation of the fin energy with respect to free parameter α_* to the corresponding work of external forces.

If $r \sim \sqrt{hR}$ we can't replace the expression (14) by (15), though it is possible to make other simplifications which are due to the fact that the deformed region is localized in the vecinity of the pole, so we can take α and θ to be small enough in the boundary-layer which appears. For the measures of strain and bending we have simplified expressions

$$\gamma = u' + \frac{1}{2}\left(\theta^2 - \alpha^2\right), \gamma_1 = \frac{u}{\alpha}, \quad \Delta k = \frac{1}{R}\left(\theta' - 1\right), \quad \Delta k_1 = \frac{1}{R}\frac{\theta - \alpha}{\alpha}.$$

For the integral it is possible to take again the infinity limits. It is convenient to introduce function φ instead of $\theta(\alpha)$: $\theta = \alpha + \varphi$. Therefore the energy is given by the formula

$$\mathcal{E} = \pi E h R^2 \int_{-\infty}^{\infty} \left[\alpha\left(u' + \varphi(\alpha + \varphi)\right)^2 + \frac{u^2}{\alpha} + \varepsilon^2\left(\alpha\varphi'^2 + \frac{\varphi^2}{\alpha}\right)\right] d\alpha$$

Here ε is determined by $\varepsilon^2 = h^2/12R^2$; we have also put $\nu = 0$ for simplicity. Calculation of the energy comes now to the solving of the system of two ordinary differential equations for $u(\alpha)$ and $\varphi(\alpha)$.

Now we consider the case of spherical shell subjected to a uniformly distributed normal load p. The potential energy due to the load p is given by

$$A = \pi p R^3 \int_0^{\alpha_0} \sin^2\alpha \ \sin\theta \ d\alpha$$

For simplicity we limit ourselves with the case of shallow shell $(\alpha_0 \ll 1)$.

For this case let us introduce the new quantity $\varrho(r)$ by the formulae

$$\sin\theta \simeq \left(\varrho + \frac{r}{R}\right), \text{ so } \cos\theta \simeq 1 - \frac{1}{2}\left(\varrho + \frac{r}{R}\right)^2$$

and also

$$\theta' \simeq \left(\varrho' + \frac{1}{R}\right), \quad \gamma = R u' + \frac{\varrho^2}{2} + \varrho\frac{r}{R}$$

Here $r(\alpha) = R\sin\alpha \simeq R\alpha$ - is radial cylindrical coordinate, prime denotes differentiation with respect to r. It is clear that $\varrho(r) = -W'(r)$, where $W(r)$ is vertical deplacement of the shell. The problem of determining equilibrium configuration comes to the finding of the extremum of energy functional

$$I = \frac{Eh}{1-\nu^2}\int_0^{r}\Big[\Big(\gamma^2 + \frac{u^2}{r^2}R^2 + \frac{2\nu u\gamma}{r}R\Big)r + \frac{h^2}{12}\Big(\varrho'^2 + \Big(\frac{\varrho}{r}\Big)^2 + \frac{2\nu\varrho\varrho'}{r}\Big)r +$$

$$+ \frac{2(1-\nu^2)}{Eh}\, p\frac{r^2}{2}\Big(\varrho + \frac{r}{R}\Big)\, dr \qquad (30)$$

where $r_0 = r(\alpha_0)$. The trivial equilibrium state is given by a constant radial and vanishing tangential displacement:

$$u_0 = -\frac{p(1-\nu)}{2Eh}\frac{r}{R}, \quad \varrho_0 = 0$$

To investigate other states of equilibrium it is convenient to put $u = u_0 + \tilde{u}$, $\varrho = \varrho_0 + \tilde{\varrho}$. In terms of new variables $\tilde{u}(r)$, $\tilde{\varkappa}(r) = \tilde{\varrho}(r)\frac{R}{r}$ the energy can be rewritten in the form:

$$I = \frac{\pi Eh}{1-\nu^2}\Big[\int_0^\infty \Big(\tilde{u}' - \frac{\tilde{u}}{r}\Big)^2 \frac{r\,dr}{R^{-2}} + 2\int_0^\infty\Big(\frac{\tilde{\varkappa}^2}{2} + \tilde{\varkappa}\Big)\Big(\tilde{u}' + \nu\frac{\tilde{u}}{r}\Big)\frac{r^3}{R}\,dr +$$

$$+ \frac{h^2}{12R^2}\int_0^\infty \tilde{\varkappa}'^2 r^3 dr + \int_0^\infty \frac{r^5}{R^4}\Big(\frac{\tilde{\varkappa}^2}{2} + \tilde{\varkappa}\Big)^2 dr + \frac{p(\nu^2-1)}{2EhR}\int_0^\infty \tilde{\varkappa}^2 r^3 dr \qquad (31)$$

where we have extended the upper integral limit to infinity, assuming that buckling is concentrated near the pole and boundary effects are not essential.

In order to determine the structure of the energy we use the simplifying assumption going back to von Karman and Tsien [11]: the deflection is vertical, i.e. parallel to the axis of symmetry. Such an assumption can be regarded as a kinematic restriction. Karman-Tsien assumption is incorrect, because of doubling the Euler critical load [12]. Nevertheless we think it to be instructive to clarify the energy structure and to obtain at least qualitative results.

Thus, setting $u = 0$ in (31), we find

$$I = \frac{Eh}{1-\nu^2}\int_0^\infty \big(\Phi(\tilde{\varkappa}, r) + \varepsilon^2 r\tilde{\varkappa}'^2\big)\,dr$$

where $\varepsilon^2 = \dfrac{h^2}{12R^2}$ is small parameter and

$$\Phi(\tilde{\varkappa}, r) = \Big(\frac{\tilde{\varkappa}^2}{2} + \tilde{\varkappa}\Big)^2\frac{r^5}{R^4} + \frac{p(\nu^2-1)}{2ERh}r^3\tilde{\varkappa}^2$$

is nonconvex function of $\tilde{\varkappa}$, parametrized with the load p .

It can be seen that the energy has a familiar structure, containing nonconvex algebraic term and quadratic term in derivatives. "Kink"-solution corresponds to a dimple with edge far from the shell axis; the "solitary wave"-solution corresponds to a dimple in the vicinity of the shell pole.

The method of dimple edge energy calculation can be generalized in a natural way for nonspherical shells as well.

INTERFACIAL BOUNDARIES

In the theory of rods and shells the bending energy, which contains second derivatives of field variables, is important in a number of problems, so the necessity of the including it into the energy expression is commonly accepted. Alternatively, the theory of interfacial boundaries, which we are going to consider, seems to be the only example of essential application of the material models with higher derivatives.

In classical theory of heterophase equilibrium surfaces of material characteristics discontinuities deviding the coexisting phases are introduced. It is suggested, that a surface of discontinuity posseses some energy. The surface position in space is considered as an additional independent degree of freedom. Alternative way of reasoning,introduced for the first time by van der Waals [4], considers interfacial boundary as a thin layer of continuous change of parameters. In order to describe such a layer van der Waals used the weakly nonlocal model of continuum with the mass density of free energy of the form:

$$\rho \Phi(\rho, \nabla\rho, T) = \rho F(\rho, T) + \frac{1}{2}\varepsilon^2 (\nabla\rho)^2 \qquad (33)$$

where ρ is the mass density, T is the temperature, ε is a small parameter. For onephase liquids $\rho F(\rho,T)$ is a convex function of ρ , so the second term in (33) appears to be nonessential. If the liquid exists in two different phase states, the function $\rho F(\rho,T)$ has 2 convex branches, determined, generaly speaking, on a noninteracting intervals of the ρ variable. Van der Waals was the first who suggested to use single nonconvex function $\rho F(\rho,T)$ on the whole interval of densities. For example, in the case of van der Waals' gas (liquid), F(ρ , T) has the form

$$F(\varrho, T) = f(T) - a\varrho - RT\ln(\frac{1}{\varrho} - b) \quad (34)$$

where a, b, R - are positive constants. Qualitative appearance of the function F (34) - when temperature is sufficiently small - is shown in Fig. 5.

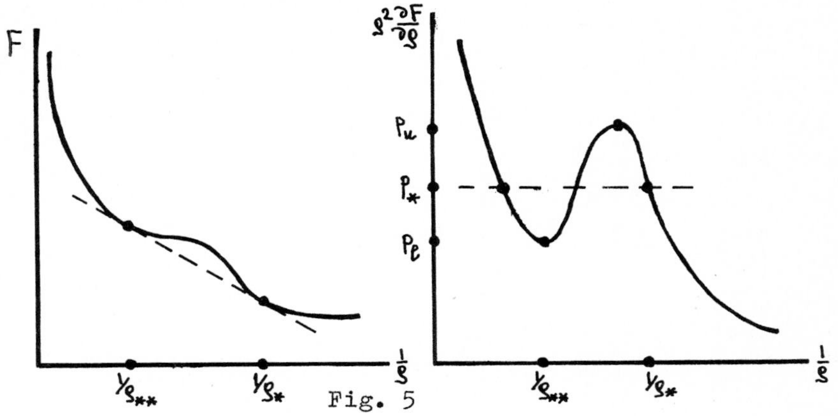

Fig. 5

Van der Waals' theory contains all the ingredients of the class of theories being considered: nonconvexity + high derivatives.

The states of heterophase equilibrium in the domain V under a fixed external pressure p can be found from the equality of variations of free energy and the work of external pressure:

$$\delta \int_V (\varrho F(\varrho, T) + \varepsilon^2 (\nabla\varrho)^2) dV = - p\,\delta V \quad (35)$$

Here it is supposed that the temperature T is also fixed.

It is seen from (35) that equilibrium states are the stationary points of the functional

$$\int_V \left[\varrho F(\varrho, T) + \frac{p}{\varrho} + \frac{1}{2}\varepsilon^2 (\nabla\varrho)^2 \right] dV \quad (36)$$

We call the function $\mu(\varrho, p, T) = F(\varrho, T) + p/\varrho$ chemical po- potential*). For a liquid with function $F(\varrho, T)$, shown in

*) This function has meaning of nonequilibrium chemical potential; usually its equilibrium analog: $\mu(p, T) = \min_{\varrho} \mu(\varrho, p, T)$ is named "chemical potential".

Fig. 5, it can be find such $p = p^*$ that the chemical potential $\mu(\varrho, p^*, T)$ would have two local minimums on ϱ with equal values of μ :

$$\mu(\varrho_*, p^*, T) = \mu(\varrho_{**}, p^*, T) = \mu_0$$

It is convinient to change the functional (36), subtracting the constant $\mu_0 M$, where M is the mass, contained in the volume V , $M = \int_V \varrho \, dv$. In the case of unbounded volume $V = R^3$ the integral

$$\int_{R^3} \left[\varrho(\mu(\varrho, p^*, T) - \mu_0) + \tfrac{1}{2}\varepsilon^2 (\nabla\varrho)^2 \right] dv$$

converges under the condition that the density tends to the values ϱ_* or ϱ_{**} at infinity.

If the density depends on the single coordinate $x_1 = x$ the equilibrium configurations are the stationary points of the functional

$$\int_{-\infty}^{\infty} \left[\varrho(\mu(\varrho, p^*, T) - \mu_0) + \frac{1}{2}\left(\frac{d\varrho}{dx}\right)^2 \right] dx \tag{37}$$

which is just the same as one being considered above. Note, that for the functional (37) conditions (4) and (6) take the form:

$$\varrho \frac{\partial F}{\partial \varrho}\Big|_{\varrho = \varrho_*, \varrho_{**}} = p^*, \qquad \left(F + \frac{p^*}{\varrho}\right)\Big|_{\varrho = \varrho_*, \varrho_{**}} = \mu_0$$

and correspond to the equality of pressures and chemical potentials in coexisting phases.

Now we turn to spherical equilibrium configurations, being the stationary points of the functional

$$4\pi \int_0^{\infty} \left[\varrho(\mu(\varrho, p, T) - \mu_0) + \frac{1}{2}\left(\frac{d\varrho}{dr}\right)^2 \right] r^2 dr \tag{38}$$

For the convergency of the integral (38) we put $\mu_0 = \mu(\varrho_\infty, p, T)$ where $\varrho_\infty = \varrho(\infty)$.

It is interesting to consider the stationary points of the functional (38) for all positive values of external pressure p . There can be defined three singular values of p : lower critical pressure p_l, critical pressure p_* and upper critical pressure p_u, $p_l < p_* < p_u$. They are determined by following conditions: for $p < p_l$ the function $\mu(\varrho, p, T)$ has the only minimum (at the point ϱ_*), for $p = p_l$ it appears an additional stationary point ϱ_{**} for $p > p_l$ the function $\mu(\varrho, p, T)$ has two local minima at the points ϱ_* and ϱ_{**}, for $p = p_*$ the values of μ at the points ϱ_* and ϱ_{**} are equal, $\mu(\varrho_*) < \mu(\varrho_{**})$ for $p > p_*$, and the local minimum ϱ_* dissappears for $p > p_u$ (Fig. 6a-g).

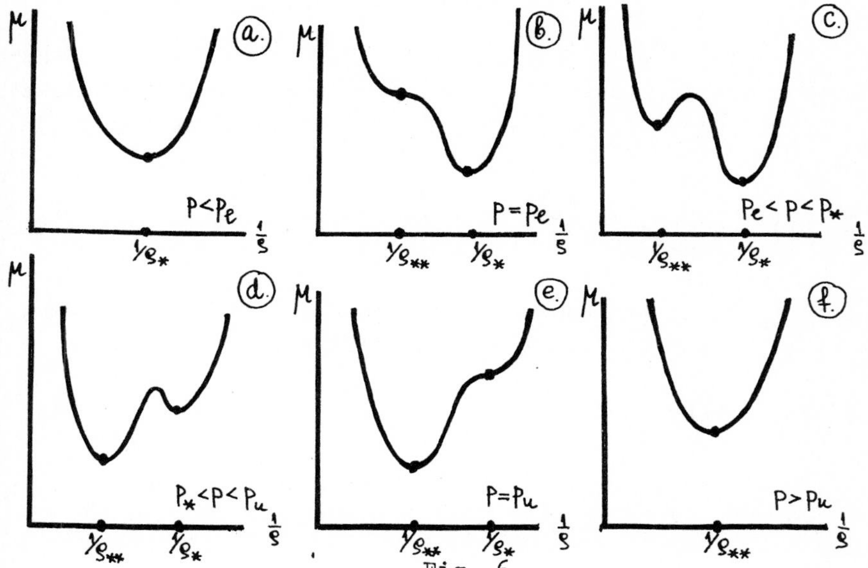

Fig. 6

Although the integrands in (38) and (2) distinguish by the multiplier r^2, the stationary points of (38) and (2) are similar. So, we have:

for $p < p_l$ there is onephase state $\varrho(r) = \varrho_*$

for $p_l < p < p_*$ there is nontrivial stationary point with $\varrho_\infty = \varrho_{**}, \varrho_0 = \varrho(0) > \varrho_*$
(nucleus of **vapour**)

for $p_* < p < p_u$ there is nontrivial stationary point with $\varrho_\infty = \varrho_*$, $\varrho_\infty < \varrho_0$
(**liquid** nucleus)

for $p_* > p_u$ there is onephase state $\varrho(r) = \varrho_{**}$

To show that we need to investigate the equation

$$\frac{\partial}{\partial \varrho}\left[\varrho(\mu - \mu_0)\right] = \varepsilon^2\left(\frac{d^2\varrho}{dr^2} + \frac{2}{r}\frac{d\varrho}{dr}\right) \qquad (39)$$

with the natural boundary conditions $d\varrho/dr_{r=0} = d\varrho/dr_{r=\infty} = 0$. Because $d^2\varrho/dr^2|_\infty = 0$ as well, the equation (39) yields $\partial[(\mu(\varrho, p, T) - \mu_0)]/\partial\varrho = 0$ for $r = \infty$. It can be seen, that this condition connects ϱ_∞ and p:

$$\varrho^2 \frac{\partial F}{\partial \varrho}\bigg|_{\varrho_\infty} = P$$

We have to omit the investigation of (39) so as the **demonstra**tion of solutions instability because of restricted paper volume; for details see [13, 14]. We note only that integration of (39) gives

$$p(0) - p(\infty) = 2\int_0^\infty \varepsilon^2 \frac{d}{dr}\frac{dr}{r} \tag{40}$$

where $p(r) = \varrho^2 \partial F/\partial\varrho|_{\varrho=\varrho(r)}$. For states, localized near $r = R$, the equation (40) is similar to the Laplace **formula**

$$p(0) - p(\infty) = \frac{2\sigma}{R} \tag{41}$$

From (40) and (41) one can find the expression for interfacial tension

$$\sigma = R\int_0^\infty \varepsilon^2 \frac{d}{dr}\frac{dr}{r}$$

It is obvious that the existence of interfacial surface tension is a manifestation of nonhydrostatic nature of stress tensor. It can be shown that the equation (39) is just an integral of the standard tensor equation of elastic equilibrium [13]. We have to omit the discussion because it needs too much technic of nonlocal continuum mechanics (see [13-16]).

NECKING IN FIBERS

If a long polymeric or metallic fiber is stretched along the axis, a simple homogeneous extension may evolve into a nonhomogeneous state in which the fiber thins down in one or more short regions, i.e. necks appear.

It seems that the formation of the neck can be also described by the theory of energy structure which we are discussing. It is clear that the neck expansion is accompanied by a number of quite different physical processes, nevertheless we suppose that the first appearence of the neck is due to nonconvexity of the energy.

Let us consider a circular slender cylindrical rod, having radius a and length l in initial nondeformed state. We assume that the deformation can be expressed in terms of the radial displacement $u(x)$ and the displacement $W(x)$ along the axis:

$$\hat{x} = x + W(x), \quad \hat{r} = r(1 + u(x)), \quad \hat{\theta} = \theta$$

Here x, r, θ are cylindrical Lagrangian coordinates in the reference configuration and $\hat{x}, \hat{r}, \hat{\theta}$ - coordinates of the rod points in deformed state; displacement $u(x)$ appears to be dimensionless.

Components of the strain tensor ε_{ij} depend on r, u, u', W', thus the density of the elastic energy can be thought as a known function of r, u, u', W'. Averaging the energy along the cross-section, we obtain

$$\Phi(u, u', W') = \frac{2\pi}{\pi a^2} \int_0^a V(r, u, u', W') r\,dr$$

We suppose, that one end of the rod is fastened: $u(0) = 0$, $W(0) = 0$, the other end is loaded with a tensile force P, and the lateral surface is free.

Then the equilibrium configurations are the stationary points of the functional

$$\int_0^l \Phi(u, u', W')dx - PW(l) = \int_0^l \left[\Phi(u, u', W') - PW'\right] dx \qquad (42)$$

It is obvious that in order to find stationary points with respect to W' we have to minimize the integrand in (42) with W'. We denote through $\Phi(u, u')$ the function

$$\Phi(u, u') = \min_{W'}(\Phi(u, u', W') - PW')$$

So we need now to find the stationary point of the functional

$$\int_0^l \Phi(u, u')dx$$

with respect to functions $u(x)$, $u(0) = 0$.

It is reasonable to suggest the following approximation for $\Phi(u, u')$

$$\Phi(u, u') = F(u) + \varepsilon^2 u'^2$$

where $F(u) = \Phi(u, 0)$. In linear elasticity

$$\Phi(u, u', W') = \frac{\lambda}{2}(2u + W)^2 + \mu(2u^2 + W'^2) + \frac{a^2}{4}\mu u'^2 \quad (43)$$

Here λ and μ are the Lame's moduli. We see that the function $F(u)$ is convex, so the localization is absent. To obtain the desired effect we need to consider nonlinear elastic material. The situation, studied above is reached, if the energy $V(\varepsilon_{ij})$ and the load P are such that the function $F(u)$ has two local minima. The formation of the "nucleus" of the new phase in a homogeneous state $u(x) \equiv u_{**}$ is possible when the values of F in stationary points u_* and u_{**} are different and $F(u_*) < F(u_{**})$ for $u_* < u_{**}$. Such configurations are characterized by the region of order where the radius of the rod will be less than u_{**}.

Let us note, that the considered model predicts the first unstable stage of neck formation, afterwards another physical mechanisms come forward (plasticity for metals), which make such a simplified conservative model inadequate.

The idea of taking into consideration nonconvex functions $F(u)$ is due to Ericksen [18], higher derivatives were introduced recently by Coleman [19]. Autman [20], investigating bifurcations of the trivial solution, obtained an analog of the upper critical load for this problem.

SOLITONS

The analogy between solitons and nuclei of a new phase is not sufficiently complete because the features which characterize solitons are essentially connected with dynamics. Nevertheless in the theory of solitons we meet the same structure of energy. To make this clear, we consider the well-known Korteweg-de Vries equation [2] describing, for instance, surface water waves

$$\eta_t + 6\eta\eta_x + \eta_{xxx} = 0$$

After the change of variables $\eta = \varphi_x$ it obtains the variational structure

$$\varphi_{xt} + 6\varphi_x \varphi_{xx} + \varphi_{xxxx} = 0$$

with the Lagrangian function [21]

$$\Lambda = -\frac{1}{2}\varphi_t\varphi_x - \varphi_x^3 + \frac{1}{2}\varphi_{xx}^2$$

To find the stationary solutions, moving with the constant speed c (we take $c > 0$) $\varphi = V(x - ct)$ we have to solve the equation with the Lagrangian function

$$\Lambda = -V_x{}^3 + \frac{1}{2}cV_x^2 + \frac{1}{2}V_{xx}{}^2$$

The last one can be rewritten in terms of $u = V_x$

$$\Lambda = F(u) + \frac{1}{2}u_x{}^2, \quad F(u) = \frac{1}{2}cu^2 - u^3$$

We see, that Λ has again just the same structure as we have discussed above with the function $F(u)$ presented in Fig. 1a.

DISLOCATIONS

Let us consider the infinite cubic atomic lattice. We can shift atoms of half-space along the crystal plane at a distance b (lattice parameter). It is obvious that the displaced lattice will coincide with itself so our transformation is a symmetry. Assume now that we shift only part of the half-space. Then a dislocation appear - a region of transition from the domain, where displacement takes place to the domain where displacement is absent. We denote by $u(x)$ relative displacement of atoms on both sides of the sliding plane. The elastic energy of distortion of the upper and lower atom rows is due to interaction of both sides along the sliding plane. It is clear that this is a periodic function of displacement $u(x)$ with period b .

The equilibrium distribution of displacement $u(x)$ can be found from the equation, suggested by Peierls [8]:

$$\frac{2\varkappa}{b}\int_{-\infty}^{\infty} \frac{du/d\xi}{\xi - x}d\xi + \frac{2\pi\sigma}{G} = \sin\frac{2\pi u}{b} \qquad (44)$$

where σ is exterior tangential stress along the sliding plane, $\varkappa = \frac{b}{2(1-\nu)}$, G is shear modulus, ν is Poisson's ratio. The integral in the l.h.s. of (44) is due to long-range elastic interaction of a distorted region with the other part of the crystal. The r.h.s. is related with nonlinear interaction of rows of atoms, which is a short range and is localized in the vicinity of the sliding plane.

The analysis of the equation (44) shows that we have again the same energetical structure: algebraic term connected with periodic (nonconvex) interaction energy and elastic energy, represented by nonlocal term. Contrary to the examples examined above, here we meet for the first time a strong nonlocality, therefore Euler's equation appears to be integro-differential.

The simplest nontrivial solution of (44), the kink, (in the case of vanishing load)was given by Peierls [8]:

$$u = \frac{b}{\pi} \operatorname{arctg} \frac{x}{\varkappa}$$

The size of the dislocation core(of the energy localization) is of order $\varkappa$

When $\sigma \neq 0$ there is a familiar nonstable solution of "solitary wave" type [22], which corresponds to the pair of dislocations with opposite signs

$$u = -\frac{b}{\pi}\left(\operatorname{arctg}\frac{x-\eta}{\mu} - \operatorname{arctg}\frac{x+\eta}{\mu}\right) + \frac{b}{2\pi}\beta$$

where $2\eta = \frac{\varkappa G}{\pi\sigma}$ is the distance between dislocations. The parameter β is related with the exterior load $\sin\beta = \frac{2\pi\sigma}{G}$, we have, also, $\mu = \varkappa/\cos\beta$, $\eta = \varkappa/\sin\beta$. The solution collapses if $\cos\beta \to 0$. When $\sigma = G/2\pi$ spontaneous formation of dislocation pair takes place. This stress value correspond to the theoretical shear regidity and appears to be an analog of the upper critical load. We note, that the application of the critical shear stress comes to metastability of the trivial configuration, thus we can interpret this solution as a nucleus of a "new phase".

The equation (44) permits us to describe in a qualitative manner the structure of the dislocation core. Being integrodifferential it is too complicated to analyze. Therefore there have been attempts to restrict to considerations with only weak nonlocality. The most successfull is the model of Frienkel and Kontorova

[23]. In their model the equilibrium configurations are extremals of the functional (for = 0):

$$\int_{-\infty}^{\infty} \left(\frac{du}{dx}\right)^2 - \frac{b^2}{2l^2}\left(1 - \cos\frac{2\pi u}{b}\right) \, dx$$

where $l = \left(\frac{E}{G(1-\nu^2)}\right)^{1/2}$, E is Young's modulus. The Euler's equation coincides with the well-known Sin-Gordon equation. Kink solution in the framework of this model can be expressed through the elementary functions

$$tg\left(\frac{\pi}{4} + \frac{\pi u}{2b}\right) = \exp\left(-\frac{\pi x}{l}\right)$$

This is not the place to discuss the advantages of the considered models of dislocations. We have to point, however, that there is another interpretation of the Frienkel-Kontorova model which is related with the concept of a kink on a dislocation line [24] and, of course, with the analogous behavior of this objects.

ELEMENTARY PARTICLES

It is well-known that the equations of electrostatic scalar field can be obtained from the variational equation

$$\delta \int_V \Lambda \, dV = 0 \tag{45}$$

where $\Lambda = (\nabla\varphi)^2$, φ is a field variable. Particles are connected with singular solutions: spherical solution of (45) is given by $\varphi \sim 1/r$ where r is radial coordinate. The energy of the considered singular solutions is infinite thus there arise apparent difficulties in physical interpretations of such solutions. One way to overcome this is to add the term $V(\varphi)$ to the Lagrangian. We can't achieve success trying to preserve linear equations, i.e. taking $V = m^2\varphi^2$; the solutions continue to be singular: $\varphi \sim \frac{1}{r}\exp(-mr)$ though "long range" action gives place to "short range" action. To obtain particle-like solutions with finite energy we need to add an essentially nonlinear term, for example, $V = \varphi^2(\varphi - \varphi_0)^2$. As we know "solitary wave" like solutions are unstable, so the problem can not be solved in this way and there have been attempts to construct models of elementary particles in terms of high order tensor fields [3]. Nevertheless the model of scalar field theory with the Lagrangians, similar to

$\Lambda = (\nabla\varphi)^2 + (\varphi^2 - a^2)$, are widely used for qualitative considerations of such phenomena as spontaneous symmetry breaking.

MECHANISM OF PHASE TRANSFORMATION

We have begun with considering of the function $F(u)$ having local minima. However, from the presented examples it is clear that $F(u)$ usually depends on some parameter p and obtains local minima only under special values of p. As a rule, energy functional has the following structure

$$I = \int_a^b F(u)dx + p\int_a^b u dx$$

The function $\mu(u, p) = F(u) + p \cdot u$ has the above property if the condition of the convexity of $F(u)$ is violated and the derivative d^2F/du^2 changes its sign twice *).

The most typical dependence of $\mu(u,p)$ on the parameter p is the following. We have three different intervals on p axis: $p < p_1$, $p_1 < p < p_u$, $p > p_u$. When $p < p_1$ the function has the only minimum, at $p = p_1$ there appears the additional stationary point. In the interval $p_1 < p < p_u$ the function $\mu(u,p)$ has already two local minima and their depths are equalized at $p = p_*$, where $p_1 < p_* < p_u$. When $p > p_u$ we have again only one stationary point, the second vanishes at $p = p_u$. The sequence of transformations of the function $\mu(u,p)$ (Fig. 6) can be characterized in terms of cusp catastrophe. In mechanics the quantities p_1 and p_u are called upper and lower critical loads respectively, in thermodynamics they are called spinodal points.

We come to the following physical picture. If $p < p_1$ the system can be found only in the phase state u_*. When $p = p_1$ there appears another phase state u_{**} which is unstable. After $p > p_1$ the newly formed phase becomes stable with respect to infinitesimal disturbances, being unstable with respect to finite disturbances. Such states are called metastable. Absolute minimum is achieved at $u = u_*$ up to the value of $p = p_*$, when the energies of phases become equal. The subsequent increasing of p leads to the shallowing of the minimum, corresponding to $u = u_*$

*) This let us formulate the main conception in the concise form: nonconvexity + higher derivatives.

and it vanishes at $p = p_u$. It is obvious that after $p > p_u$ the only $u = u_{**}$ phase can exist.

If we increase p, the phase transformation from state u_* to the state u_{**} will occur. We shall consider the process of phase transformation in some detail. Let us assume that the system is in homogeneous state $u \equiv u_*$ when $p < p_1$. After $p > p_1$ it appears additional ability to jump into another phase, though, in the absence of fluctuations the transformation into u_{**} phase can occur only after we go beyond the upper critical load p_u, i.e. after the state u_* becomes unstable (strategy of maximal delay). Classical thermodynamics, assuming fluctuations, accept as a transformation pressure the value $p = p_*$, which corresponds to the equality of phases' energies (Maxwell's strategy).

Let us turn now to continuum, characterized by the function u(x). It is obvious that simultaneous transformation of all particles from one state to another is enegretically unfavorable: the necessity to overcome the energetical barrier causes significant energy losses. It seems that this fact was understood for the first time in connection with the investigation of phase transitions, though analogous ideas can be met in the theory of dislocations. The transformation from one phase state to another realizes through the formation of critical nuclei - the localized states of the new phase. Corresponding nonhomogeneous soliton configurations can be modeled by the saddle point of the energy functional.Such solutions of the Euler's equations describe the minimal energy of fluctuation, which transforms the system from metastable state to the stable state. Critical nuclei apparently unstable: the subcritical nuclei have a tendecy to collaps while for the overcritical nuclei the forgoing growth is energetically favorable: such growing nuclei provide the transformation of the system into the new phase state.

Critical nuclei correspond to the localized solutions of "solitary wave" type. It can be shown that by use of suggested analogy we can describe on equal grounds such different processes as the formation of vapour bubbles in overheated liquid, the formation of dimples on shells under pressure, the formation of dislocations in stressed crystal etc. We shall limit ourselves with: a) liquid-vapour transition, b) shells buckling.

In the theory of liquid-vapour transformation the specific volume V corresponds to variable u, F is specific free energy, μ - is specific Gibbs' energy or chemical potential, parameter

p coincides with the exterior pressure. The analysis of the structure of critical nuclei [6, 13] provides the following picture of transformation. When $p < p_l$ only one phase 1 **(vapour)** is stable and there are no nontrivial solutions of our variational problem. When $p_l \leq p < p_*$ there appears metastable phase 2 (**liquid**) and we have nontrivial **solutions** of "solitary wave" type corresponding to **vapour** nuclei in **liquid** phase. Phase 1 stay stable and corresponds to the absolute minimum. When $p = p_*$ the indifferent equilibrium liquid-vapour with flat interfacial boundary (kink solution) is possible thus the transition from the phase 1 to the phase 2 occurs through the formation of the critical nucleus of infinite radious. When $p_* < p < p_u$ phase 2 becomes metastable. In this interval of pressure nontrivial "solitary wave" solutions that model **liquid** bubbles in **vapour** are possible. Effective bubble "radious" so as the energy of critical nucleus diminish **to zero value** when pressure increases from $p = p_*$ to $p = p_u$. The value of pressure $p = p_u$ **corresponds** to the state of absolute instability of phase 1. After $p > p_u$ only phase 2 is stable and nontrivial solutions are absent.

Thus, phase transformation is possible in the following interval: $p_l < p < p_u$. Having calculated for all of these pressures the energy of the critical nucleus we can estimate the degree of stability of the equilibrium states.

Let us turn now to the dimple formation on the shell under the compressive exterior pressure p . The dependence of the deflection on the load is presented in Fig. 7. When $p > p_l$ there exists buckled equilibrium state, which differs significantly from homogeneously deformed initial state. When $p = p_*$ both equilibrium states have the same energy and at $p = p_u$ the initial state becomes absolutely unstable.

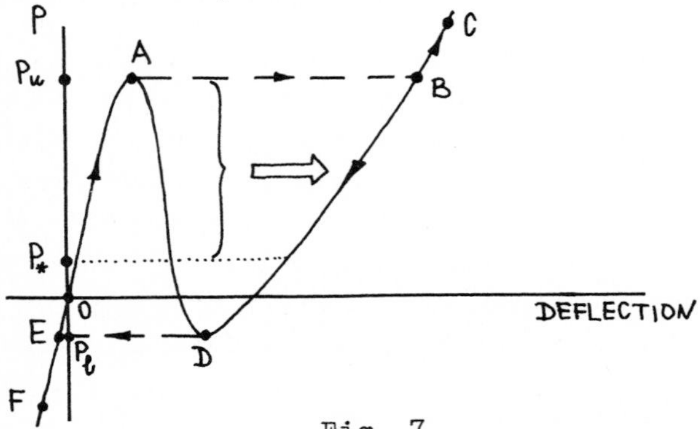

Fig. 7

The pressure p_u is called the upper critical load and it can be calculated in a standard way from the linear theory (see [25]).

In the literature on elastic stability it is usually assumed that, when the load p increases its value, the system evolve to the buckled states through the OABC parth (Fig. 7) and when p decreases - through the CBDEF parth. As we have already seen, the most important are not upper and lower critical loads but rather $p = p_*$: the loss of stability occurs not after $p = p_u$, but after $p = p_*$ (see also [12]).

The estimation of the lower critical load p_l so as the p_* seems to be one of the most important problems in the theory of shells. The analogy of the buckling phenomena with the process of nuceation makes it possible to suggest the following buckling procedure. The transition to the buckled state is possible when $p_* < p \leq p_u$ and occurs through the formation of the critical uncleus i.e. the dimple, corresponding to the saddle point of the energy functional. The energy of such dimple characterize the energetical level of finite disturbances bringing out of the initial state of equilibrium. If we have $p \sim p_*$ the estimation of the "critical" energy can be made by use of Pogorelov's theory. The equilibrium configurations corresponding to reflection of some spherical segment appear to be unstable, thus dimples with the overcritical radius continue to grow and subcritical dimples vanishes. This coincides with the picture of phase transition in liquid and we obtain analogy between the dimple edge energy and the energy of the interfacial tension. This analogy permits to suggest the new understanding of upper and lower critical loads and to yield the investigation of degree of stability to the problem of saddle points of the energy functional.

CONCLUDING REMARKS

The examples presented above show the identity of the mathematical structure of the energy functional leading to the localization of the energy in quite different physical situations. The main conception concerning energy structure can be repeated in a concise form: nonconvexity + higher derivatives. If the energy of homogeneous states has a number of minima, then the equilibrium configuration is not unique. Regularization can be achieved by the introduction to the functional of additional terms with the derivatives [26]; in all of our examples this was done according to the additional reasons of physical character. The existance of higher derivatives in the equations equilibrium leads to the appearance of states with localized energy. Thus it can be formulated the following proposal: if a nonconvexity appear, then one mustlook for higher derivatives.

REFERENCES

1 A. Einstein, L.Infeld, B. Hoffman, Ann. Math. 39 (1938) 65-100.
2 D.J. Korteweg, G. de Vries, Philos. Mag. 39 (1895), 422-443.
3 R. Jackiv, Rev. Mod. Phys. 49 (1977) 681-706.
4 J.D. van der Waals, Verhandel. Konink. Akad. Weten., Amsterdam (Sect. 1) 1 (1893).
5 L.D. Landau, E.M. Lifshits, Electrodynamics of Continuum, Nauka, Moscow, 1982 (in Russian).
6 J.W.Cahn, J.E.Hilliard, J. Chem. Phys., 28 (1958) 258-267.
7 A.Pogorelov, Geometrical Methods in Nonlinear Theory of Shells, Nauka, Moscow, 1967 (in Russian).
8 R. Peierls, Proc. Phys. Soc., 52 (1940).
9 J.W. Gibbs, Trans. Connect. Acad., 3 (1876) 108-248; 3 (1878) 343-524.
10 V. Berdichevskii, Variational Principles of Continuum Mechanics, Nauka, Moscow, 1983 (in Russian).
11 T. Karman, H.S. Tsien, J. Aeronaut. Sci. 7 (1939) 43.
12 K.O. Friedrichs, in T. Karman Anniversary Volume, 1941, p.258.
13 L. Truskinovskii, Dokl. AN SSSR, 269 (1983).
14 L. Truskinovskii, Dokl. AN SSSR, 265 (1982).
15 L.I. Sedov, Continuum Mechanics, v. 1, 2, Nauka, Moscow, 1976 (in Russian).
16 R.D. Mindlin, Int. J. Solids and Struct. 1 (1965) 417-436.
17 V. Berdichevskii, J. Appl. Math. and Mech. (PMM) 30 (1966) 510-530.
18 J.L. Ericksen, J. Elast. 5 (1975).
19 B. Coleman, Arch. Rat. Mech. Anal., 83 (1983) 115-137.
20 S.S. Antman, J. Math. Anal. Appl., 44 (1973) 333.
21 J. Whitham, Linear and Nonlinear Waves, 1973.
22 F.R.N. Nabarro, Proc. Phys. Soc. 59 (1947).
23 I.Frienkel, T. Kontorova, JETP (1938) 1340.
24 A. Seeger, D. Schiller, in W.P.Mason (ed.), Physical Acoustics v.3, p.361.
25 R.Joelly, Ueber ein Knickungsproblem an der Kugelschale, Zürich, 1915.
26 J.M. Ball, J.C. Currie, J. Func. Anal., 41 (1981).

CHAPTER 2:

EDGE EFFECTS IN COMPOSITE STRUCTURES

EDGE EFFECTS IN ROTATIONALLY SYMMETRIC COMPOSITE SHELLS

MAHIR SAYIR
Swiss Federal Institute of Technology, Zurich, Switzerland

ABSTRACT

Fiber-reinforced rotationally symmetric shells are usually much stiffer along the meridian than across the transverse direction. Thus, their bending resistance is much larger than their resistance against shear deformation. Strong shear coupling results even for thin shells if they are subjected to transverse loads. Engineering approaches have to be reassessed and may prove to be inaccurate, particularly in regions close to the edges. We start from the three dimensional equations of linear elasticity and present an asymptotic approach based on a systematic discussion of the relative magnitude of two geometrical and one material parameters. Cases of moderately and strongly anisotropic behaviour are compared to isotropy, and simplified sets of equations are derived and solved for the edge zones, where shear coupling becomes decisive in understanding some of the anomalies encountered here in both numerical and experimental analyses of such structures. A transversely isotropic cylindrical shell is used to illustrate the general procedure and its physical consequences.

1. INTRODUCTION

The mechanical behaviour of thin elastic structures with one "preferred" direction along which the stiffness is much larger than in perpendicular directions, may differ quite considerably from the isotropic case (see for example ref.1, 2, 3). In isotropic structures subjected to bending, simplifying assumptions (attributed to Navier, Kirchhoff, Love) about"cross sections remaining plane and perpendicular to the middle surface" are perfectly justifiable in sufficient distance from the"edges" (discontinuities in load or geometry). In strongly anisotropic structures such assumptions may prove to be unrealistic even at large distances from "edges". The reason is essentially due to strong shear coupling resulting from the fact that the structure behaves "soft" in shear and "stiff" in bending. Most of the composite structures exhibit such strongly anisotropic features, at least near the edges. In the following, a general method of quantitative analysis based on "multiple scale asymptotic expansions" will be presented. It has the advantage of correlating explicitly and transparently equations to the corresponding main physical mechanisms.

The essential assumptions are

- linear elasticity with "small" deformations and displacements,
- rotational symmetry in geometry, loading and anisotropic material properties,

- "thin structures" such that their thickness is much smaller than other characteristic dimensions (radius of curvature, length, etc).

A circular cylindrical shell will be chosen to illustrate the method with reduced formalism. Its special geometry is not essential for the applicability of the method and for the physical implications of the results. We therefore qualify as "auxiliary assumptions" the following restrictions which are simply introduced in order to reduce formalism and to make equations more transparent:

- circular cylindrical shell,
- transverse loading only, distributed or applied at "ends",
- transversely isotropic behaviour with the "preferred" (stiff) direction along the axis of the cylinder.

The assumption of "transversely isotropic behaviour" is a mild restriction indeed, since the results obtained in the following can easily be adapted with some minor algebraic changes to orthotropy, provided that the modulus of elasticity E along the axis is at least an order of magnitude larger than the transverse modulus E_I (perpendicular to the shell surface) and the shear modulus G. Such conditions can be realized for example in fiber-reinforced composite shells with stiff fibers arranged symmetrically with respect to the axis and parallel to the shell surface.

2. BASIC EQUATIONS

The geometry of the shell, loading, coordinates and the five elasticity constants E, ν, E_I, G, ν_I of transversely isotropic behaviour can be seen in Fig.1.

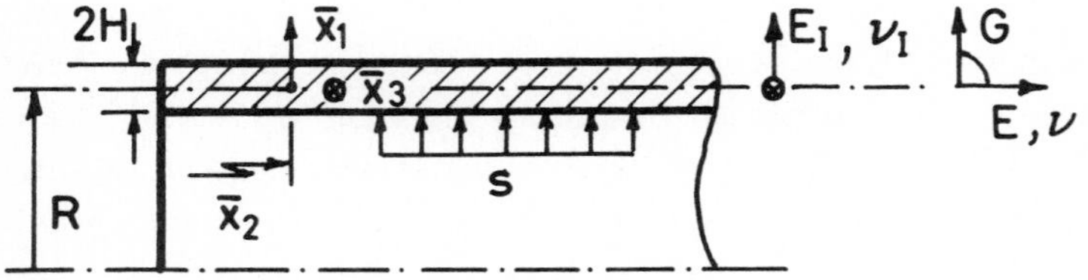

Fig.1. Geometry, loading, coordinates and elasticity constants of transverse isotropy.

We introduce the following dimensionless quantities:

$$\sigma_{kl} := \bar{\sigma}_{kl} / \sigma_0 \quad ; \quad u_\alpha := \bar{u}_\alpha . E_I/\sigma_0 H \quad ; \quad x_1 := \bar{x}_1/H \quad ; \quad x_2 := \bar{x}_2/L \quad ;$$
$$\varepsilon_R := H/R \quad ; \quad \varepsilon_L := H/L \quad ; \quad \varepsilon_M := E_I/E \quad ; \quad \mu := E_I/G \ , \tag{2.1}$$

where latin indices k,l run through 1, 2, 3, greek indices α through 1, 2, σ_0 is the maximum of the load $|s(x_2)|$ and quantities with a bar designate the ones with dimension. L is a characteristic length along the x_2-direction. If L is of the same order of magnitude as H, we will call the corresponding domain "edge zone"; if L is of the order of magnitude R, we will call the corresponding domain "interior zone". Two examples in Fig.2a and 2b illustrate the situation.

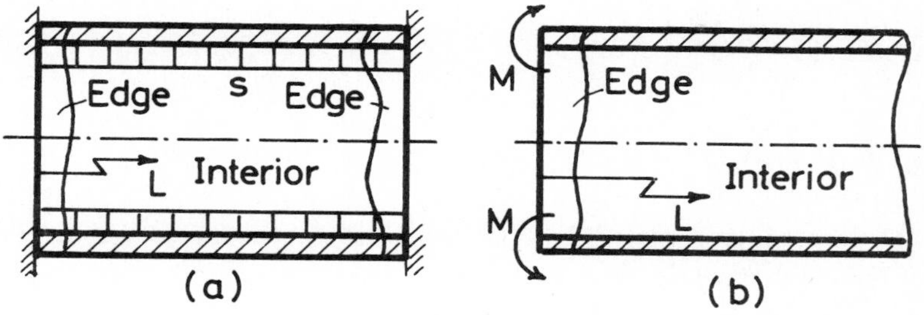

Fig. 2a. Built-in shell under internal pressure.
Fig. 2b. Long shell under bending moments at the free edge.
Edge and interior zones.

The basic three dimensional constitutive relations and equilibrium equations for transversely isotropic, axisymmetrical behaviour can now be written as follows in dimensionless stresses and displacements:

$$\begin{aligned}
&u_{1,1} = \sigma_{11} - \nu\varepsilon_M\sigma_{22} - \nu_I\sigma_{33} \; , \\
&\varepsilon_L u_{2,2} = \varepsilon_M\,[\sigma_{22} - \nu(\sigma_{33} + \sigma_{11})] \; , \\
&\varepsilon_R u_1 = (1 + x_1\varepsilon_R)[\sigma_{33} - \nu_I\sigma_{11} - \nu\varepsilon_M\sigma_{22}] \; , \\
&u_{2,1} + \varepsilon_L u_{1,2} = \mu\sigma_{12} \; , \\
&\sigma_{11,1} + \varepsilon_L\sigma_{12,2} + \varepsilon_R(\sigma_{11} - \sigma_{33})/(1 + \varepsilon_R x_1) = 0 \; , \\
&\sigma_{21,1} + \varepsilon_L\sigma_{22,2} + \varepsilon_R\sigma_{12}/(1 + \varepsilon_R x_1) = 0 \; ,
\end{aligned} \tag{2.2}$$

where $(\;)_{,\alpha} = \partial(\;)/\partial x_\alpha$.

Simplified approximate equations for thin shells may be obtained from (2.2) by considering the limit $\varepsilon_R \to 0$ and expanding stresses and displacements in terms of ε_R. But since we want to consider both interior and edge zones and the case of highly anisotropic behaviour, parameters ε_L and ε_M may also take "small" values comparable with ε_R. A relative evaluation of these three parameters ε_R, ε_L, ε_M with respect to each other becomes necessary, so that also ε_L and $\varepsilon_M \to 0$ in a certain manner connected with $\varepsilon_R \to 0$. This will allow us a systematic classification of different physically relevant cases discussed in the next sections. Only first order approximations corresponding to the first terms of expanded stresses and displacements will be considered.

3. INTERIOR ZONE AND MEMBRANE STATE

The interior zone is defined by

$$\varepsilon_L = 0(\varepsilon_R) \ , \quad L = 0(R) \ , \tag{3.1}$$

where we use 0( ) in the usual sense, i.e. $f(x,\varepsilon_R) = 0(\varepsilon_R^n)$ if and only if

$$\lim_{\varepsilon_R \to 0} \frac{[f(x,\varepsilon_R)]}{\varepsilon_R^n} = f(x) < \infty \ . \tag{3.2}$$

3.1. Isotropy, $\varepsilon_M = 1$

The special case of isotropy will help us introduce some useful ideas. Proceeding as in ref.4 we first remark that due to the scaling of stresses by $\sigma_0 = \max|s(x_2)|$, the transverse normal stress

$$\sigma_{11} = 0(1) \ . \tag{3.3}$$

Then, we obtain from (2.2) the evaluations

$$\sigma_{33} = 0(\varepsilon_R^{-1}) \quad ; \quad u_1 = 0(\varepsilon_R^{-2}) \quad ; \quad u_2 = 0(\varepsilon_R^{-2}) \tag{3.4}$$

and

$$\sigma_{22} = 0(1) \quad ; \quad \sigma_{12} = 0(\varepsilon_R) \ . \tag{3.5}$$

Thus, in a first step of approximation (2.2) may be reduced to the well-known relations of a "membrane state":

$$\begin{aligned} & u_1,_1 = 0(u_1\varepsilon_R) \cong 0 \ , \\ & \varepsilon_R u_1 - \sigma_{33} = 0(\sigma_{33}\varepsilon_R) \cong 0 \ , \\ & \sigma_{11},_1 - \varepsilon_R\sigma_{33} = 0(\sigma_{11}\varepsilon_R) \cong 0 \ . \end{aligned} \tag{3.6}$$

The remaining three relations are irrelevant. Equations (3.6) lead to the classical engineering formulas

$$\bar{\sigma}_{33} = s.R/2H \quad ; \quad \bar{u}_1 = s.R^2/2EH \tag{3.7}$$

for long tubes or cylindrical containers by using the appropriate boundary conditions at the surfaces $x_1 = \pm 1$.

3.2. Moderate anisotropy, $\varepsilon_M = 0(\varepsilon_R)$

Consider now a cylindrical shell which is stiffer in the axial direction, so that the ratio $\varepsilon_M := E_I/E$ defined in (2.1) has a numerical value comparable with $\varepsilon_R := H/R$. We write

$$\varepsilon_M = O(\varepsilon_R) \tag{3.8}$$

and obtain by a similar evaluation procedure as in subsection 3.1 a reduced set of equations of a first step of approximation from (2.2). This set turns out to be identical to (3.6), so that in the interior zone L= O(R), for moderate anisotropy defined by (3.8), a membrane state prevails. From this point of view there is no difference with respect to the isotropic case discussed in 3.1.

3.3. Strong anisotropy, $\varepsilon_M = O(\varepsilon_R^{\alpha})$, $\alpha \geq 2$

In the case of strong anisotropy defined by

$$\varepsilon_M = O(\varepsilon_R^{\alpha}) \ , \quad \alpha \geq 2 \ , \tag{3.9}$$

one can show that there is no consistent reduced set of equations leading to a membrane state. The reason becomes clear if in both the isotropic case and the case of moderate anisotropy we consider an intermediate zone lying between the edge zone L= O(H) and the interior zone L= O(R). We call this zone the "bending range" and discuss it in the next section.

4. BENDING RANGE

4.1. Isotropy, $\varepsilon_M = 1$

It is well-known that an isotropic cylindrical shell shows bending effects under transverse loading only in the range

$$L = O[(RH)^{\frac{1}{2}}] \ , \quad \varepsilon_L = O(\varepsilon_R^{\frac{1}{2}}) \ . \tag{4.1}$$

Evaluating orders of magnitude of stresses and displacements on the basis of (3.3) from (2.2), one obtains now

$$\begin{aligned} &\sigma_{33} = O(\varepsilon_R^{-1}) \quad ; \quad \sigma_{12} = O(\varepsilon_R^{-\frac{1}{2}}) \quad ; \quad \sigma_{22} = O(\varepsilon_R^{-1}) \quad ; \\ &u_2 = O(\varepsilon_R^{-3/2}) \quad ; \quad u_1 = O(\varepsilon_R^{-2}) \quad . \end{aligned} \tag{4.2}$$

Thus, with respect to the membrane state with evaluations (3.4), (3.5), both shear stresses σ_{12} (factor $\varepsilon_R^{-3/2}$) and bending normal stresses σ_{22} (factor ε_R^{-1}) become much larger, whereas axial displacements u_2 are reduced by a factor $\varepsilon_R^{\frac{1}{2}}$.

The following set of equations for the first step of approximation can now be derived from (2.2):

$$\begin{aligned} &u_{1,1} = O(u_1 \varepsilon_R) \cong 0 \quad , \\ &\varepsilon_L u_{2,2} - (\sigma_{22} - \nu\sigma_{33}) = O(\sigma_{22}\varepsilon_R) \cong 0 \quad , \\ &\varepsilon_R u_1 - (\sigma_{33} - \nu\sigma_{22}) = O(\sigma_{33}\varepsilon_R) \cong 0 \quad , \end{aligned} \tag{4.3}$$

$$u_{2,1} + \varepsilon_L u_{1,2} = O(\varepsilon_R u_2) \cong 0 \quad ,$$

$$\sigma_{11,1} + \varepsilon_L \sigma_{12,2} - \varepsilon_R \sigma_{33} = O(\sigma_{11}\varepsilon_R) \cong 0 \quad ,$$

$$\sigma_{21,1} + \varepsilon_L \sigma_{22,2} = O(\sigma_{12}\varepsilon_R) \cong 0 \quad .$$

The first relation shows that transverse displacement u_1 is independent of the thickness coordinate x_1, the next two are constitutive relations of plane stress, the fourth one leads to the well-known assertion that "cross sections remain plane and perpendicular to the deformed middle surface". The last two relations are simplified equilibrium equations. Eliminating all quantities except u_1 and using the appropriate boundary conditions at the surfaces $x_1 = \pm 1$, one obtains the classical differential equation for an isotropic cylindrical shell

$$\varepsilon_R^{-2} \varepsilon_L^4 u_{1,2222} + 3(1 - \nu^2) u_1 = \frac{3}{2} s (1 - \nu^2) \varepsilon_R^{-2} \quad . \tag{4.4}$$

Considering Φ as a quantity representing all four stresses and two displacements, one obtains a solution uniformly valid both in the interior zone and in the bending range by setting $\bar{x}_2 = 0$ at the edge and writing in a general way

$$\Phi = \Phi_{Membrane} + \Phi_{Bending} \quad , \tag{4.5}$$

where

$$\Phi_{Bending} \sim \exp[-k\bar{x}_2/(RH)^{\frac{1}{2}}] \quad , \quad k \cong 0{,}9 \quad , \tag{4.6}$$

and decreases hence exponentially out of the bending range towards the interior zone.

4.2. Moderate anisotropy, $\varepsilon_M = O(\varepsilon_R)$

The "small" material parameter ε_M will now alter the definition of the bending range. Indeed, (4.1) must be modified by setting

$$\varepsilon_L = O(\varepsilon_R^{\frac{1}{2}} \varepsilon_M^{\alpha})$$

with α to be determined from (2.2) to obtain a consistent set of reduced equations. We find $\alpha = 1/4$, hence

$$\varepsilon_L = O(\varepsilon_R^{\frac{1}{2}} \varepsilon^{1/4}) \quad , \quad L = O[(RH)^{\frac{1}{2}} (E/E_I)^{1/4}] \tag{4.7}$$

defines the bending range in moderate anisotropy. This means that with respect to isotropic behaviour, the bending range of moderately anisotropic composite shells tends to widen towards the interior zone. Besides, evaluations (4.2) have to be replaced by

$$\sigma_{33}= O(\varepsilon_R^{-1}) \quad ; \quad \sigma_{12}= O(\varepsilon_R^{-\frac{1}{2}}\varepsilon_M^{-1/4}) \quad ; \quad \sigma_{22}= O(\varepsilon_R^{-1}\varepsilon_M^{-\frac{1}{2}}) \quad ;$$
$$u_2= O(\varepsilon_R^{-3/2}\varepsilon_M^{1/4}) \quad ; \quad u_1= O(\varepsilon_R^{-2}) \ . \tag{4.8}$$

One sees that both shear stress σ_{12} (factor $\varepsilon_M^{-1/4}$) and bending normal stress σ_{22} (factor $\varepsilon_M^{-\frac{1}{2}}$) are larger than in the isotropic case, whereas axial displacement u_2 is reduced by a factor $\varepsilon_M^{1/4}$.

Equations of the first step of approximation are now

$$\begin{aligned}
&u_{1,1}= O(u_1\varepsilon_R)\cong 0 \ , \\
&\varepsilon_M^{-1}\varepsilon_L u_{2,2} - \sigma_{22}= O(\sigma_{22}\nu\varepsilon_M^{\frac{1}{2}})\cong 0 \ , \\
&\varepsilon_R u_1 - \sigma_{33}= O(\sigma_{33}\nu\varepsilon_M^{\frac{1}{2}})\cong 0 \ , \\
&u_{2,1} + \varepsilon_L u_{1,2}= O(u_2\varepsilon_R\varepsilon_M^{-\frac{1}{2}})\cong 0 \ , \\
&\sigma_{11,1} + \varepsilon_L\sigma_{12,2} - \varepsilon_R\sigma_{33}= O(\sigma_{11}\varepsilon_R)\cong 0 \ , \\
&\sigma_{21,1} + \varepsilon_L\sigma_{22,2}= O(\sigma_{12}\varepsilon_R)\cong 0 \ .
\end{aligned} \tag{4.9}$$

They can be interpreted along similar lines to those discussed in Subsection 4.1. The main point is here the order of magnitude of the neglected shear stress term on the right hand side of the fourth equation. It is larger by a factor $\varepsilon_M^{-\frac{1}{2}}$ and shows definitely that shear coupling is getting more important as the degree of anisotropy increases. Elimination of all quantities except u_1 by appropriate use of the boundary conditions at the surfaces $x_1= \pm 1$ leads to the main differential equation

$$\varepsilon_M^{-1}\varepsilon_R^{-2}\varepsilon_L^{4}u_{1,2222} + 3u_1= \frac{3}{2}s\varepsilon_R^{-2} \ . \tag{4.10}$$

Writing again a uniformly valid solution in the general form (4.5), one finds that

$$\Phi_{Bending} \sim \exp[-k(E_I/E)^{1/4}.(RH)^{-\frac{1}{2}}.\bar{x}_2] \ , \quad k\cong 0{,}9 \ , \tag{4.11}$$

so that the exponential decay of the bending effects slows down by a factor $\varepsilon_M^{1/4}$ in the exponent with respect to the isotropic case.

4.3. Strong anisotropy, transitional case $\varepsilon_M= O(\varepsilon_R^2)$

The denomination "transitional" for the case

$$\varepsilon_M= O(\varepsilon_R^{2}) \tag{4.12}$$

will become clear in the next subsection. Proceeding as in subsection 4.2 we find again that the bending range is characterized by (4.7). But now (4.7) means that

$$\varepsilon_L = O(\varepsilon_R) \tag{4.13}$$

in the bending range, if one considers (4.12). Thus, the bending range reaches the interior zone and merges with "membrane" type behaviour. Evaluation of orders of magnitude leads to

$$\begin{aligned} &\sigma_{33} = O(\varepsilon_R^{-1}) \quad ; \quad \sigma_{12} = O(\varepsilon_R^{-1}) \quad ; \quad \sigma_{22} = O(\varepsilon_R^{-2}) \quad ; \\ &u_2 = O(\varepsilon_R^{-1}) \quad ; \quad u_1 = O(\varepsilon_R^{-2}) \ . \end{aligned} \tag{4.14}$$

Hence, shear stresses are as large as hoop stresses σ_{33}. The consequence is strong shear coupling and ensuing shear deformation even for the first step of approximation. This becomes very clear by studying the corresponding reduced set of equations

$$\begin{aligned} &u_{1,1} = O(u_1 \varepsilon_R) \cong 0 \quad , \\ &\varepsilon_M^{-1} \varepsilon_L u_{2,2} - \sigma_{22} = O(\nu \sigma_{22} \varepsilon_L^2 \varepsilon_R^{-1}) \cong 0 \quad , \\ &\varepsilon_R u_1 - \sigma_{33} = O(\nu \sigma_{33} \varepsilon_R) \cong 0 \quad , \\ &u_{2,1} + \varepsilon_L u_{1,2} = \mu \sigma_{12} \quad , \\ &\sigma_{11,1} + \varepsilon_L \sigma_{12,2} - \varepsilon_R \sigma_{33} = O(\sigma_{11} \varepsilon_R) \cong 0 \quad , \\ &\sigma_{21,1} + \varepsilon_L \sigma_{22,2} = O(\sigma_{12} \varepsilon_R) \cong 0 \quad . \end{aligned} \tag{4.15}$$

Accordingly, no reduction is possible in the fourth equation and "cross sections cannot remain plane and perpendicular to the deformed middle surface". Neither are the bending stresses linearly distributed over the thickness, since now, by appropriate elimination from (4.15),

$$\sigma_{22,11} + \varepsilon_L^2 \varepsilon_M^{-1} \sigma_{22,22} = 0 \quad , \tag{4.16}$$

so that bending stress distributions over the thickness are strongly coupled with distributions along the axis. In particular, a given thickness distribution of σ_{22} at $x_2 = 0$, irrespective of the global force and moment to which it may lead, will influence the behaviour deep in the interior zone L= O(R). Saint Venant's principle seems to fail in this transitional case of strong anisotropy characterized by (4.12).

Even though most of the real structures do not reach as strong an anisotropy as (4.12), the strong shear coupling giving rise to the complication (4.16) becomes particularly important in the neighbourhood of the edges of moderately anisotropic shells (see Section 5).

4.4. Strong anisotropy, limiting case $\varepsilon_M = O(\varepsilon_R^\alpha)$, $\alpha > 2$

We will now see that for

$$\varepsilon_M = O(\varepsilon_R^\alpha) \ , \quad \alpha > 2 \tag{4.17}$$

a new type of behaviour arises in the interior zone L= O(R), which is here identical to the bending range according to the results of the last subsection. In fact, setting

$$\varepsilon_L = O(\varepsilon_R) \ , \quad L = O(R) \tag{4.18}$$

we find with

$$\sigma_{33} = O(\varepsilon_R^{-1}) \ ; \quad \sigma_{12} = O(\varepsilon_R^{-1}) \ ; \quad u_1 = O(\varepsilon_R^{-2}) \tag{4.19}$$

that the reduction of (2.2) for the first step of approximation reads

$$\begin{aligned} &u_{1,1} = O(u_1 \varepsilon_R) \cong 0 \ , \\ &\varepsilon_R u_1 - \sigma_{33} = O(\nu \sigma_{33} \varepsilon_R) \cong 0 \ , \\ &\varepsilon_L u_{1,2} - \mu \sigma_{12} = O(\sigma_{12} \varepsilon_M \varepsilon_R^{-2}) \cong 0 \ , \\ &\sigma_{11,1} + \varepsilon_L \sigma_{12,2} - \varepsilon_R \sigma_{33} = O(\sigma_{11} \varepsilon_R) \cong 0 \ , \\ &\sigma_{12,1} = O(\sigma_{12} \varepsilon_R) \cong 0 \ . \end{aligned} \tag{4.20}$$

According to the last equation, shear stress σ_{12} should be constant over the thickness. In fact, transverse displacement u_1 causes solely shear deformation and obeys the differential equation

$$\mu^{-1} \varepsilon_R^{-2} \varepsilon_L^2 u_{1,22} - u_1 = -\tfrac{1}{2} s \varepsilon_R^{-2} \tag{4.21}$$

which can be derived by appropriate elimination from(4.20). But then, boundary conditions along the surfaces $x_1 = \pm 1$ requiring the shear stresses to vanish cannot be fulfilled, so that these surfaces become singular with respect to the regular asymptotic expansion. Hence, we must introduce boundary layers in the neighbourhood of $x_1 = \pm 1$ and a corresponding singular perturbation scheme.

Considering a boundary layer along the outer surface $x_1 = +1$, we define a boundary layer coordinate

$$\xi := (-1 + x_1) \varepsilon_L \varepsilon_M^\beta$$

and find $\beta = -\tfrac{1}{2}$ from the requirement that for a first step of approximation a consistent set of equations follows from (2.2) in the boundary layer. Setting

$$\xi := (-1 + x_1) \varepsilon_L \varepsilon_M^{-\frac{1}{2}} \ , \tag{4.22}$$

one obtains from (2.2) in addition to (4.19)

$$\sigma_{22} = O(\varepsilon_R^{-1}\varepsilon_M^{-\frac{1}{2}}) \quad ; \quad u_2 = O(\varepsilon_R^{-2}\varepsilon_M^{\frac{1}{2}}) \tag{4.23}$$

and the set of equations

$$\begin{aligned} & u_{1},_{\xi} = O(u_1\varepsilon_M^{\frac{1}{2}}) \cong 0 \quad , \\ & \varepsilon_M^{-1}\varepsilon_L u_{2},_{2} - \sigma_{22} = O(\sigma_{22}\varepsilon_M^{\frac{1}{2}}) \cong 0 \quad , \\ & \varepsilon_M^{-\frac{1}{2}}\varepsilon_L u_{2},_{\xi} + \varepsilon_L u_{1},_{2} = \mu\sigma_{12} \quad , \\ & \sigma_{12},_{\xi} + \varepsilon_M^{\frac{1}{2}}\sigma_{22},_{2} = O(\sigma_{12}\varepsilon_M^{\frac{1}{2}}) \cong 0 \quad . \end{aligned} \tag{4.24}$$

According to (4.17) and (4.23), as $\alpha \to \infty$ bending normal stress σ_{22} also $\to\infty$ and axial displacement $u_2 \to 0$. We then obtain a case similar to the ones discussed by Spencer in Ref.1 with inextensible layers along the lateral surfaces $x_1 = \pm 1$ carrying finite tensile or compressive forces and giving rise to a finite bending moment, whereas the core of the shell remains under constant shear.

The more realistic case $2<\alpha<\infty$ has been illustrated in Fig.3 where a "thickness" of the boundary layers

$$h := H\varepsilon_M^{\frac{1}{2}}\varepsilon_L^{-1} \sim L(G/E)^{\frac{1}{2}} \tag{4.25}$$

consistent with (4.22) is defined. In this boundary layer, bending stresses σ_{22} rise sharply towards the lateral surfaces to large values and shear stresses σ_{12} reach the constant value of the core exponentially towards the core according to the differential equation

$$\sigma_{12},_{\xi\xi} + \sigma_{12},_{22} = \varepsilon_L\, u_{1},_{222} \tag{4.26}$$

derived by appropriate elimination from (4.24).

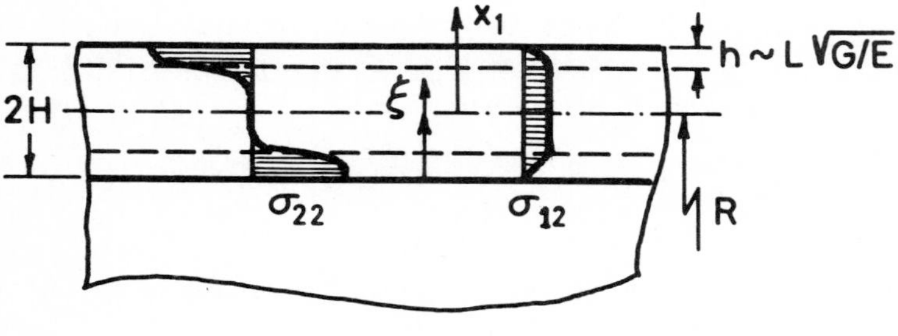

Fig.3 : Boundary layers and shear core in the strongly anisotropic composite shell for $\varepsilon_M = O(\varepsilon_R^{\alpha})$, $\alpha>2$.

Thus, a "natural" sandwich structure emerges, the core of which "naturally" refuses to carry normal bending stresses σ_{22} and shifts them to the boundary layers, while accepting constant shear stresses in a given cross section (ref.5, see also ref.2).

A uniformly valid solution over the thickness in the interior zone may be obtained by adding the solution of (4.20) to solutions in both boundary layers along $x_1 = \pm 1$ derived from (4.24), i.e.

$$\Phi = \Phi_{Regular} + \Phi_{Layer\ 1} + \Phi_{Layer\ 2} ,$$

where Layer 1 corresponds to the vicinity of $x_1 = +1$ and Layer 2 to $x_1 = -1$.

Both $\Phi_{Layer\ 1}$ and $\Phi_{Layer\ 2}$ decay exponentially inwards, towards the core, with exponents proportional to $\varepsilon_L \varepsilon_M^{-\frac{1}{2}}$ (the reciprocal value of (4.25)).

5. EDGE ZONE

Solutions of differential equations like (4.4), (4.10), (4.16), etc. require knowledge of boundary conditions at the edges, for example at $x_2 = 0$. Here, both displacement components, or one displacement and one stress component or both stress components must be given. In the following, only the cases of isotropy ($\varepsilon_M = 1$) and moderate anisotropy ($\varepsilon_M = O(\varepsilon_R)$) will be considered. Edge zone problems in strong anisotropy will be reported elsewhere. Both in isotropy and in moderate anisotropy, solutions in the bending range discussed in Section 4 impose restrictions on the thickness distributions of u_1, u_2, σ_{22} and σ_{12}. Thus, u_1 must be constant over the thickness (for the first step of approximation), u_2, σ_{22} linear in x_1 and σ_{12} quadratic. We might easily imagine or encounter in practical cases boundary conditions at $x_2 = 0$ which do not meet these requirements. For example, in Fig.2b bending moments M at $x_2 = 0$ might be induced with a bending normal stress distribution given by

$$\sigma_{22}(x_1;0) = -a.\sin(\pi x_1/2) \quad , \tag{5.1}$$

instead of the linear distribution required in the interior zone, so that

$$M = -\int_{-1}^{+1} \sigma_{22} x_1 dx_1 = 8a/\pi^2 \tag{5.2}$$

(M dimensionless, per unit length).

Thus, the first step of approximation becomes singular at $x_2 = 0$ and an edge layer allowing the passage from (5.1) to a linear distribution becomes necessary. We will show that this edge layer is much more complicated in the case of moderate anisotropy as compared to the isotropic case.

5.1. Isotropy, $\varepsilon_M = 1$

We obtain for the isotropic case equations in the edge layer region by simply setting it identical to the edge zone defined in Secion 2, explicitly

$$\varepsilon_L = O(1) \ , \quad L = O(H). \tag{5.3}$$

We then find out that all dimensionless stresses and displacements in the edge zone are of the order of the quantity which has to be corrected to match the boundary conditions at $x_2 = 0$. From (2.2) we obtain the following reduced set of equations:

$$\begin{aligned}
&u_{1},_{1} = \sigma_{11} - \nu(\sigma_{22} + \sigma_{33}) \ , \\
&\varepsilon_L u_{2},_{2} = \sigma_{22} - \nu(\sigma_{11} + \sigma_{33}) \ , \\
&\sigma_{33} - \nu(\sigma_{11} + \sigma_{22}) = O(\sigma_{33}\varepsilon_R) \cong 0 \ , \\
&u_{2},_{1} + \varepsilon_L u_{1},_{2} = 2(1 + \nu)\sigma_{12} \ , \\
&\sigma_{11},_{1} + \varepsilon_L \sigma_{12},_{2} = O(\sigma_{11}\varepsilon_R) \cong 0 \ , \\
&\sigma_{21},_{1} + \varepsilon_L \sigma_{22},_{2} = O(\sigma_{12}\varepsilon_R) \cong 0 \ .
\end{aligned} \tag{5.4}$$

Obviously, (5.4) are the basic equations for a simple problem in plane strain. They lead to a solution which we may represent generally with the notation Φ_{Edge}. If we want a solution uniformly valid in the interior zone, bending range and edge zone, we write

$$\Phi = \Phi_{Membrane} + \Phi_{Bending} + \Phi_{Edge} \tag{5.5}$$

and choose for Φ_{Edge} at $x_2 = 0$ boundary conditions which correspond to the difference of distributions required by $\Phi_{Bending}$ and the one imposed at the boundary. For example, in the case (5.1), the proper (self-equilibrated) boundary conditions for Φ_{Edge} would be

$$\begin{aligned}
&\sigma_{12}(x_1;0) = 0 \ , \\
&\sigma_{22}(x_1;0) = a(12x_1/\pi^2 - \sin(\pi x_1/2)) \ , \\
&\sigma_{12}(x_1;\infty) \to 0 \ , \\
&\sigma_{22}(x_1;\infty) \to 0 \ .
\end{aligned} \tag{5.6}$$

A plane strain problem for the semi-infinite block of Fig.4 has now to be solved either numerically or by conformal mapping (ref.6) on the basis of (5.4) and (5.6).

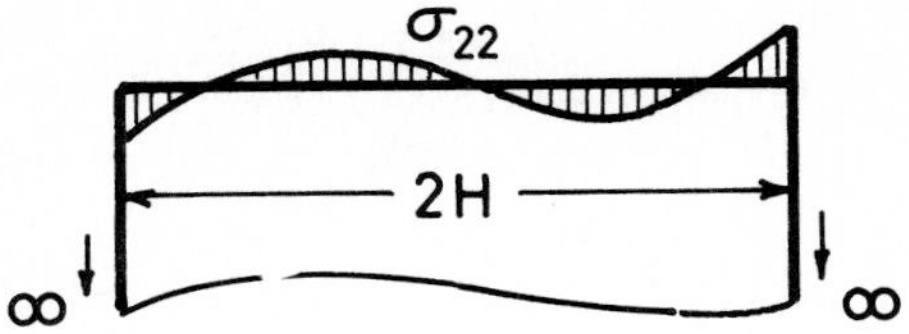

Fig.4. Semi-infinite block in plane strain characterizing situation in edge zone for isotropic behaviour.

5.2. Moderate anisotropy, $O(\varepsilon_R) = O(\varepsilon_M)$

- Edge layer for bending stresses

We assume now that the bending stress distribution over the thickness has to be corrected in an "edge layer". We expect the "extent" of this edge layer to depend on ε_M. We thus set

$$\varepsilon_L = O(\varepsilon_M^{\alpha})$$

and determine α so, that a consistent set of equations in the edge layer is derivable from (2.2). The result is $\alpha = \frac{1}{2}$, hence

$$\varepsilon_L = O(\varepsilon_M^{\frac{1}{2}}) \quad , \quad L = O[H(E/E_I)^{\frac{1}{2}}] \quad . \tag{5.7}$$

The following evaluations follow from (2.2):

$$\sigma_{21} = O(\sigma_{22}\varepsilon_M^{\frac{1}{2}}) \quad ; \quad \sigma_{11} = O(\sigma_{22}\varepsilon_M) \quad ; \quad \sigma_{33} = O(\sigma_{22}\varepsilon_M) \quad ;$$
$$u_2 = O(\sigma_{22}\varepsilon_M^{\frac{1}{2}}) \quad ; \quad u_1 = O(\sigma_{22}) \tag{5.8}$$

and the reduced set of equations read

$$u_{1,1} = O(u_1\varepsilon_R) \cong 0 \quad ,$$
$$\varepsilon_M^{-1}\varepsilon_L u_{2,2} - \sigma_{22} = O(\sigma_{22}\nu\varepsilon_M) \cong 0 \quad ,$$
$$\varepsilon_R u_1 - \sigma_{33} + \nu\sigma_{11} + \nu\varepsilon_M\sigma_{22} = O(\sigma_{33}\varepsilon_R) \cong 0 \quad ,$$
$$u_{2,1} + \varepsilon_L u_{1,2} = \mu\sigma_{12} \quad , \tag{5.9}$$
$$\sigma_{11,1} + \varepsilon_L\sigma_{12,2} = O(\sigma_{11}\varepsilon_R) \cong 0 \quad ,$$
$$\sigma_{21,1} + \varepsilon_L\sigma_{22,2} = O(\sigma_{12}\varepsilon_R) \cong 0 \quad .$$

Equations (5.9) for the edge layer in moderate anisotropy are almost identical to (4.15) for the transitional case of strong anisotropy in the interior zone. Only the third and fifth equations exhibit irrelevant differences connected with σ_{33}. In fact, appropriate elimination from (5.9) leads exactly to

(4.16). This is due to the fact that if, starting from a case of moderate anisotropy, the degree of anistropy is increased, the "thickness" of the edge layer characterized by (5.7) grows and reaches eventually the interior zone as $\varepsilon_M \to 0(\varepsilon_R^2)$, the condition for which we derived (4.15).

Solving the harmonic differential equation (4.16) with the requirement that σ_{22}, $\sigma_{12} \to 0$ as $x_2 \to \infty$ (that is, out of the range of the edge layer), we see (Section 6) that the general solution is of the form

$$\Phi_1 = \sum_{k=1}^{\infty} f_k(x_1).\exp[-\ \omega_k \bar{x}_2 (G/E)^{\frac{1}{2}}/H] \quad , \tag{5.10}$$

where ω_k is the solution of $\tan\omega_k = \omega_k$. Because of the factor $(G/E)^{\frac{1}{2}}$ in the exponent, the decay is not as strong as in the isotropic case.

- Edge layer for shear stresses

Solution (5.10) leads to shear stresses which do not vanish at the edge boundary $x_2 = 0$. Thus, a new edge layer is required to correct the "self-equilibrated" shear stresses arising from Φ_1(or from any independent shear stress distribution imposed as boundary condition). It can easily be shown that this edge layer coincides with the "edge zone" defined by

$$\varepsilon_L = 0(1) \quad , \quad L = 0(H) \tag{5.11}$$

as in the isotropic case discussed in subsection 5.1.

Evaluation of stresses and displacements in terms of σ_{12} leads to the result that they are of the same order with respect to each other, so that (2.2) reduces to

$$\begin{aligned}
&u_{1,1} - \sigma_{11} + \nu_I \sigma_{33} = 0(u_1 \varepsilon_M) \cong 0 \quad , \\
&u_{2,2} = 0(u_2 \varepsilon_M) \cong 0 \quad , \\
&\sigma_{33} - \nu_I \sigma_{11} = 0(\sigma_{33} \varepsilon_M) \cong 0 \quad , \\
&u_{2,1} + \varepsilon_L u_{1,2} = \mu \sigma_{12} \quad , \\
&\sigma_{11,1} + \varepsilon_L \sigma_{12,2} = 0(\sigma_{11} \varepsilon_R) \cong 0 \quad , \\
&\sigma_{21,1} + \varepsilon_L \sigma_{22,2} = 0(\sigma_{12} \varepsilon_R) \cong 0 \quad .
\end{aligned} \tag{5.12}$$

By appropriate elimination from (5.12) we obtain the differential equation

$$\sigma_{11,11} + (1 - \nu_I^2)\mu^{-1} \varepsilon_L^2 \sigma_{11,2} = 0 \quad . \tag{5.13}$$

Both σ_{11} and σ_{12} turn out to be of the form (see Section 6)

$$\Phi_2 = \sum_{n=1}^{\infty} g_n(x_1).\exp[-n\pi\mu^{\frac{1}{2}}(1 - \nu_I^2)^{-\frac{1}{2}} \bar{x}_2/H] \quad . \tag{5.14}$$

Unfortunately, this solution does not fulfill boundary condition $\sigma_{12} = 0$ at the lateral surfaces $x_1 = \pm 1$, so that a boundary layer zone along the lateral surfaces of the shell, similar to the one encountered in the limiting case of strong anisotropy (subsection 4.4, Fig.4), but here confined to the edge zone should be considered.

The"layer thickness", for example along $x_1 = +1$, is found by setting

$$\xi = (-1 + x_1)\varepsilon_M^{-\alpha}\mu^{\alpha}\varepsilon_L$$

and determining α to obtain a consistent set of equations. Again $\alpha = \frac{1}{2}$, so that the boundary layer coordinate becomes

$$\xi = (-1 + x_1)\varepsilon_M^{-\frac{1}{2}}\mu^{\frac{1}{2}}\varepsilon_L \quad . \tag{5.15}$$

Proceeding straightforwardly as in the previous cases, we end up with a harmonic differential equation of the form

$$\sigma_{12},_{\xi\xi} + \sigma_{12},_{22} = 0 \quad . \tag{5.16}$$

The boundary value problem corresponds to the "quarter-space" illustrated in Fig.5, loaded by a strongly decaying exponential distribution of shear stresses σ_{12} along the face $\xi = 0$. Of course, all quantities should vanish both for $\xi \to \infty$ and $x_2 \to \infty$.

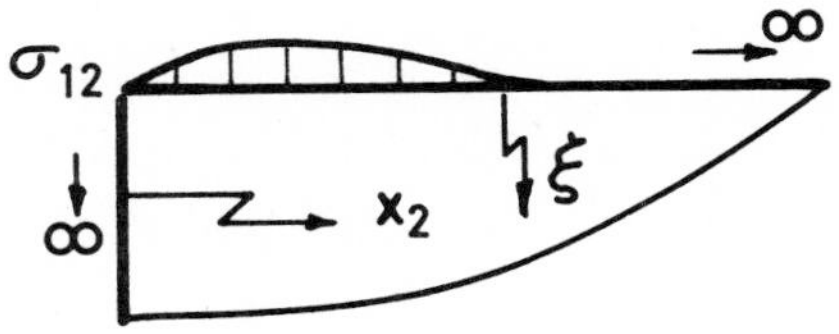

Fig.5. Correction of shear stresses at faces $x_1 = \pm 1$ in a "quarter-space".

In practical cases, the error on shear stresses induced by Φ_2 at the faces $x_1 = \pm 1$ is expected to be small (see Section 6), so that a general solution of the form

$$\Phi = \Phi_{Memebrane} + \Phi_{Bending} + \Phi_1 + \Phi_2 \tag{5.17}$$

may prove to be sufficient to obtain an acceptable approximation. Here, Φ_1 corresponds to (5.9) - correction of bending normal stresses σ_{22} along $x_2 = 0$ - and Φ_2 to (5.12) - correction of shear stresses σ_{12} along $x_2 = 0$ -.

The different domains have been visualized in Fig.6. Table 1 shows the orders of magnitude for dimensionless stresses in the different zones and their typical decay characteristics with corresponding layer thicknesses.

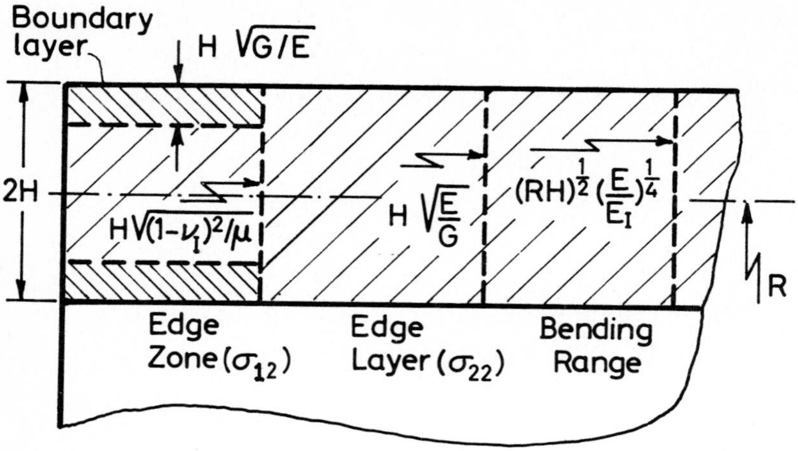

Fig.6. Bending range, edge layer, edge zone and boundary layers in the edge vicinity of a moderately anisotropic shell.

TABLE 1

Dimensionless stresses in the bending range, edge layer, edge zone and boundary layer.

	σ_{22}	σ_{12}	σ_{11}	Layer thickness and decay
$\Phi_{Bending}$	$O(\varepsilon_R^{-1}\varepsilon_M^{-\frac{1}{2}})$	$O(\varepsilon_R^{-\frac{1}{2}}\varepsilon_M^{-\frac{1}{4}})$	$O(1)$	$L_B = (RH)^{\frac{1}{2}}(E/E_I)^{\frac{1}{4}}$ $\sim \exp[-k\bar{x}_2/L_B]$, $k \cong 0{,}9$
Φ_1	$O(\varepsilon_R^{-1}\varepsilon_M^{-\frac{1}{2}})$	$O(\varepsilon_R^{-1})$	$O(\varepsilon_R^{-1}\varepsilon_M^{\frac{1}{2}})$	$L_1 = H(E/G)^{\frac{1}{2}}$ $\sim \exp[-\omega_k\bar{x}_2/L_1]$, $\omega_k = \tan\omega_k$
Φ_2	$O(\varepsilon_R^{-1})$	$O(\varepsilon_R^{-1})$	$O(\varepsilon_R^{-1})$	$L_2 = H(1 - \nu_I^2)^{\frac{1}{2}}\mu^{-\frac{1}{2}}$ $\sim \exp[-n\pi\bar{x}_2/L_2]$, $n = 1, 2, \ldots$
Φ_{Layer}	$O(\varepsilon_R^{-1}\varepsilon_M^{-\frac{1}{2}})$	$O(\varepsilon_R^{-1})$	$O(\varepsilon_R^{-1}\varepsilon_M^{\frac{1}{2}})$	$h = L(G/E)^{\frac{1}{2}}$, $L = O(H)$ $\sim \exp[-\omega(H - \lvert\bar{x}_1\rvert)/h]$, $\omega > 0$

Remark : It might be highly interesting to use the above decay characteristics in order to improve the efficiency of finite element schemes with elements of higher order.

6. AN EXAMPLE

The relatively simple example of fig.2b with the bending normal stress distribution (5.1) at $x_2 = 0$ will now be solved for moderate transverse isotropy. Further examples illustrating both different shell geometry and full orthotropic behaviour will be reported elsewhere.

In this example, $\Phi_{Membrane} \equiv 0$, since only loading by end moments is considered. Setting

$$y := \varepsilon_M^{\frac{1}{4}} \varepsilon_R^{\frac{1}{2}} \varepsilon_L^{-1} (3/4)^{\frac{1}{4}} x_2 \quad , \tag{6.1}$$

the full solution of (4.9) and (4.10) corresponding to $\Phi_{Bending}$ in (5.17) can be shown to be

$$\begin{aligned}
\sigma_{22} &= -\frac{3}{2} M x_1 e^{-y}(\cos y + \sin y) \quad , \\
\sigma_{21} &= \frac{3}{2} \varepsilon_M^{\frac{1}{4}} \varepsilon_R^{\frac{1}{2}} (3/4)^{\frac{1}{4}} M (1 - x_1^2) e^{-y} \sin y \quad , \\
\sigma_{33} &= \varepsilon_M^{\frac{1}{2}} (3/4)^{\frac{1}{2}} M e^{-y} (\cos y - \sin y) \quad , \\
u_2 &= \frac{3}{2} \varepsilon_M^{\frac{3}{4}} \varepsilon_R^{-\frac{1}{2}} (4/3)^{\frac{1}{4}} M x_1 e^{-y} \cos y \quad , \\
u_1 &= \varepsilon_M^{\frac{1}{2}} \varepsilon_R^{-1} (3/4)^{\frac{1}{2}} M e^{-y} (\cos y - \sin y) \quad ,
\end{aligned} \tag{6.2}$$

where M is the dimensionless bending moment at the end.

If the stress distribution at $x_2 = 0$ is not linear as required by (6.2) but given as in (5.1), boundary conditions (5.6) must be considered and (5.9), (4.16) solved.

We now set

$$\eta := \varepsilon_M^{\frac{1}{2}} \mu^{-\frac{1}{2}} \varepsilon_L^{-1} x_2$$

and obtain the solution corresponding to Φ_1 in (5.17) :

$$\begin{aligned}
\sigma_{22} &= \sum_{k=1}^{\infty} A_k \exp(-\omega_k \eta).\sin(\omega_k x_1) \quad , \\
\sigma_{21} &= \mu^{-\frac{1}{2}} \varepsilon_M^{\frac{1}{2}} \sum_1^{\infty} A_k \exp(-\omega_k \eta).(\cos\omega_k - \cos(\omega_k x_1)) \quad , \\
\sigma_{11} &= \mu^{-1} \varepsilon_M \sum_1^{\infty} A_k \exp(-\omega_k \eta).(\omega_k x_1 \cos\omega_k - \sin(\omega_k x_1) \quad , \\
u_1 &= -\mu^{-1} \sum_1^{\infty} A_k \omega_k^{-1} \exp(-\omega_k \eta).\cos\omega_k \quad , \\
u_2 &= -\varepsilon_M^{\frac{1}{2}} \mu^{\frac{1}{2}} \sum_1^{\infty} A_k \omega_k^{-1} \exp(-\omega_k \eta).\sin(\omega_k x_1) \quad , \\
\sigma_{33} &= \sum_1^{\infty} A_k \exp(-\omega_k \eta).[-\mu^{-1} \varepsilon_R \omega_k^{-1} \cos\omega_k + \nu\mu^{-1} \varepsilon_M (x_1 \sin\omega_k - \sin(\omega_k x_1)) \\
&\quad + \nu \varepsilon_M \sin(\omega_k x_1)] \quad ,
\end{aligned} \tag{6.3}$$

where numbers ω_k are solutions of

$$\tan\omega_k = \omega_k \tag{6.4}$$

($\omega_1 = 4.493$; $\omega_2 = 7.725$; $\omega_3 = 10.904$; ... ; $\omega_n \cong (n + \frac{1}{2})\pi$). Set (6.3) represents a general solution to match any bending normal stress distribution imposed at $x_2=0$. For the particular case given in (5.6) we find

$$A_k = M\pi^2/\sin\omega_k/4(\omega_k^2 - \pi^2/4) \ . \tag{6.5}$$

As can be seen from (6.3)

$$\sigma_{21}(x_1;0) = \mu^{-\frac{1}{2}}\varepsilon_M^{\frac{1}{2}} \cdot \sum_{k=1}^{\infty} A_k(\cos\omega_k - \cos(\omega_k x_1)) \quad , \tag{6.6}$$

although this quantity should vanish identically. To correct this distribution, we solve (5.11) and (5.12) with "self-equilibrated" loading corresponding to (6.6) but of opposite sign. The result is listed below :

$$\begin{aligned}
\sigma_{12} &= \sum_{n=1}^{\infty} B_n \exp(-n\pi z).\cos(n\pi x_1) \quad , \\
\sigma_{11} &= \mu^{\frac{1}{2}}(1 - \nu_I^2)^{-\frac{1}{2}} \cdot \sum_1^{\infty} B_n \exp(-n\pi z).\sin(n\pi x_1) \quad , \\
\sigma_{22} &= \mu^{-\frac{1}{2}}(1 - \nu_I^2)^{\frac{1}{2}} \cdot \sum_1^{\infty} B_n [1-\exp(-n\pi z)]\sin(n\pi x_1) \quad , \\
\sigma_{33} &= \nu_I \mu^{\frac{1}{2}}(1 - \nu_I^2)^{-\frac{1}{2}} \cdot \sum_1^{\infty} B_n \exp(-n\pi z).\sin(n\pi x_1) \quad , \\
u_1 &= -[\mu(1 - \nu_I^2)]^{\frac{1}{2}} \cdot \sum_1^{\infty} B_n n^{-1} \pi^{-1} \exp(-n\pi z).\cos(n\pi x_1) \quad , \\
u_2 &\equiv 0 \quad ,
\end{aligned} \tag{6.7}$$

where

$$z := (\mu/(1 - \nu_I^2))^{\frac{1}{2}} \varepsilon_L^{-1} x_2 \tag{6.8}$$

Again, the B_n in (6.7) can be adjusted to satisfy any "self-equilibrated" shear stress distribution imposed at $x_2 = 0$. For (6.6) and (6.5) we find

$$B_n = \mu^{-\frac{1}{2}}\varepsilon_M^{\frac{1}{2}}.M\pi^2 . \sum_{k=1}^{\infty} \omega_k (-1)^k/2(\omega_k^2 - n^2\pi^2)/(\omega_k^2 - \pi^2/4) \ , \tag{6.9}$$

with ω_k defined according to (6.4).

The"erroneous" shear stress distribution at $x_1 = \pm 1$ is now

$$\sigma_{12}(\pm 1;z) = \mu^{-\frac{1}{2}}\varepsilon_M^{\frac{1}{2}}.M\pi^2 . \sum_{k=1}^{\infty} \omega_k/2(\omega_k^2 - \pi^2/4) \sum_{n=1}^{\infty} \exp(-n\pi z)/(\omega_k^2 - n^2\pi^2) \ . \tag{6.10}$$

Both (6.7) and (6.10) corresponding to corrections of the shear stress distributions lead in this particular case to extremely small numerical values for both stresses and displacements. The maximum value of σ_{12} resulting from the sum of solutions (6.2), (6.3) and (6.7) is 0.045 at $\bar{x}_2$= 5.5H if we set M= 2/3. Compared with σ_{22max}= 0.92 at $\bar{x}_2$= H,this is a small value indeed. The "erroneous" shearing stress (6.10) takes maximum values of 0.006 at $\bar{x}_2$= 0.15H, so that results (6.2) and (6.3) are largely sufficient to describe the uniformly valid solution with satisfactory accuracy.

Distributions of bending normal stress σ_{22} over the thickness at various distances from the edge and the decay of σ_{22max} along the axis of the shell have been visualized in Fig.7. Typical values of ε_M= 0.05, ε_R= 0.1, μ= 2.7, ν_I= 0.3 have been chosen and M has been set equal to 2/3.

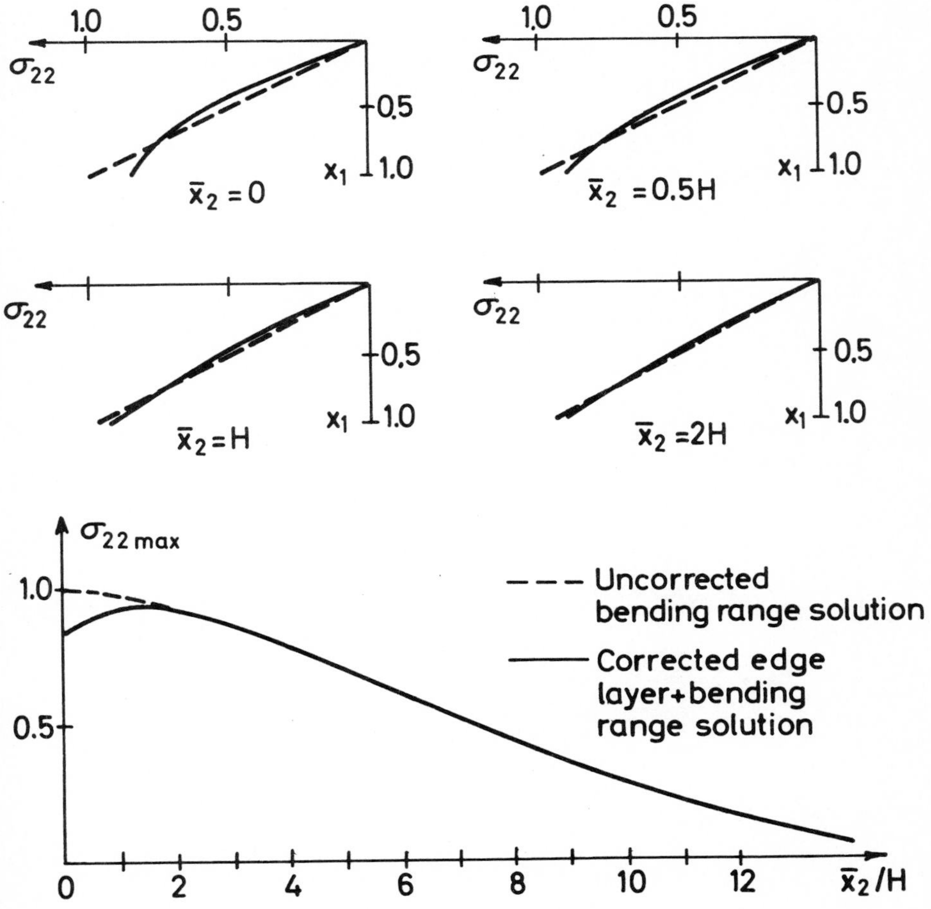

Fig.7. Bending normal stress distributions

It is interesting to note that, whereas corrections due to the sinusoidal distribution at $x_2 = 0$ vanish practically at distances $\bar{x}_2 \cong 3H$ (edge layer for bending stresses), the bending range extends quite a long way into the interior zone. The rapid vanishing of the corrections is due to the fact that the sinusoidal distribution (5.1) chosen here as a first simple illustration, represents a particularly "mild" perturbation with respect to the linear distribution required in the bending range. Other more significant special cases of edge perturbation will be reported elsewhere.

Note : Part of this work has been based on ideas developped in some detail by author's former PhD student C.S.Mitropoulos in his dissertation cited as ref.5.

REFERENCES

1 A.J.M. Spencer, Deformation of Fibre Reinforced Materials, Oxford 1972.
2 M.Sayir, Flexural Vibrations of Strongly Anisotropic Beams, Ing.Arch. 49(1980) 309-323.
3 T.G.Rogers, A.C. Pipkin, Small Deflection of Fibre Reinforced Beams or Slabs, J.Appl.Mech. 38(1971) 1047.
4 M.Sayir, C.S.Mitropoulos, On elementary Theories of Linearelastic Beams, Plates and Shells, J.Appl.Math.Phys.(ZAMP) 31(1980) 1-55.
5 C.S.Mitropoulos, Zur Theorie der schwach und starkanisotropen faserverstaerkten Rotationsschalen, Dissertation No.6317, ETH Zuerich, 1978.
6 N.I.Mushkelishvili, Some Basic Problems of the Mathematical Theory of Elasticity, Noordhoff, Groningen, Holland, 1953.

SOME THEORETICAL ASPECTS IN THE MODELLING OF DELAMINATION FOR MULTILAYERED PLATES

J.L. DAVET Ph. DESTUYNDER Th. NEVERS
Ecole Centrale des Arts et Manufactures
Grande Voie des Vignes - 92290 CHATENAY MALABRY (FRANCE)

KEYWORDS : Delamination, Multilayered plates, Composite-Materials, Stress singularities near the edges.

ABSTRACT

A new approach for the modelling of delamination in a composite multilayered plate is suggested in this paper. We prove that for a free edge logarithmic singularities appear and that after delamination one has a new kind of singular behaviour. The latter being stronger than the former. The influence of the fiber orientation is pointed out on the coefficients of these singularities for [0-90°] plies.

INTRODUCTION TO THE MODELLING OF DELAMINATION FOR PLATES

Since laminate composite materials have been used in aircraft industry many problems have arisen concerning edge effects and delamination. First of all one discovered that stress concentrations could appear near the edges - even for free edges - involving damage growth (decohesion between fibers and matrix, rupture of fibers, bubble growth etc) and then a delamination.

Several important technical difficulties have slowed down studies on this problem. At the moment there are too few experimental data for a comparison with theoretical models. Hence all the numerical results deduced from theoretical models have to be interpreted very carefully. Furthermore a basic difficulty appears in the accuracy of numerical methods which could be used to solve these models ; specially when bending is concerned. As a matter of fact transverse shearing stresses are often used in the prediction of delamination and finite element schemes lead to a poor approximation of this quantity.

The purpose of this article is to give some information concerning the stress field near the edges of multilayered composite plates. The framework is the linear elasticity. There are four sections. The first one is a brief recall of the three dimensional model. The so-called asymptotic method is recalled in the second section and this leads to boundary layer theory. A local solution of this effect is performed in section three where the logarithmic behaviour of the transverse shearing stress is proved. As an application, the effect of the fiber orientation on the coefficient of this singularity is shown. Finally the stress field at the crack tip of a delamination is computed in section four and stress intensity factor with respect to fiber orientation discussed.

I - THE THREE DIMENSIONAL MODEL FOR PLATES

Let us consider a plate which occupies in space the open set $\Omega = \omega \times]-\varepsilon,\varepsilon[$, the medium surface of which is ω and its thickness 2ε. It is submitted to both body and surface forces. The latter are applied on a part of the lateral boundary, say Γ_1^ε, or on the upper and lower faces $\Gamma_\pm^\varepsilon$. The plate is assumed to be fixed on another part Γ_o^ε of the lateral boundary. This plate is made of stacked layers each one containing unidirectional fibers. But obviously the fiber orientation can be different from one layer to the other. The thickness of each one is εh_i and number of layers is N_c. Hence :

$$2\varepsilon = \sum_{i=1,N_c} \varepsilon h_i .$$

The plate is built symmetrically around its medium surface. This assumption permits to decouple membrane and bending effects. The displacement field is denoted by $u = (u_i)$ and the stress field $\sigma = (\sigma_{ij})$. Then if R_{ijkl} is the stiffness tensor and S_{ijkl} its inverse - the compliance - the constitutive relationship is :

$$\sigma_{ij} = R_{ijkl}\, \gamma_{kl}(u) \text{ or else } \quad \gamma_{ij}(u) = S_{ijkl}\, \sigma_{kl}, \tag{1}$$

where

$$\gamma_{ij}(u) = 1/2(\partial_i u_j + \partial_j u_i).$$

The Principle of Virtual Work can be written :

$$\forall v \in \underset{\sim}{V}, \int_{\Omega^\varepsilon} \sigma_{ij} \partial_i v_j = \int_{\Omega^\varepsilon} f_i v_i + \int_{\Gamma_\pm^\varepsilon} g_i^\pm v_i + \int_{\Gamma_1^\varepsilon} h_i v_i , \tag{2}$$

where $\underset{\sim}{V}$ is the space of admissible displacement fields.

The purpose of asymptotic method in plate theory is to construct an approximation of the solution (σ,u) of (1)-(2) considering that the thickness is small compared to the other dimensions (length or width). An important point in asymptotic methods is that only energy is well approximated. Hence the obtained results - through asymptotic methods - are only justified from the energy point of view. For instance it would be hazardous to draw conclusions concerning the local behaviour of the three dimensional solution. As a matter of fact asymptotic methods rest on a zoom technique which consists in a dilatation of a small volume. Obviously, there are several ways to consider that a volume is small. This leads to different kinds of approximation of the three dimensional model. First of all, if we consider that the thickness is small we obtain classical plate theories for which the boundary conditions are not always satisfied. Hence the approximation near the edge is not appropriate and one has to consider a better approximation of three dimensional elasticity near the edges. This is obtained by a zoom technique - in the vicinity of the edges - with respect to both thickness and width of the neighbourhood surrounding the edge. This is the way to derive boundary layer theory from the three dimensional.

II - ASYMPTOTIC THEORY OF PLATES AND BOUNDARY LAYERS EFFECTS

Let us consider the open set $\Omega = \omega \times]-1,1[$ and the mapping from Ω into Ω^{ε} defined by :

$$X = (x_1,x_2,x_3) \in \Omega \to X^{\varepsilon} = F^{\varepsilon}(X) = (x_1,x_2,x_3) \in \Omega^{\varepsilon}. \tag{5}$$

To this change of coordinates we associate a change of functions as follows :

$$\left\{\begin{array}{llll} \sigma^{\varepsilon}_{\alpha\beta}(X) = \sigma_{\alpha\beta}\circ F^{\varepsilon}(X) & , & u^{\varepsilon}_{\alpha}(X) = u_{\alpha}\circ F^{\varepsilon}(X) & , \\ \sigma^{\varepsilon}_{\alpha\beta}(X) = \varepsilon^{-1}\sigma_{\alpha 3}\circ F^{\varepsilon}(X) & , & u^{\varepsilon}_{3}(X) = \varepsilon u_{3}\circ F^{\varepsilon}(X) & , \\ \sigma^{\varepsilon}_{33}(X) = \varepsilon^{-2}\sigma_{33}\circ F^{\varepsilon}(X) & . & & \end{array}\right. \tag{6}$$

<u>Remark 1</u>. The powers of ε in the relations (6) seem to be arbitrary. As a matter of fact this choice simplifies very much the computation. But it can also be interpreted as a change in tangent and cotangent space to the manifold Ω^{ε}, induced by the mapping F^{ε}.

⊂⊃

Then a simple computation leads to an equivalent problem to (3)-(4), but set over Ω instead of Ω^{ε}. The unknowns are $(\sigma^{\varepsilon},u^{\varepsilon})$ such that for arbitrary elements τ,v (respectively virtual stress and displacement fields) :

$$\left\{\begin{array}{l} a_o(\sigma^{\varepsilon},\tau) + \varepsilon^2 a_2(\sigma^{\varepsilon},\tau) + \varepsilon^4 a_4(\sigma^3,\tau) + B(\tau,u^{\varepsilon}) = 0, \\ B(\sigma^{\varepsilon},v) = F(v) \ , \end{array}\right. \tag{7}$$

where the bilinear forms $a_o(.,.)$, $a_2(.,.)$, $a_4(.,.)$ and $B(.,.)$ are defined as follows :

$$\left\{\begin{array}{l} a_o(\sigma,\tau) = \int_{\Omega} S_{\alpha\beta\mu\nu}\sigma_{\alpha\beta}\tau_{\mu\nu}, \ a_4(\sigma,\tau) = \int_{\Omega} S_{3333}\sigma_{33}\tau_{33} \\ a_2(\sigma,\tau) = 4\int_{\Omega} S_{\alpha 3\alpha 3}\sigma_{\alpha 3}\tau_{\beta 3} + \int_{\Omega} S_{\alpha\beta 33}\{\sigma_{\alpha\beta}\tau_{33} + \sigma_{33}\tau_{\alpha\beta}\} \\ B(\tau,\nu) := -\int_{\Omega} \tau_{ij}\partial_i v_j, \ F(v) = -\{\varepsilon^{-2}\int_{\Gamma\pm} g_3 v_3 + \varepsilon^{-1}\{\int_{\Gamma_1} h_3 v_3 + \int_{\Gamma\pm} g_{\alpha} v_{\alpha} + \int_{\Omega} f_3 v_3 \\ \qquad + \int_{\Gamma_1} h_{\alpha} v_{\alpha} + \int_{\Omega} f_{\alpha} v_{\alpha}\} \end{array}\right. \tag{8}$$

Remark 2. As a matter of fact the problem that we are dealing with, is linear. Hence, using superposition principle it is always possible to combine the right handside multiplied by any scalar ; the solution is then modified in the same way. Therefore one can assume without loss of generality that there is no power of ε in the expression of the right hand side (it means that we can choose $\varepsilon = 1$). The general solution will then be deduced by linear combination of the solutions obtained for different right hand sides.

⊂⊃

Remark 3. In the formulation of problem (7) it has been assumed that for any α, β and μ in the set $\{1,2\}$, we have :

$$S_{\alpha 333} = S_{\alpha\beta\mu 3} = 0.$$

This is the case of a monoclinic material and each layer satisfies this property.

⊂⊃

The asymptotic method consists then to set :

$$(\sigma^{\varepsilon}, u^{\varepsilon}) = (\sigma^{o}, u^{o}) + \varepsilon^{2}(\sigma^{2}, u^{2}) + \dots$$

which implies :

$$\begin{cases} \forall \tau, \ a_{o}(\sigma^{o},\tau) + B(\tau,u^{o}) = 0 \ , \\ \forall v, \ B(\sigma^{o},v) = F(v) \ , \end{cases} \tag{9}$$

etc.

Equations (9) are the plate model. Obviously it has to be explicited in each case. Let us just recall how (σ^{o}, u^{o}) can be computed for an inplane loading ($f_3 = g_3^{\pm} = g_a^{\pm} = f_{\alpha} = 0$ for instance).

The displacement u_{α}^{o} is independent on the coordinate x_3 and u_3^{o} is null. The stress field is such that on the one hand :

$$\sigma_{33}^{o} = \sigma_{\alpha 3}^{o} = 0 \qquad (\alpha = 1,2)$$

and on the other hand

$$\sigma_{\alpha\beta}^{o} = \overline{R}_{\alpha\beta\mu\nu} \, \gamma_{\mu\nu}(u^{o}) \qquad (\alpha,\beta,\mu,\nu \in \{1,2\}) \ ;$$

where $R_{\alpha\beta\mu\nu}$ is the inverse of the restriction of the compliance tensor to planar stresses. It is worth noticing that $\overline{R}_{\alpha\beta\mu\nu} = R_{\alpha\beta\mu\nu}$. The equilibrium equations are :

$$\left\{\begin{array}{ll} \partial_\beta n^o_{\alpha\beta} = 0 & \text{on } \omega \\ n^o_{\alpha\beta} b_\beta = \int_{-1}^{+1} h_\alpha & \text{on the boundary } \gamma_1 \end{array}\right. \tag{10}$$

where

$$n^o_{\alpha\beta} = \int_{-1}^{+1} \sigma^o_{\alpha\beta}$$

and the displacement field is null on γ_o.

Remark 4. The three dimensional model is such that on the loaded (or free) edge one has :

$$\sigma_{\alpha\beta} b_\beta = h_\alpha \qquad (\text{or } \sigma_{\alpha\beta} b_\beta = 0). \tag{11}$$

The solution $\sigma^o_{\alpha\beta}$ of (10) doesn't satisfy these conditions. It is only true for the mean value across the thickness of the plate.

Hence, it seems necessary to improve the approximation of the three dimensional solution near the free edges (at least). A similar phenomenon arises near clamped edges ; but it appears in the computation of the second term of the asymptotic expansion.

⊃

Let us focus our attention on the free edge problem. First of all we use a translation :

$$(\sigma^\varepsilon, u^\varepsilon) = (\sigma^o, u^o) + (y^\varepsilon, u^\varepsilon), \tag{12}$$

where (σ^o, u^o) is solution of (9) and $(\sigma^\varepsilon, u^\varepsilon)$ is the one of (7). Hence $(y^\varepsilon, \mu^\varepsilon)$ is a solution of :

$$\left\{\begin{array}{l} \forall\tau,\ a_o(y^\varepsilon,\tau) + \varepsilon^2 a_2(y^\varepsilon,\tau) + \varepsilon^4 a_4(y^\varepsilon,\tau) + B(\tau,u^\varepsilon) \\ \qquad = -\varepsilon^2 a_2(\sigma^o,\tau) - \varepsilon^4 a_4(\sigma^o,\tau), \\ \forall v \in \underset{\sim}{V} \quad B(y^\varepsilon, v) = -B(\sigma^o, v) \end{array}\right. \tag{13}$$

It is then possible to restrict the preceding relations to a neighbourhood β of the boundary Γ_1 of Ω ($\Gamma_1 = \gamma_1 \times]-1,1[$) such that :

$$\beta = \{(s,\xi x_3) \mid \in \gamma_1 x]0,L[x]-1,1[\} \tag{14}$$

where (s,ξ) is a system of local coordinates on the boundary γ of ω (see figure 2).

The basic idea of the boundary layer theory is to use a zoom technique in the direction normal to the boundary, say ξ, such that we set :

$$\eta = \frac{\xi}{\varepsilon}$$

Then the neighbourhood β is mapped on β^{ε} with:

$$\beta^{\varepsilon} = \{(s,\eta,x_3) \Big| \in \gamma_1 \times]0,\tfrac{L}{\varepsilon}[x]-1,1[\} .$$

Equations (13) are also transformed because of this change of coordinates. If we set a priori :

$$(y^{\varepsilon},\mu^{\varepsilon}) = (y^{o},\mu^{o}) + \varepsilon\ (y^{1},\mu^{1}) + \ldots \text{ etc}$$

one can characterize the model (y^{o},μ^{o}) is solution of which. The computations are quite easy but a little bit voluminous. Only the results obtained are mentioned here. For sake of simplicity we assume that each layer is orthotropic in the same system of axes- say (s,η). It is also admitted in the sequel that this configuration can be deduced up to a rotation - the angle of which is θ - around the axis x_3 (see figure 2).

Then the component φ of the displacement μ^{o} along the axis "s" is solution of the following set of equations over the semi infinite strip β^{∞} (β^{ε} for $\varepsilon = 0$) :

$$R_{s3s3} \frac{\partial^2 \varphi}{\partial x_3^2} + R_{\eta 3\eta 3} \frac{\partial^2 \varphi}{\partial \eta^2} = 0 \text{ on each layer} \tag{15}$$

$$R_{s3s3} \frac{\partial \phi}{\partial x_3} \text{ and continuous at each interface} \tag{16}$$

$$R_{s3s3} \frac{\partial \varphi}{\partial x_3} = 0 \quad \text{at } x_3 = \pm 1 \tag{17}$$

$$R_{s3s3}\frac{\partial \varphi}{\partial \eta} = - T^{o}_{s\eta} = - [\sigma^{o}_{12} \cos\theta + \sigma^{o}_{22} \sin\theta] \text{ at } \eta = 0 , \tag{18}$$

(where x_1 and x_2 are the tangential and normal coordinates to the boundary of ω (see figure 2).

<u>Remark 5</u>. In order to obtain a solution φ of equations (15)-(18), which is confined in the vicinity of the lateral boundary, we prescribe the additional condition on φ :

$$\lim_{\eta \to \infty} \varphi(s,\eta,x_3) = 0.$$

⊂⊃

<u>Remark 6</u>. It is clear that s appears as a parameter in the boundary layer model. But φ depends on s because of $T^{o}_{s\eta}$.

⊂⊃

Remark 7. For sake of simplicity the analysis is limited to the computation of φ. We shall see - at the end of section 3 - that this is sufficient to determine the transverse shearing stresses because of the hypothesis concerning the orthotropy of each layer in the same system of axes (s,η).

⊂⊃

3 - COMPUTATION OF THE TRANSVERSE SHEARING STRESS IN THE BOUNDARY LAYER

In order to analyze the stress field in the vicinity of the interface between two layers it is sufficient to solve the system (15)-(18) in a neighbourhood of the interface. If "h" is the coordinate of a boundary between two layers, we set :

$$
\begin{aligned}
z &= \sqrt{\frac{R^{+}_{s\eta s\eta}}{R^{+}_{s3s3}}}\,(x_3 - h) \qquad \text{for } x_3 \geqslant h \\
z &= \sqrt{\frac{R^{-}_{s\eta s\eta}}{R^{-}_{s3s3}}}\,(x_3 - h) \qquad \text{for } x_3 \leqslant h
\end{aligned}
\tag{19}
$$

where R^{+}_{ijkl} (respectively R^{-}_{ijkl}) is the stiffness tensor in the upper layer (respectively lower layer) to the interface at $x_3 = h$.

If φ is expressed as a function of s, η and z, the model (15)-(18) can be formulated locally (in the strip surrounding the line $x_3 = h$) :

$$\frac{\partial^2\varphi}{\partial z^2} + \frac{\partial^2\varphi}{\partial \eta^2} = 0 \qquad \text{on } B_1 \text{ and } B_2, \quad \text{(see figure 3)} \tag{20}$$

$$\sqrt{R_{s3s3}R_{s\eta s\eta}}\,\frac{\partial\varphi}{\partial z} \text{ and } \varphi \text{ continuous at } z = 0 , \tag{21}$$

$$R_{s\eta s\eta}\,\frac{\partial\varphi}{\partial\eta} = -\,T^{o}_{s\eta} \qquad \text{at } \eta = 0 , \tag{22}$$

$$\lim_{\eta\to\infty} \varphi(s,\eta,z) = 0. \tag{23}$$

As a matter of fact boundary conditions on the lines $z = \pm a$ should be prescribed in order to determine completely φ. Let us arbitrarily choose $\partial\varphi/\partial z = 0$. Then, the solution $\underset{\sim}{\varphi}$ of (20)-(23) satisfying this additional condition is not the restriction of φ to the strip $B_1 \cup B_2$. But if we set :

$$\varphi = \underset{\sim}{\varphi} + \varphi^{c}$$

the complementary term φ^{c} is solution of an homogeneous Neuman problem in the vicinity of z = 0. Hence, using a reflexion procedure, one knows that φ^{c} is regular in the vicinity of z = 0. As a consequence, we can claim that the singularities of φ are those of $\underset{\sim}{\varphi}$.

In order to compute $\underset{\sim}{\varphi}$ let us first notice that the general solution of (20) can be written :

$$\varphi = \sum_{n \geq 0} A_n \cos\left(2\frac{n}{a}z\right) e^{-\frac{2n}{a}\eta} + B_n \sin\left(\frac{(2n+1)}{a}z\right) e^{-\frac{(2n+1)}{a}\eta}$$

(because of the condition at $z = \pm a$). There is no reason to have the same coefficients A_n and B_n for $z > 0$ or $z < 0$. As a matter of fact the conditions (21) imply that :

$$z \geq 0, \ \varphi = \sum_{n \geq 0} A_n \cos\left(\frac{2nz}{a}\right) e^{-\frac{2n}{a}\eta} + \frac{X_n}{\sqrt{R^+_{s\eta s\eta} R^+_{s3s3}}} \sin\left(\frac{(2n+1)z}{a}\right) e^{-\frac{(2n+1)\eta}{a}}$$

$$z \leq 0, \ \varphi = \sum_{n \geq 0} A_n \cos\left(\frac{2nz}{a}\right) e^{-\frac{2n}{a}\eta} \ \frac{X_n}{\sqrt{R^-_{s\eta s\eta} R^-_{s3s3}}} \sin\left(\frac{(2n+1)z}{a}\right) e^{-\frac{(2n+1)\eta}{a}} .$$

The coefficients A_n and X_n can be deduced from condition (22) as follows. Let us observe that (22) is equivalent to :

$$- \left[R_{s\eta s\eta} \frac{\partial \varphi}{\partial \eta}\right]^a = \sum_{n \geq 0} (2n+1) \frac{Xn}{2a} (G^+ + G^-) \sin\left(\frac{(2n+1)z}{a}\right) \quad (24)$$

$$- \left[R_{s\ s} \frac{\partial \varphi}{\partial \eta}\right]^s = \sum_{n \geq 0} 2n\frac{An}{2a} (R^+_{s\eta s\eta} + R^-_{s\eta s\eta}) \cos\left(\frac{2nz}{a}\right) \quad (25)$$

$$(2n+1) \frac{Xn}{2a} (G^+ - G^-) \sin\left(\frac{(2n+1)}{a}|z|\right) .$$

where we have set :

$$G^{\pm} = \sqrt{\frac{R^{\pm}_{s\eta s\eta}}{R^{\pm}_{s3s3}}} ,$$

and $[f]^a$ (respectively $[f]^s$) is the odd (respectively even) component of f with respect to z. Furthermore the assumption that each layer is orthotropic in the same system of axes (s,η) implies $R^+_{s\eta s\eta} = R^-_{s\eta s\eta}$. Hence :

$$- \left[R_{s\eta s\eta} \frac{\partial \varphi}{\partial \eta}\right]^s = \sum_{n \geq 0} 2n \frac{An}{a} R_{s\eta s\eta} \cos\left(\frac{2nz}{a}\right) + 2n+1\ X_n \left(\frac{G^+ - G^-}{2a}\right) \sin\left(\frac{(2n+1)}{a}|z|\right) . \quad (26)$$

Let us set :

$$T^{oa}_{s\eta} = [T^o_{s\eta}]^a \text{ and } T^{os}_{s\eta} = [T^o_{s\eta}]^s ,$$

such that equation (24) implies :

$$X_n = \frac{2a}{(2n+1)\Pi(G^+ + G^-)} \int_{-a\frac{\Pi}{2}}^{a\frac{\Pi}{2}} T^{oa}_{s\eta} \sin\left(\frac{(2n+1)}{a}z\right)$$

and from (26)

$$A_n = - \frac{(G^+ - G^-)}{2an\Pi R_{s\eta s\eta}} \sum_{p \geq 0} (2p+1)\ X_p \int_{-a\frac{\Pi}{2}}^{a\frac{\Pi}{2}} \sin\left(\frac{(2p+1)}{a}|z|\right) \cos\left(\frac{2nz}{a}\right)$$

The plate is assumed to be equilibrated (the same number of layers at 0° and 90°). The meanvalue of $T^{o}_{s\eta}$ is null and therefore $T^{os}_{s\eta} = 0$. Hence we have $A_o = 0$.

A simple computation leads to the following expressions for X_n and A_n :

$$\begin{cases} A_n = \dfrac{2\, T^{oa}_{s\eta}}{n\,\Pi} \left(\dfrac{G^+ - G^-}{G^+ + G^-}\right) \dfrac{a}{R_{s\eta s\eta}} \sum\limits_{p>0} \left[\dfrac{1}{2p+1} \left(\dfrac{1}{2(p+n)+1} + \dfrac{1}{2(p-n)+1}\right)\right] \\ X_n = \dfrac{4\, a^2\, T^{oa}_{s\eta}}{(2n+1)^2 \Pi (G^+ + G^-)} \end{cases} \qquad (27)$$

The transverse shearing stress Y_{s3} is given by :

$$y_{s3} = R_{s3s3} \frac{\partial \varphi}{\partial x_3} = \sqrt{R_{s3s3} R_{s\eta s\eta}}\ \frac{\partial \varphi}{\partial z}$$

or else for $z \geqslant 0$:

$$y_{s3} = \sum_{n>1} \left[- \frac{2n}{a} A_n \sqrt{R^+_{s3s3} R^+_{s\eta s\eta}}\ \sin\left(\frac{2n}{a} z\right) e^{-\frac{2n}{a}\eta}\right] + \sum_{n\geqslant 0} \left[\frac{2n+1}{a} X_n \cos\left(\frac{(2n+1)}{a} z\right) e^{-\frac{2n+1}{a}\eta}\right] \qquad (28)$$

From (27), one can deduce that the series $\Sigma\, n\, A_n$ is convergent and only the second one - in the above expression - can lead to a singularity. From (27) and (28) one obtains for the singular part of Y_{sz} :

$$y_{s3} = \frac{T^{oa}_{s\eta}}{(G^+ + G^-)} \operatorname{Log} \left| \frac{1 + 2\cos\left(\frac{z}{a}\right) e^{-\frac{\eta}{a}} + e^{-\frac{2\eta}{a}}}{1 - 2\cos\left(\frac{z}{a}\right) e^{-\frac{\eta}{a}} + e^{-\frac{2\eta}{a}}} \right|$$

The summ of the series is obtained with the help of the following identity :

$$\sum_{n\geqslant 0} \frac{1}{2n+1} \cos((2n+1)\, z) e^{-(2n+1)\eta} = \frac{1}{4} \operatorname{Log} \left| \frac{1 + 2\cos(z)\, e^{-\eta} + e^{-2\eta}}{1 - 2\cos(z)\, e^{-\eta} + e^{-2\eta}} \right| .$$

As a matter of fact the stress vector $T^{oa}_{s\eta}$ depends on the angle θ of the fibers with respect to the axis x_1 (see figure 2 and formula 18).

From practical point of view it is very convenient to analyze the dependence of the transverse shearing stress along the boundary. Following the notations of figure 2, one has :

$$y_{13} = [\, y^{\theta}_{s3} \cos(\theta) - y^{\theta}_{\eta 3} \sin(\theta)]$$

The component $y_{\eta 3}$ is out of reach, up to now. But just at the interface between two layers the continuity of the normal stress implies :

$$y_{3} = y_{s3}^{\theta + \frac{\Pi}{2}} \qquad \underline{(\text{at } z = 0 \text{ !})}$$

hence :

$$y_{13} = [y_{s3}^{\theta} \cos(\theta) - y_{s3}^{\theta + \frac{\Pi}{2}} \sin(\theta)] \qquad (\text{at } z = 0 \text{ !}) \tag{30}$$

In order to check the influence of the fiber orientation on the coefficient of the logarithmic singularity we have drawn on figures 4-1 and 4-2. The evolution of this coefficient in both cases : prescribed stress or strain. It appears that the maximum is obtained for two values of the angles between the fibers and the normal to the boundary.

4 - INITIATION OF A CRACK AT THE INTERFACE BETWEEN TWO LAYERS

Because of the influence of the stress singularity described in section 3, a delamination can occur just at the interface between two layers. The purpose of this section is to analyze the stress field which can arise near the crack tip. The study is limited to the transverse shearing stress. We prove that the logarithmic singularity is transformed into another one, similar - but different - to the one met in Fracture Mechanics.

It is assumed in the sequel that the crack length is of the same order that the thickness of the plate. This allows to consider that the membrane state of stress is still solution of the classical model where the crack does not appear.

Only the boundary layer is modified as follows :

$$\begin{array}{l} R_{s3s3} \dfrac{\partial^2 \varphi}{\partial x_3^2} + R_{s\eta s\eta} \dfrac{\partial^2 \varphi}{\partial \eta^2} = 0 \qquad \text{on each layer} \\ R_{s3s3} \dfrac{\partial \varphi}{\partial x_3} \text{ and } \varphi \text{ continuous at each interface} \\ R_{s3s3} \dfrac{\partial \varphi}{\partial x_3} = 0 \quad \text{on the lips of the crack} \\ R_{s\eta s\eta} \dfrac{\partial \varphi}{\partial \eta} = - T^{o}_{s\eta} \quad \text{at } \eta = 0. \qquad \lim\limits_{\eta \to \infty} \varphi(s, x_3, \eta) = 0. \end{array} \tag{31}$$

Remark 7. The difference between equations (31) and (15)-(18) is limited to the boundary condition on the lips of the crack

In order to compute the solution of (31) at the vicinity of the crack tip, it is convenient to use a local system of polar coordinates around the crack tip - say (r,ξ) - but associated to the cartesian coordinates (z,η) (and not (x_3,η) !) where z has been introduced in (19). Then the system to be solved can be written as follows :

$$\begin{cases} \frac{1}{r^2}\frac{\partial^2\varphi}{\partial\xi^2} + \frac{1}{r}\frac{\partial}{\partial r}\left(r\frac{\partial\varphi}{\partial r}\right) = 0 \quad \text{on each layer} \\ \frac{\partial\varphi}{\partial\xi} = 0 \quad \text{at } \xi = \pm\Pi \\ R_{s3s3}\frac{\partial\varphi}{\partial\xi} \text{ and } \varphi \text{ continuous at } \xi = 0. \end{cases} \tag{32}$$

The general solution of (32) is :

$$= A_n\, r^n \cos(n\xi) + \frac{X_n}{\sqrt{R^{\pm}_{s3s3}R^{\pm}_{s3s3}}}\, r^{\frac{2n+1}{2}} \sin\left(\frac{2n+1}{2}\xi\right).$$

Among all these functions, one is singular (i.e leads to infinite stress field). It is obtained for n = 0 and is given by :

$$S_o(r,\xi) = \frac{X_o}{\sqrt{R^{\pm}_{s\eta s\eta}R^{\pm}_{s3s3}}}\sqrt{r}\quad \sin\left(\frac{\xi}{2}\right).$$

If we express S_o with respect to the cartesian coordinates (x_3,η) we have :

$$S_o(x_3,\eta) = \frac{X_o}{\sqrt{2\,R^{\pm}_{s3s3}R^{\pm}_{s\eta s\eta}}}\left[\left[(\eta-\ell)^2 + \frac{R_{s\eta s\eta}}{R^{\pm}_{s3s3}}(x_3-\varepsilon h)^2\right]^{1/2} - (\eta-\ell)\right]^{1/2} \tag{33}$$

where ℓ is the crack length. From Stokes formula it is possible to express X_o with the help of a path independent integral. Let $\mathscr{C}$ be a curve surrounding the crack tip and $b = (b_\eta, b_z)$ the components of the unit normal to $\mathscr{C}$ as shown on figure 5. Then Stokes formula permits to establish that the following expression doesn't depend on $\mathscr{C}$:

$$J = -\frac{1}{2}\oint_{\mathscr{C}} R_{s\eta s\eta}\left[\left(\frac{\partial\varphi}{\partial z}\right)^2+\left(\frac{\partial\varphi}{\partial\eta}\right)^2\right]b_\eta + \int_{\mathscr{C}} R_{s\eta s\eta}\left[\left(\frac{\partial\varphi}{\partial\eta}\right)^2 b + \frac{\partial\varphi}{\partial\eta}\frac{\partial\varphi}{\partial x_3}b_z\right]. \tag{34}$$

Hence, it is possible to choose particular curves $\mathscr{C}$ in order to compute J. First of all let us choose a very small circle surrounding the crack tip. As the radius tends to zero one deduces that:

$$J = \frac{\Pi X_o^2}{8}\left[\sqrt{\frac{R^{+}_{s\eta s\eta}}{R^{+}_{s3s3}}} + \sqrt{\frac{R^{-}_{s\eta s\eta}}{R^{-}_{s3s3}}}\right] \tag{35}$$

Another choice for $\mathscr{C}$ is the external boundary of the semi-infinite strip β^∞ on which the model (31) is set, (see figure 5). Using the fact that on the one hand $\partial\varphi/\partial z = 0$ at $x_3 = \pm\varepsilon$ and $x_3 = \varepsilon h$ $(\eta \leq \ell)$, and on the other hand $b_z = 0$ at $\eta = 0$, one obtains from (34) :

$$X_o^2 = \frac{8}{\Pi}\,\frac{\left[\sum_{i=1,Nc}\frac{\varepsilon h_i}{R_{s\eta s\eta}}|T^{oi}_{s\eta}|^2 + \int_{-\varepsilon}^{\varepsilon} R_{s3s3}\left(\frac{\partial\varphi}{\partial x_3}\right)^2\right]}{\sqrt{\frac{R^{+}_{s\eta s\eta}}{R^{+}_{s3s3}}} + \sqrt{\frac{R^{-}_{s\eta s\eta}}{R^{-}_{s3s3}}}}$$

where εhi (respectively $T^{oi}_{s\eta}$) is the thickness (respectively the inplane shearing stress) of the layer i.

The components $T^{oi}_{s\eta}$ can be deduced from the normal stress vector on the boundary, say $(\sigma^o_{12}, \sigma^o_{22})$ by a rotation around the axe x_3 such that if θ is the angle of this rotation :

$$T^{oi}_{s\eta} = - [\sin(\theta)\, \sigma^o_{22} + \cos(\theta)\, \sigma^o_{12}] .$$

For physical reasons it seems interesting to focus our attention on the component Y_{13} of the stress field. It is given by :

$$y_{13} = [y_{s3} \cos(\theta) - y_{\eta 3} \sin(\theta)] .$$

As a matter of fact $Y_{\eta 3}$ has not been computed. But the continuity of the transverse shearing stress implies :

$$y^{\theta}_{\eta 3} = y^{\theta + \frac{\Pi}{2}}_{s3}$$

Hence :

$$y_{s3} = R_{s3s3} \frac{\partial \varphi}{\partial x_3} = \frac{X_o(\theta)}{2} \frac{1}{\sqrt{\eta - \ell}} + \ldots\ldots \quad (z = 0)$$

therefore :

$$y^{\theta}_{13} = [X_o(\theta) + X_o(\theta + \frac{\Pi}{2})]_2 \frac{1}{\sqrt{(\eta - \ell)}} + \ldots\ldots \quad (z = 0) .$$

The computation fo X_o needs the one of $\frac{\partial \varphi}{\partial x_3}$ at η = 0. But in particular cases this therm can be neglected. It corresponds to a three layers plate with a crack initiated from the boundary at each interface (hence there is no more logarithmic singularity). Then X_o^2 - given par (36) - represents twice the value of X_o^2 for one crack (because there are two cracks on β (see figure 5).

Furthermore, $\frac{\partial \varphi}{\partial x_3}$ can be neglected only if the crack length is large enough compared to the thickness of the layers. But it should be noticed that if it is too long the plate model leading to the determination of $\sigma^o_{\alpha\beta}$ is no longer valid. Anyway the variation of X_o with such assumptions with respect to the orientation of the fibers is given on figure 6.

5 - CONCLUSIONS

The study given in this paper is certainly insufficient for at least two reasons. First of all only [0,90°] plies have been discussed. This assumption has been mainly used for decoupling the torsion effect and the bending and stretching effects in the boundary layer. Then, the second assumption is the elasticity of the material which is certainly not acceptable as far as local behaviour is concerned. It would be necessary to introduce damage theory to explain what is happening in the edge effect. This is another but quite more difficult step. It has also to be mentioned that elasticity may be reasonable for large delamination but in such cases the classical plate theory has to be improved in order to take into account the discontinuity of the material.

REFERENCES

[1] L. Anquez, A. Bern, Calcul des singularités dans les effets de bord. La recherche Aérospatiale, ONERA, Avril 1985.
[2] R. Barsoum, C. Freese, An iteractive approach for the evaluation of delamination stresses in laminated composites, Int. J. Num. Meth. Eng. Vol. 20, 1415-1431 (1984)
[3] P.G Ciarlet, Ph. Destuynder, A justification of two dimensional linear plate model, J. Mecan., Vol. 128, n°2 - 1979.
[4] J.L Davet, Ph. Destuynder, Singularités logarithmiques dans les effets de bord d'une plaque en matériaux composites, Jour. Méca. Theo. Appl., Vol. 4, n° 3, p. 54-71 (1985)
[5] Ph. Destuynder, Th. Nevers, Une modélisation du délaminage d'une plaque mince, C.R. Acad. Sc. Paris T299, Série II n ° 8 (1984)
[6] K.O. Friedrichs, R.F. Dressler, A boundary layer theory for elastic plates, CPAM-XIV, 1961, p. 1-33
[7] R. Spilker, S. Chou, Edge effect in symmetric composite laminates : Importance of satisfying the traction-free edge condition. J. Composite Materials vol. 14, p. 2-20 (1980)
[8] R. Zwiers T. Ting, R. Spilker, On the logarithmic singularity of free edge stress in laminated composite under uniform extension, J. Appl. Mechs., Vol. 49, p. 561-569 (1982)

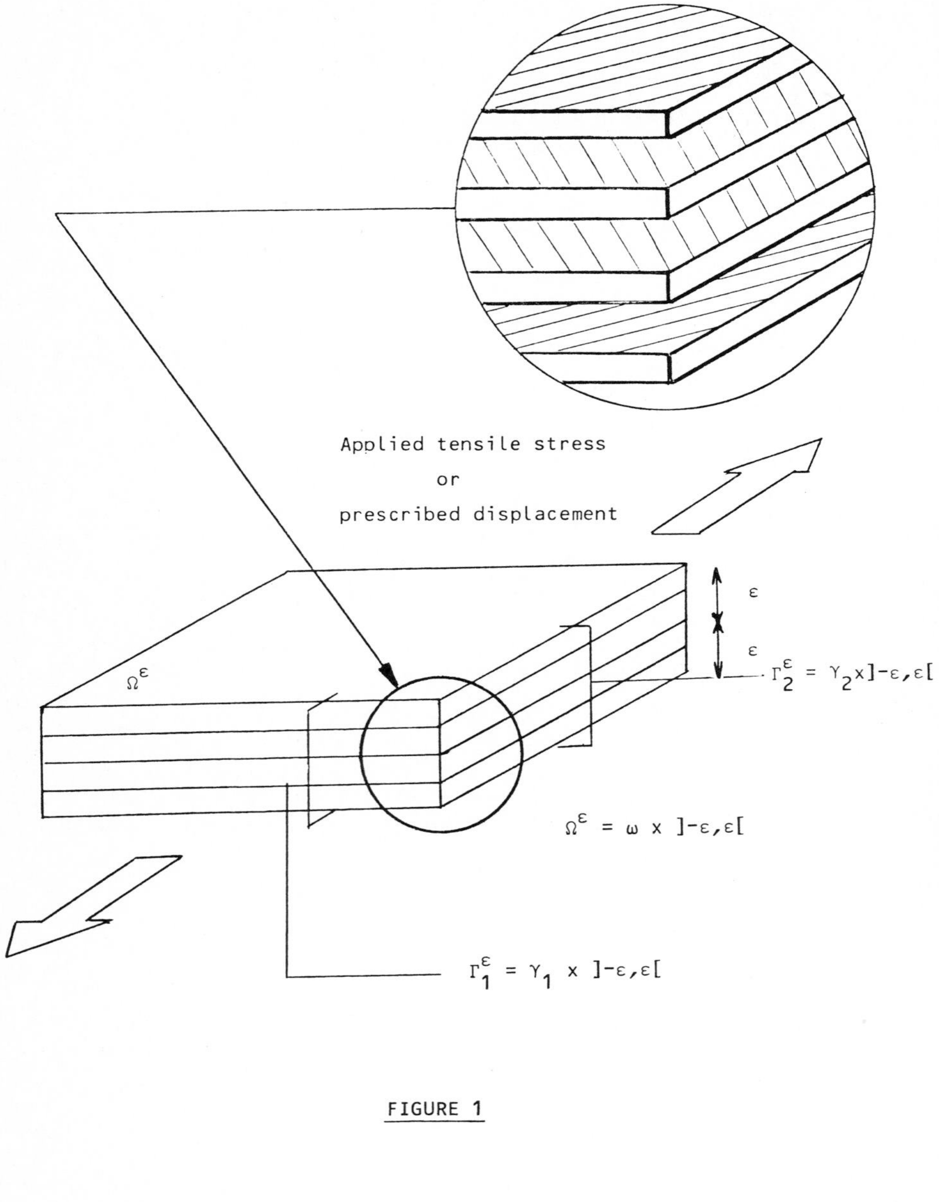
Applied tensile stress
or
prescribed displacement
ε
ε
$\Gamma_2^\varepsilon = \gamma_2 x]-\varepsilon,\varepsilon[$
Ω^ε
$\Omega^\varepsilon = \omega \times]-\varepsilon,\varepsilon[$
$\Gamma_1^\varepsilon = \gamma_1 \times]-\varepsilon,\varepsilon[$

FIGURE 1

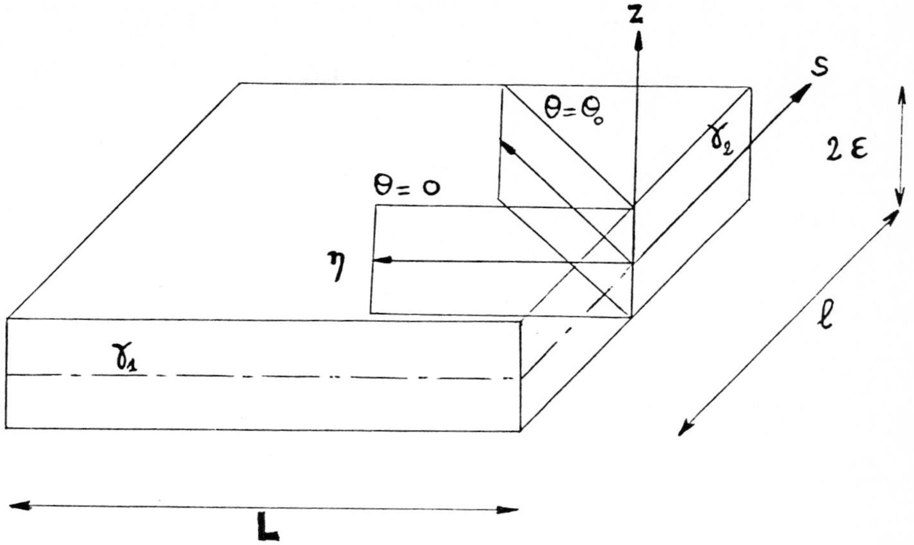

FIGURE 2

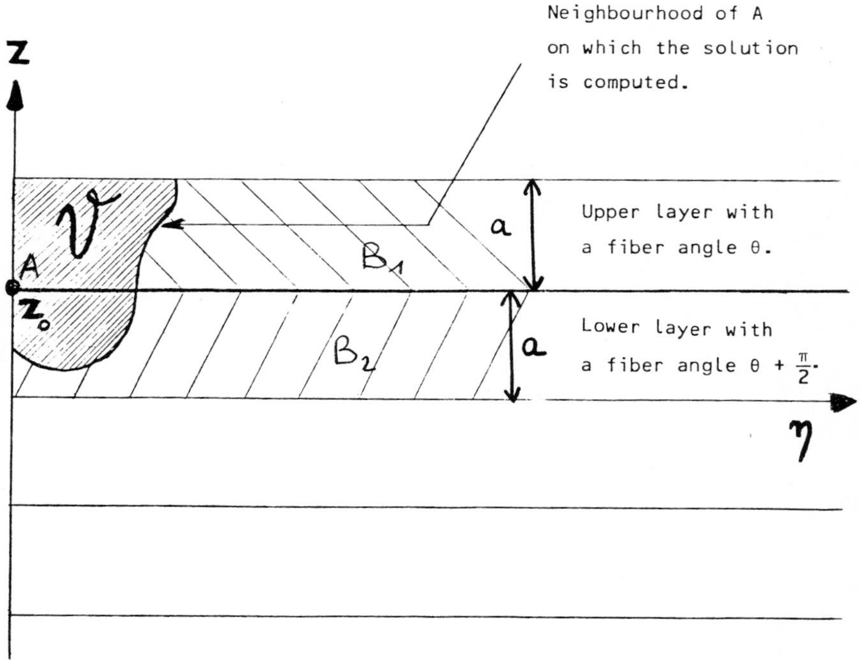

Figure 3. : Semi-infinite strip on which the boundary layer problem is set.

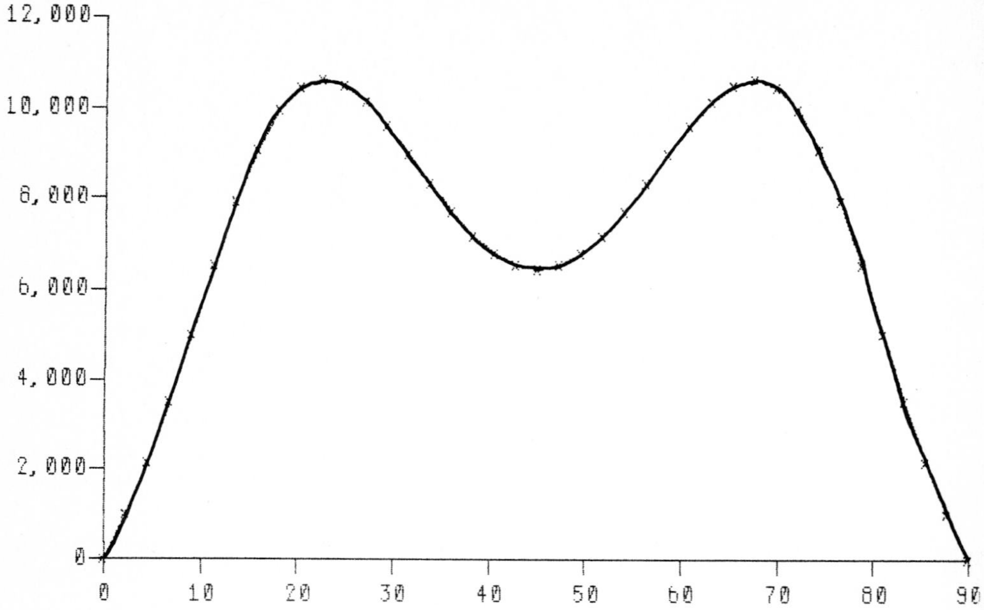

Figure 4-1 variation of the stress intensity factor with the fiber orientation (prescibed stresses)

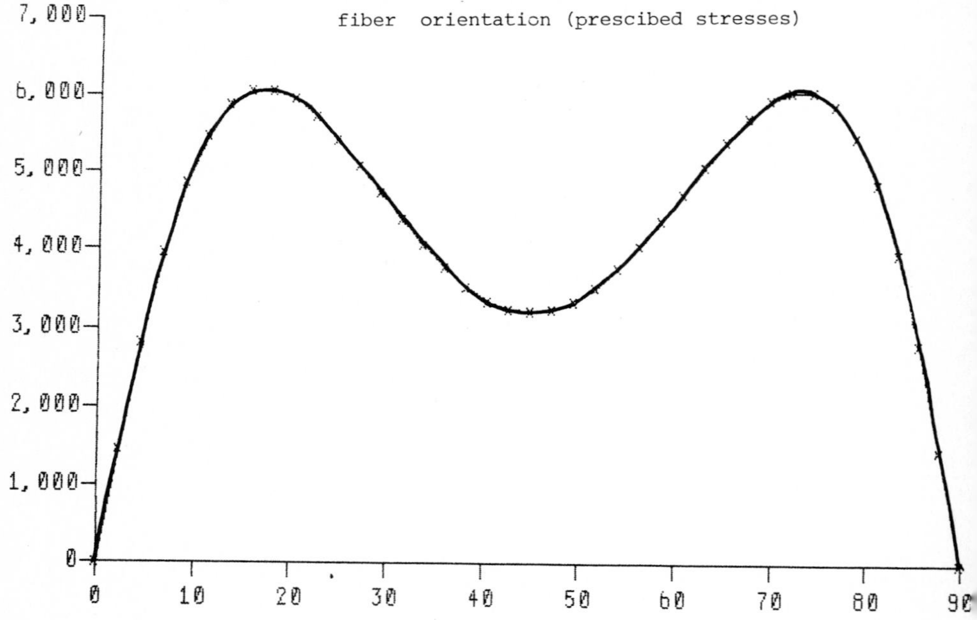

Figure 4-2 variation of the stress intensity factor with the fiber orientation (prescibed strains)

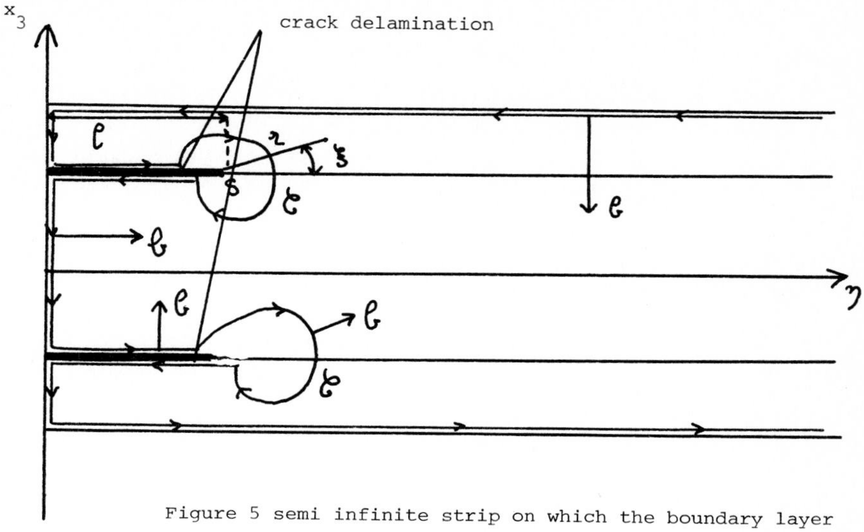

Figure 5 semi infinite strip on which the boundary layer problem is set (with cracks)

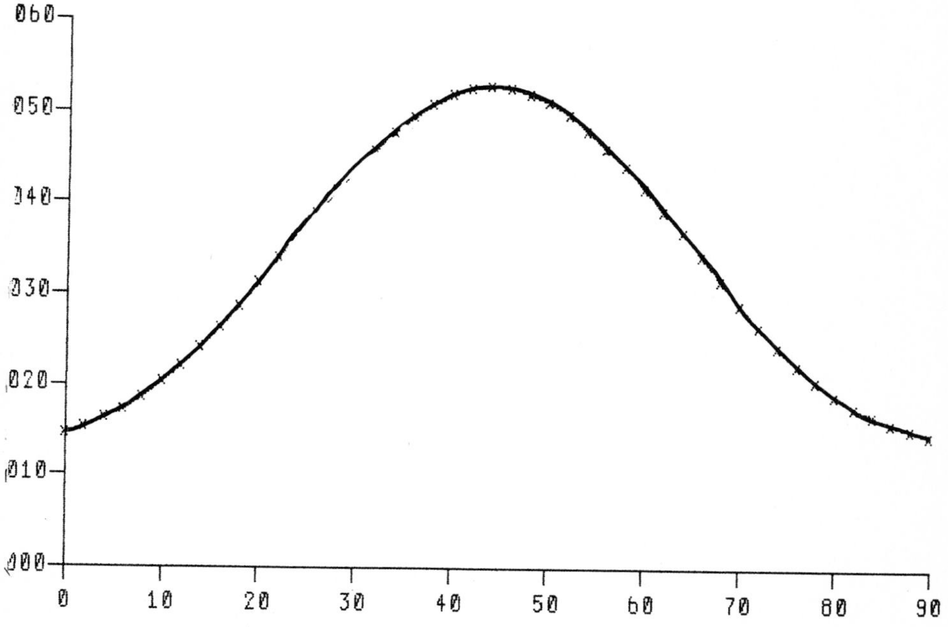

Figure 6 variation of the stress intensity factor at the crack tip (prescibed stresses)

LOCAL EFFECTS CALCULATIONS IN COMPOSITE PLATES BY A BOUNDARY LAYER METHOD

D. ENGRAND

BERTIN & Cie, B.P. 3 - 78370 PLAISIR (France)

ABSTRACT

In composite structures, local effects can be of particular importance for they are responsible for local damages like delamination due to locally three dimensional stress fields.

Such effects can be accounted for by means of a boundary layer analysis that uncouple local calculations from overall usual (finite elements) analysis.

In this paper, we present a very simple approximate solution to the boundary layer equations, that can be applied to various local effects. It is based on the variational principle of complementary energy, using the simplest possible admissible stress field.

Amongst a large number of applications, (e·g· various free edge effects due to inplane loadings) we show here two rather unusual ones related to adhesively bonded joints and to the bending of composite plates, illustrating the effectiveness and versatility of this method.

INTRODUCTION

Various types of boundary layer problems can occur in structural mechanics, but are mainly concerned here with linear plate and shell theories.

In layered composite structures, such problems can be of particular importance, for they are responsible for locally three dimensional stress fields, possibly resulting in delamination.

Independently of theoretical considerations on possible stress singularities (ref. 1,2) it is important, from an engineering point of view, to account for such effects in composite plates and shells by means of specific methods that uncouple as much as possible local calculations from overall(F.E.M.) analysis, with no need for very fine and expensive mesh refinements.

As was earlier pointed out (ref. 4), this can be achieved through a boundary layer approach, associated with particular methods of solution.

In such a way it is possible to carry out classical laminate analysis, and to add, by linear superposition, a correcting stress field obtained separately by boundary layer analysis, when necessary. In the present work, we develop a very simple approximate method of solution to the complicated boundary layer equations, that can be applied to a wide range of local problems in composite plates, including : free edge problems for extensional loads, influence of in-plane edge loads, free or loaded holes, influence of temperature and hygrometry, curing (residual) stresses, free edge stresses due to bending loads ...

For the sake of completeness, we first recall the boundary layer equations in terms of stresses or stress functions (ref. 5), the displacement field in the B.L. being of little interest. Then, considering different sets of boundary and loading conditions, we derive a method of solution based on the variational principle of complementary energy, together with assumed stress functions that satisfy exactly all the boundary and continuity conditions.

It must be underlined that such a method cannot account for possible stress singularities. However, it provides a quite efficient and cost effective tool for investigation of dangerous stress concentrations, with a large number of possible applications in strutural mechanics.

BOUNDARY LAYER EQUATIONS

Notations

Standard notations are used throughout. In Cartesian axes, we denote the coordinates in the mid-plane of the plate by x, y, and the coordinate normal to this plane by z ($-h \leqslant z \leqslant +h$). In order to highlight the influence of the small parameter h (thickness = 2h) we put :

$$\eta = \frac{z}{h} \quad (-1 \leqslant \eta \leqslant +1)$$

The material is taken to be heterogeneous in the thickness direction, so that we can define the matricial compliances Sij as functions of η. In the case of layered plates, these functions are piecewise constant.

As the laminate is assumed to be symmetric, we impose the relationship :
$Sij(-\eta) = Sij(\eta)$

The non-zero coefficients in the 6x6 symmetric matrix S are S_{11}, S_{12}, S_{13}, S_{16}, S_{22}, S_{23}, S_{26}, S_{33}, S_{36}, S_{44}, S_{45}, S_{55}, S_{66}, and the tangential part of S is denoted by S :

$$\overline{S} = \begin{bmatrix} S_{11} & S_{12} & S_{16} \\ S_{12} & S_{22} & S_{26} \\ S_{16} & S_{26} & S_{66} \end{bmatrix}$$

The stress tensor for the interior solution is denoted by σ, and the additional tensor in the boundary layer is denoted by F. The engineering components are used for the strains.

Boundary conditions and classical laminate theory (CLT)

Various types of boundary and loading conditions may be applied to a plate. For the sake of simplicity, we first consider here the case where only inplane loads are present, and, in addition, we restrict the local analysis to force-type boundary conditions.

Following a widely used notation, the edge is defined by $y = o$. The local coordinates are then : x, tangential to the intersection of the edge and the mid-plane, z (or η) normal to the mid plane, y being positive in the interior direction of the plate.

Two components of edge loads may be considered :

$p_{yy}(x,\eta)$ normal to the edge

$p_{xy}(x,\eta)$ tangential to the edge

In classical laminate theory (CLT), the plane part $\overline{\sigma}$ of the stress tensor σ is a function of η (piecewise constant since the S_{ij} are piecewise constant), and is computed from the generalized plane stress N, by the equation :

$$\overline{\sigma} = C(\eta)\, N \qquad (1)$$

With N defined by

$$N = \frac{1}{2}\int_{-1}^{+1} \overline{\sigma}(\eta)\, d\eta \qquad (2)$$

and $C(\eta)$ being a stress concentration tensor.

Within the framework of CLT, it is generally not possible to satisfy exactly the boundary conditions, either for free or loaded edges. The only conditions that may be verified are :

$$N_{yy}(x,o) = \frac{1}{2}\int_{-1}^{+1} p_{yy}(x,\eta)\, d\eta \qquad (3)$$

$$N_{xy}(x,o) = \frac{1}{2}\int_{-1}^{+1} p_{xy}(x,\eta)\, d\eta$$

When interpreting CLT as an asymptotic theory of composite plates (ref. 6), this can be seen as a singular perturbation problem. Indeed, if h is a small parameter tending to zero, the CLT solution is obtained as the first term in the asymptotic expansion of the three dimensional field variables, and is only valid as an interior solution. The satisfying of boundary conditions generally involves a boundary layer that has to be analysed by specific methods. In a more physical interpretation, this can be explained as follows : the CLT tangential strain $\overline{\varepsilon}$ is independent of η (constant throughout the thickness) and strain compatibility near the edges requires the application of very special edge loads. For general loading cases, including free edge conditions, this is not the case.

It is thus necessary to analyse in more detail the stress field near the edge.

Boundary layer (B.L.) analysis

In the usual approach (ref. 6) the boundary conditions (3) associated with a CLT solution are supposed to be unknown, and appear as a final result of the boundary layer theory.

Since our aim is not to construct a theory, but rather to obtain a representation of the stress field near the edge using boundary layer equations, we first suppose that a problem has been solved by mean of CLT and conditions (3), and look for a correcting solution that can be added to the CLT solution in order to satisfy exactly the edge conditions. We thus define an additional stress tensor F such that :

$$\left.\begin{array}{l} Fyy\ (x,o,\eta) + \sigma yy\ (x,o,\eta) = pyy\ (x,\eta) \\ Fxy\ (x,o,\eta) + \sigma xy\ (x,o,\eta) = pxy\ (x,\eta) \\ Fyz\ (x,o,\eta) = o \end{array}\right\} \quad (4)$$

And, since there are no loads on the upper and lower faces :

$$\left.\begin{array}{l} Fxz\ (x,y,\pm 1) = o \\ Fyz\ (x,y,\pm 1) = o \\ Fzz\ (x,y,\pm 1) = o \end{array}\right\} \quad (5)$$

As we want to ensure that the influence of the stress field F does not extend far away from the edge, we have also to impose a rapid decrease of F with increasing y.

Typically, F must vanish at a distance δ from the edge, δ being of the order of magnetude of h.

In order to derive appropriate equations in the B.L., we first set :

$$\xi = \frac{y}{h} \quad \text{together with } \eta = \frac{z}{h}$$

so that the three dimensional equilibrium equations become :

$$\left.\begin{array}{l} h\ Fxx,x + Fxy,\xi + Fxz,\eta = o \\ h\ Fxy,x + Fyy,\xi + Fyz,\eta = o \\ h\ Fxz,x + Fyz,\xi + Fzz,\eta = o \end{array}\right\} \quad (7)$$

Similarly, the compatibility equations expressed in terms of stresses via the constitutive equations can be put into the form :

$$h^2\ E^{(2)}\ (F) + h\ E^{(1)}\ (F) + E^{(o)}\ (F) = o \quad (8)$$

where $E^{(2)}$, $E^{(1)}$, and $E^{(o)}$ are partial differential linear operators that are not specifically defined here because of their lengthy nature.

Now we can observe that in eqns (7) and (8) the higher order terms in h are precisely those involving derivatives in x.direction. We can thus simply drop these terms and try to satisfy all the equations and conditions with F independent of x (or depending only parametrically ou x), or, alternatively use the approach of (ref. 7,8) invoking matched asymptotic expansion of the total B.L. stress field. Both approaches give the same equations, but the first is simpler and does not give rise to the question of the existence of an asymptotic expansion.

The boundary layer equilibrium equations are then written :

$$\left.\begin{array}{l} Fxy,\xi + Fxz,\eta = o \\ Fyy,\xi + Fyz,\eta = o \end{array}\right. \quad (9)$$

$Fyz,_\xi + Fzz,_\eta = o$ $\}$

together with the corresponding compatibility equations :

$$E^{(o)}(F) = o \tag{10}$$

and the condition of vanishing stresses away from the edge :

$$F(x,\xi,\eta) \underset{\xi \to \infty}{\to} o \tag{11}$$

(this condition is equivalent to the "matching condition" at order zero of asymptotic expansions).

It is interesting to note that F depends on x, only in a parametric sense, via the dependence on x of pyy and pxy (see eq. (4)). Furthermore, it can be shown that the εxx partial strain associated with F through the constitutive equations is zero. In fact, three equations among the six involved in the $E^{(o)}$ group can be expressed in terms of strain, giving :

$$\varepsilon xx,_{\xi\xi} = \varepsilon xx,_{\xi\eta} = \varepsilon xx,_{\eta\eta} = o \tag{12}$$

so that we can immediately deduce :

$$\varepsilon xx = \alpha \xi + \beta\eta + \gamma \tag{13}$$

but the right hand side is zero, for it has to vanish as ξ tends to infinity.

Finally, we have $\varepsilon xx = o$, and we can eliminate Fxx from the remaining three equations in the $E^{(o)}$ group, using the relationship :

$$F\,xx = -\frac{1}{S_{11}}(S_{12}\,Fyy + S_{13}\,Fzz + S_{16}\,Fxy) \tag{14}$$

the remaining three compatibility equations now take the form :

$$(A_{22}\,Fyy),_{\eta\eta} + (A_{23}\,Fzz),_{\eta\eta} + (A_{26}\,Fxy),_{\eta\eta} + (A_{23}\,Fyy),_{\xi\xi} + (A_{33}\,Fzz),_{\xi\xi}$$
$$+ (A_{36}\,Fxy),_{\xi\xi} - (A_{44}\,Fyz),_{\eta\xi} - (A_{45}\,Fxz),_{\eta\xi} = o \tag{15}$$

$$\Delta,_\xi = o \tag{16}$$

$$\Delta,_\eta = o \tag{17}$$

with Δ given by :

$$\Delta = (A_{26}\,Fyy),_\eta + (A_{36}\,Fzz),_\eta + (A_{66}\,Fxy),_\eta - (A_{45}\,Fyz),_\xi - (A_{55}\,Fxz),_\xi \tag{18}$$

and $Aij = Sij - \dfrac{Si1\ S1j}{S_{11}}$

From (16) and (17), we deduce that Δ is a constant, k. But it is easy to see that the symmetry with respect to $\eta = o$ implies

$$2\,k = \int_{-1}^{+1} \Delta \; d\eta = o$$

then, equations (16) and (17) are reduced to :

$$\Delta = o \tag{19}$$

Now, from the first equilibrium eqn (9), it can be seen that there exits a stress function ψ such that

$$Fxy = \psi,_\eta \quad ; \quad Fxz = -\psi,_\xi \tag{20}$$

From the two remaining eqns (9), we deduce that there exists a stress function $\emptyset$ such that

$$F_{yy} = \emptyset_{,\eta\eta} \; ; \; F_{yz} = -\emptyset_{,\eta\xi} \; ; \; F_{zz} = \emptyset_{,\xi\xi} \qquad (21)$$

we are now able to express the two eqns (15) and (19), and the six boundary conditions (4) and (5) in terms of the two unknown stress functions $\emptyset(\xi,\eta)$ and $\psi(\xi,\eta)$, defined for $-1 < \eta < +1$, $0 < \xi < +\infty$. Despite their rather complex nature, the equations will be useful in deriving a variational formulation of the problem, as a basis for numerical analysis.

Variational formulation

We first express eqns (15) and (19) in terms of $\emptyset$ and ψ :

$$(A_{22}\,\emptyset_{,\eta\eta})_{,\eta\eta} + (A_{23}\,\emptyset_{,\xi\xi})_{,\eta\eta} + A_{23}\,\emptyset_{,\eta\eta\xi\xi} + A_{33}\,\emptyset_{,\xi\xi\xi\xi}$$
$$+ (A_{44}\,\emptyset_{,\eta\xi})_{,\eta\xi} + (A_{26}\,\psi_{,\eta})_{,\eta\eta} + A_{36}\,\psi_{,\eta\xi\xi} + (A_{45}\,\psi_{,\xi})_{,\eta\xi} = 0 \qquad (22)$$

$$(A_{66}\,\psi_{,\eta})_{,\eta} + A_{55}\,\psi_{,\xi\xi} + (A_{26}\,\emptyset_{,\eta\eta})_{,\eta} + (A_{36}\,\emptyset_{,\xi\xi})_{,\eta} + A_{45}\,\emptyset_{,\eta\xi\xi} = 0 \qquad (23)$$

Then, let us consider two test-functions $\delta\emptyset(\eta)$ and $\delta\psi(\eta)$ that are sufficiently regular functions of η only and that verify the admissibility conditions :

$$\delta\psi = \delta\emptyset = \delta\emptyset_{,\eta} = 0 \quad \text{for} \quad \eta = \pm 1 \qquad (24)$$

Now, we multiply eqns (22) by $\delta\emptyset$ an (23) by $\delta\psi$, and integrate through the thickness. In the resulting expressions, we then integrate by parts once or twice, depending on the order of differentiation in η, and use the well known arguments on arbitrariness of $\delta\emptyset$ and $\delta\psi$, to derive formally the following equations :

$$\int_{-1}^{+1} (A_{22}\,\emptyset_{,\eta\eta}\,\delta\emptyset_{,\eta\eta} + A_{23}\,\emptyset_{,\xi\xi}\,\delta\emptyset_{,\eta\eta} + A_{23}\,\emptyset_{,\eta\eta\xi\xi}\delta\emptyset + A_{33}\,\emptyset_{,\xi\xi\xi\xi}\,\delta\emptyset$$
$$- A_{44}\,\emptyset_{,\eta\xi\xi}\delta\emptyset_{,\eta})\,d\eta + \int_{-1}^{+1}(A_{26}\psi_{,\eta}\delta\emptyset_{,\eta\eta} + A_{36}\,\psi_{,\eta\xi\xi}\delta\emptyset$$
$$- A_{45}\,\psi_{,\xi\xi}\delta\emptyset_{,\eta})\,d\eta = 0 \;\; \forall\, \delta\emptyset \;\; \text{admissible} \qquad (25)$$

$$\int_{-1}^{+1}(A_{55}\,\psi_{,\xi\xi}\,\delta\psi - A_{66}\,\psi_{,\eta}\delta\psi_{,\eta})\,d\eta - \int_{-1}^{+1}(A_{26}\emptyset_{,\eta\eta}\delta\psi_{,\eta} - A_{36}\emptyset_{,\xi\xi}\delta\psi_{,\eta}$$
$$+ A_{45}\,\emptyset_{,\eta\xi}\delta\psi)\,d\eta = 0 \;\; \forall\, \delta\psi \;\; \text{admissible} \qquad (26)$$

We then assume a particular form of product functions for $\emptyset$ and ψ :

$$\left.\begin{aligned} \emptyset &= a(\xi)\,b(\eta) \\ \psi &= c(\xi)\,d(\eta) \end{aligned}\right\} \qquad (27)$$

with :

$$\left.\begin{aligned} & a(0) = c(0) = 1 \; \text{(arbitrary normalization)} \\ & \frac{da}{d\xi}(0) = 0 \\ & d(\pm 1) = b(\pm 1) = 0 \\ & \frac{db}{d\eta}(\pm 1) = 0 \end{aligned}\right\} \qquad (28)$$

In this way, we define a set of a admissible stress functions, and (25)(26) provide a particular variational projection formulation that can be very useful in approximating the solutions of equations (22) (23). This formulation can be used in different ways : if we know particular functions b and d satisfying the edge conditions (4), we can put $\delta\emptyset = b$, $\delta\psi = d$, into (25) and (26). Integrating through the thickness, we are led to a set of linear ordinary differential equations, the solution of which can be obtained analytically.

Alternatively, we can look for finite element approximations by unidirectional discretization of b and d. The variational formulation then leads to a set of differential equations, the solutions of which can be combined to satisfy the edge condition (4).

In the following, we use only the first approach to obtain an explicit and very simple approximation.

The simplest approximation

The edge conditions (4) can be exactly satisfied if we take :

$$\left.\begin{aligned} \delta\emptyset\ (\eta) &= b\ (\eta) = -\int_{-1}^{\eta} d\eta' \int_{-1}^{\eta'} (\sigma_{yy}\ (x,o,\tau) - p_{yy}\ (x,\tau)\ d\tau \\ \delta\psi\ (\eta) &= d\ (\eta) = -\int_{-1}^{\eta} (\sigma_{xy}\ (x,o,\tau) - p_{xy}\ (x,\tau)\ )\ d\tau \end{aligned}\right\} \quad (29)$$

(this particular choice is analogous to the one used by Horvay (ref. 9) for the biharmonic semi infinite strip problem). Furthermore, it is easy to see that all interface continuity requirements on Fxz, Fyz, Fzz are exactly satisfied with such a definition. Integrating (25) and (26) over the thickness gives the set of ordinary differential equations :

$$M \frac{d^4X}{d\xi^4} + N \frac{d^2X}{d\xi^2} + PX = o \qquad (X = \begin{Bmatrix} a \\ c \end{Bmatrix}) \qquad (30)$$

with

$$M = \int_{-1}^{+1} \begin{bmatrix} A_{33}\ b^2 & o \\ o & o \end{bmatrix} d\eta$$

$$N = \int_{-1}^{+1} \begin{bmatrix} 2\ A_{23}\ b''b - A_{44}\ b'^2 & A_{36}\ d'b - A_{45}\ db' \\ A_{36}\ d'b - A_{45}\ db' & - A_{55}\ d^2 \end{bmatrix} d\eta$$

$$P = \int_{-1}^{+1} \begin{bmatrix} A_{22}\ b''^2 & A_{26}\ d'b'' \\ A_{26}\ d'b'' & A_{66}\ d'^2 \end{bmatrix} d\eta \qquad (31)$$

Together with the edge conditions : $X\ (o) = \begin{Bmatrix} 1 \\ 1 \end{Bmatrix}$; $\frac{da}{d\xi}(o) = o$

(in all these expressions, the prime (') is used to denote η derivatives).

The solution of (30) is straightforward. It depends on the roots of a characteristic equation :

$$\det(\lambda^4 M + \lambda^2 N + P) = 0 \qquad (32)$$

these roots can be explicitly computed since this determinant is a bicubic polynomial :

$$q_0\,\mu^3 + q_1\,\mu^2 + q_2\,\mu + q_3 \qquad (\mu = \lambda^2) \qquad (33)$$

In order to satisfy the condition of vanishing stresses as $\xi \to \infty$, we have to select three roots λ_1, λ_2, λ_3, that have strictly negative real parts. This is possible only if none of the three roots μ_1, μ_2, μ_3, of (33) is negative if real. Since P is strictly positive, M positive degenerate, it is sufficient that N be strictly negative to ensure this property. However, it does not appear simple to demonstrate, though it never failed to hold for numerical applications in a large variety of problems. Finally, we obtain a general solution of the following type :

$$X(\xi) = \Re e\,(\alpha_1 X_1 e^{-\lambda_1\xi} + \alpha_2 X_2 e^{-\lambda_2\xi} + \alpha_3 X_3 e^{-\lambda_3\xi})$$

(in the case of multiple roots, it can assume different well known forms, including polynomials and exponentials).

The complex constants α_i are determined to satisfy conditions (28). Similar, but more complex (and more accurate) solutions can be obtained using the Ritz method with many more polynomial terms (ref. 3).

Some remarks

In practical applications, all coefficients Sij are piecewise constant, so that σ_{yy} and σ_{xy} are piecewise constant if the edge is free or if pxy and pyy are piecewise constant. The function b is then piecewise quadratic with a continuous first derivative, and d is piecewise linear and continuous.

More complicated edge loads can be accounted for in the same way, but piecewise constant loads have interesting applications (eg. to the problem of adhesive bonding).

- the variational formulation used here is nothing else than a particular form of the principle of complementary energy expressed in terms of stress functions. It is interesting to recall that this formulation includes as natural conditions higher order complicated interface continuity requirements that should be considered if equations (22) and (23) had to be solved analytically (note that, due to discontinuities of the Aij, these equations have to be interpreted in the sense of distributions).

- In the case of free edge conditions, the plane stresses σ_{yy} and σ_{xy} on the edge are proportional to the generalized stress Nxx on this edge (see eq.(1)) The resulting B.L. stresses are thus all proportional to Nxx. If we put Nxx = 1, we obtain a stress localization tensor in the B.L. that is a characteristic of the plate (material properties, stacking sequence, etc...). It is thus possible to characterize the susceptibility to delamination of a layered plate without any additional calculation (note that the stacking sequence is of particular importance).

- For free edges, the boundary layer thickness (BLT), i.e. the length over which the B.L. stresses are significant, appears to be of the order of magnitude of the thickness.

However, it is possible to gain more precise information on this order of magnitude : given an initial stacking sequence of different plies in a composite one can pose the question of extrapolating this composite to carry a plane load n times greater than design load with the same design stress. Is it better to bond n identical plates together, or to multiply by n the thickness of each ply ?

In the frame of CLT, the result is the same in both cases. However, with respect to the B.L. structure, the cases are quite different : in the first case, the boundary layer stresses are the same, in each of the n identical subassemblies as in the initial composite and thus the BLT is the same as for the initial composite. In the second case, the distribution of BL stresses is obtained from the initial composite with a magnification of coordinate scale, so that the maxima are conserved while the gradients are divided by n , The BLT is thus multiplied by n, and, as such, is of the order of magnitude of the new plate thickness. In both cases, the maxima are the same (although they occur n times in the first and only once in the second), and the BLT is n times greater in the second than in the first.

Indeed, in the first case, the subassemblies do not interact for they have the same inplane stiffnesses, and the same phenomenon always occur, even if the subassemblies are different,provided that they have the same inplane stiffness

EXTENSIONS

The type of analysis presented in the preceding section can be extended in a straightforward way to various problems. We only give here an outline of two such extensions that can be useful for applications of the method :

- thermal and hygroscopic stresses in a free plate
- effect of bending on free edge stresses.

Thermal and hygroscopic stresses

The general problem of thermo mechanical or hygromechanical boundary layers can be very complicated, for it involves transient analysis of thermal or hygroscopic B.L. in addition to the mechanical one . However, in the simplest case of equilibrium with a constant temperature T (or moisture concentration), the thermal (hygroscopic) B.L. disappears.

In this case, the generalized stress N is zero everywhere in the plate, and the local plane stress $\overline{\sigma}$ due to compatibility of strain can be computed in the frame of CLT (Ref. 10).

The resulting free edge plane stresses σ yy and σ xy are generally not zero, leading to a B.L. problem, the solution of which is almost identical to that of the preceding. This approach is quite useful in the analysis of residual (curing) stresses in composites.

Effect of bending on free edge stresses

Though the B.L. equations remain the same, the edge problem is rather different in nature.

The asymptotic theory of composite plates (ref. 6) leads to a generalization of the Kirchhoff theory and is thus associated with the Kirchhoff boundary conditions on the free edges, allowing the bending moment Myy to be zero, but not the twisting moment Mxy, so that the B.L. stresses will be dependent on two edge parameters : Mxx, and Mxy. All remaining steps in the simplest B.L. solution are almost analogous to those described in the preceding section. Note that every piecewise polynominal function involved in the solution has a degree increased by 1, since the edge stresses σ yy, σ xy are piecewise linear (instead of piecewise constant in the preceding). Similarly, the different stress components that were even (resp. odd) functions of η become odd (resp. even).

EXAMPLES OF APPLICATION

The method has been programmed in FORTRAN 77 on a PRIME 850 computer at BERTIN & Cie. In addition to the basic analysis and extensions, the program has a special capability to account for unloaded holes in Composite plates, using the analytical solution of LEKHNITSKII (ref. 11) for plane stress calculations around holes, and superimposing automatically the boundary layer stresses at user-specified values of the polar angle, for all possible prescribed plane stress fields at infinity.

A number of applications have been carried out, using this program (refs 5, 10,13). We only wish to show here some new applications. One is concerned with adhesively bonded joints as an application of loaded edge B.L. theory, and another with free edge stresses in a bent and twisted plate.

The results are illustrated on figures 1 to 6. It must be emphasized that the double lap joint assembly (figures 1,2) could be loaded in any mode of plane stress, although, in this example, it is only loaded in the y direction (orthogonal to the edge), and that the number of layers, orientations, and different materials can be arbitrary. Figure 2 illustrates the concept of a bonded joint edge zone as a loaded edge. Figures 3 and 4 highlight quite well the mechanism of load-transfer between different layers and adhesive (this layer is also modeled) through local sharp variations of transverse stresses.

The results on edge stresses in the plate under bending loads are interesting, for this is rather new. Indeed, the problem seems to have been disregarded by the majority of workers in Composite materials mechanics, despite its importance in various applications such as leaf springs, mainly in the field of transport (e.g automobile suspension and shock absorbers (ref. 16) where the excellent specific strain energy of Composites results in spectacular mass reductions.)

Figures 6 and 7 illustrate the local edge stresses corresponding to $M_{xx} = 10.$ and $M_{xy} = 1$. Note that the variations in the thickness coordinate on figure 6 is taken at $\xi = .7$, for σ_{yz} is zero on the edge.

CONCLUSION

Though the difficult problem of singularities was posed in a number of early theoretical and numerical investigations (ref. 14), it is still a difficult task to account for stress singularities in boundary layer calculations. Moreover, from a practical and engineering point of view, their physical meaning is questionable and still not demonstrated. In fact, they appear as "super local" effects related at least as much to the mathematical model as to the real heterogeneous structure.

In the real Composite, every layer is heterogeneous, and singularities can occur at each intersection between a fiber and the edge, and not only at layer interfaces.

Indeed, the distance over which interface singularities extend their influence is of the order of magnitude of a few fiber diameters, rendering most questionable the adequacy of a piecewise homogeneous linear elastic model for phenomena at such a small scale (ref. 3).

Thus,in a rather pragmatic attempt to solve engineering problems, we used the simplest amongst approximate rigorous methods of solution, ignoring possible singularities. This method has many advantages, such as simplicity, versatility, cost-effectiveness, extremely fast calculations even on a small computer, applicability without any restriction on materials (elastic properties), layer orientations, thicknesses, and number of layers.

Furthermore, experimental correlations (ref. 12, 15) have shown fairly good reliability of the results obtained in different cases of plane problem, even with such a crude approximation.

REFERENCES

1 S.S. Wang, I. Choi Boundary layer effects in Composite laminates , ASME jal of Applied Mechanics vol 49 (sept. 82) pp 541 - 560.

2 R.I. Swiers, T.C.T. Ting, R.L. Spilker , on the logarithmic singularity of free edge stresses in laminated Composites under uniform extension, ASME Jal of Applied Mechanics vol 49 (sept. 82) pp 561 - 569.

3 P. Bar - Yoseph, on the accuracy of interlaminar stresses calculation in laminated plates, computer Methods in Applied Mechanics and Engineering 36 (1983) pp 309 - 329 North Holland publi. Company.

4 S. Tang, A boundary layer theory - Part 1 : laminated Composites in plane stress, Jal. Compos.Mat. 9 (jan. 1975), 33 - 41.

5. D. Engrand, A boundary layer approach to the Calculation of transverse stresses along the free edges of a symmetric laminated plate, Composite Structures, I.H. Marshal ed., (1981) pp 247-261 Applied Science Publishers.

6 P. Destuynder, sur une justification des modèles de plaques par les méthodes asymptotiques, thèse de doctorat es-sciences mathematiques, Paris (1980).

7 E.L. Reiss, S. Locke, on the theory of plane stress, Quarterly of Applied Mathematics, vol 19 (1961) p 195.

8 K.O. Friedrichs, R.F. Dressler, A boundary layer theory for elastic plates, Comm. on pure Appl. Math. vol 14 (1961) pp 1 - 33.

9 G. Horvay, the end problem of rectangular strips. Jal of Appl. Mech. 20 (1953) pp 87-94.

10 D. Engrand, Calcul des contraintes de bord libre dans les plaques composites en chargement mécanique et thermique Proc. of the 3rd international symposium on Numerical Methods in Engineering 14 - 16 march 1983 Paris, editions Pluralis.

11 K.H. Sayers, C. Beaulieu, D. Engrand, Trous - calcul pratique des plaques composites, composites - plastiques renforcés - fibres de verre textile N° 2 mars avril 1984.

12 D. Engrand, calcul des contraintes de bords libres dans les plaques composites avec ou sans trou. Comparaison avec l'expérience, Troisièmes Journées Nationales sur les Composites, Paris (1982) pp 289 - 297 ed. Pluralis.

13 D. Engrand, C. Thiebault, Effets locaux dans les plaques composites, Quatrièmes Journées Nationales sur les Composites, Paris (1984) pp 265 - 279 ed. Pluralis.

14 N.J. Salamon, an assessment of the interlaminar stress problem in laminated composites, Jal of Composite Mat. supplement vol 14 (1980) p. 177

15 F. Benedic, D. Engrand, C. Thiebault, to appear.

Material	Mechanical property	Value
Carbon/epoxy	E_1	145 000 MPa
	$E_2 = E_3$	10 500 Mpa
	$G_{12} = G_{23} = G_{13}$	4 600 Mpa
	$\nu_{12} = \nu_{23} = \nu_{13}$	.29
Adhesive	E	4 000 Mpa
	ν	0,35
Aluminium	E	75 000 Mpa
	ν	.3

Table 1 : adhesive bonded joint ; mechanical characteristics

Mechanical characteristics	Value
E_1	140 000 Mpa
E_2	10 000 Mpa
$G_{12} = G_{23} = G_{13}$	4 000 Mpa
$\nu_{12} = \nu_{23} = \nu_{13}$	.29

Table 2 : Carbon/Epoxy, 8 layers plate
under bending : mechanical characteristics

Geometrical characteristics	Value
Length	300 mm
Width	100 mm
Thickness	2 mm

Table 3 : Carbon/epoxy, 8 layers plate
under bending : geometrical characteristics
Stacking sequence : $(90°, 45°, 0°, 45°)_S$

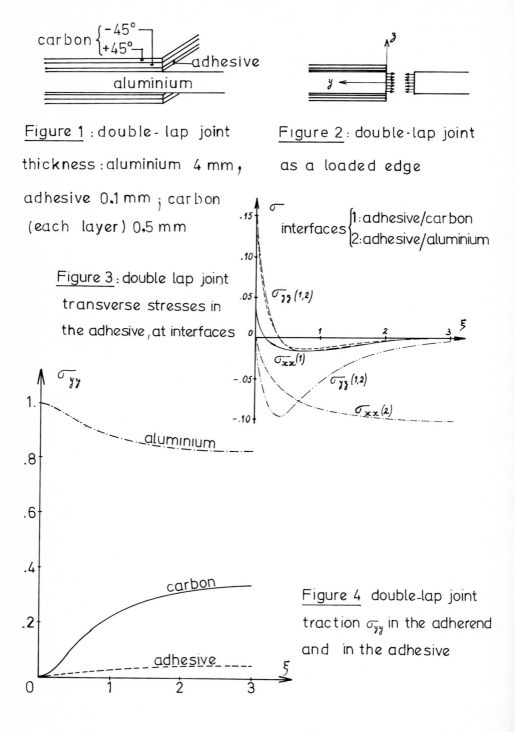

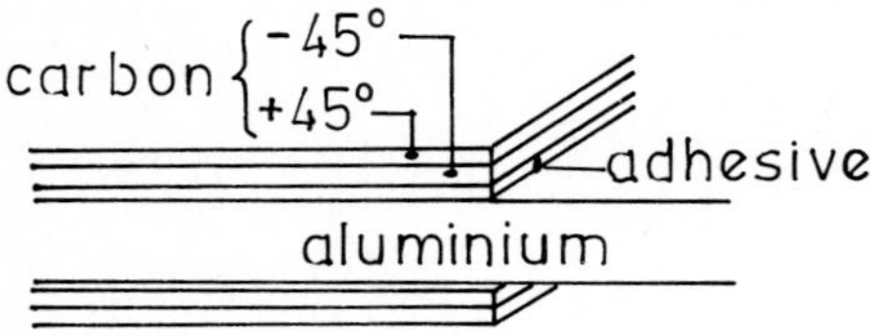

Figure 1 : double- lap joint
thickness : aluminium 4 mm, adhesive 0.1 mm ; carbon (each layer) 0.5 mm

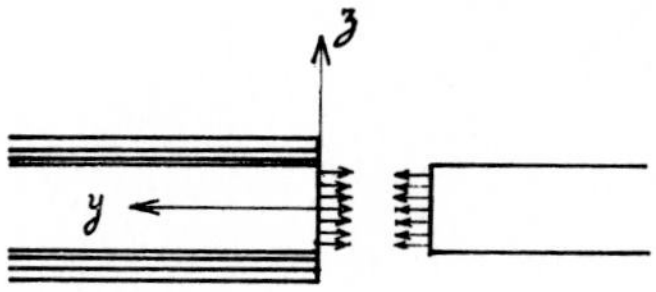

Figure 2 : double-lap joint as a loaded edge

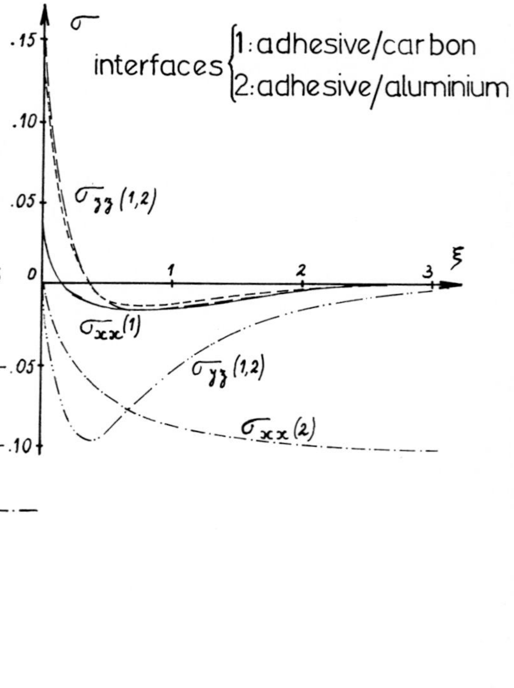

Figure 3 : double lap joint transverse stresses in the adhesive, at interfaces

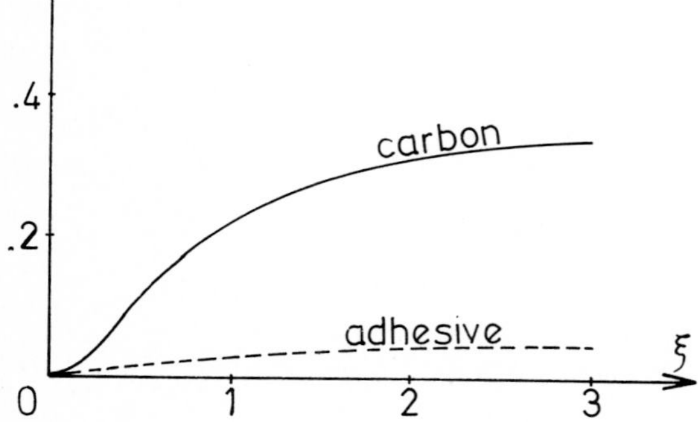

Figure 4 double-lap joint traction σ_{yy} in the adherend and in the adhesive

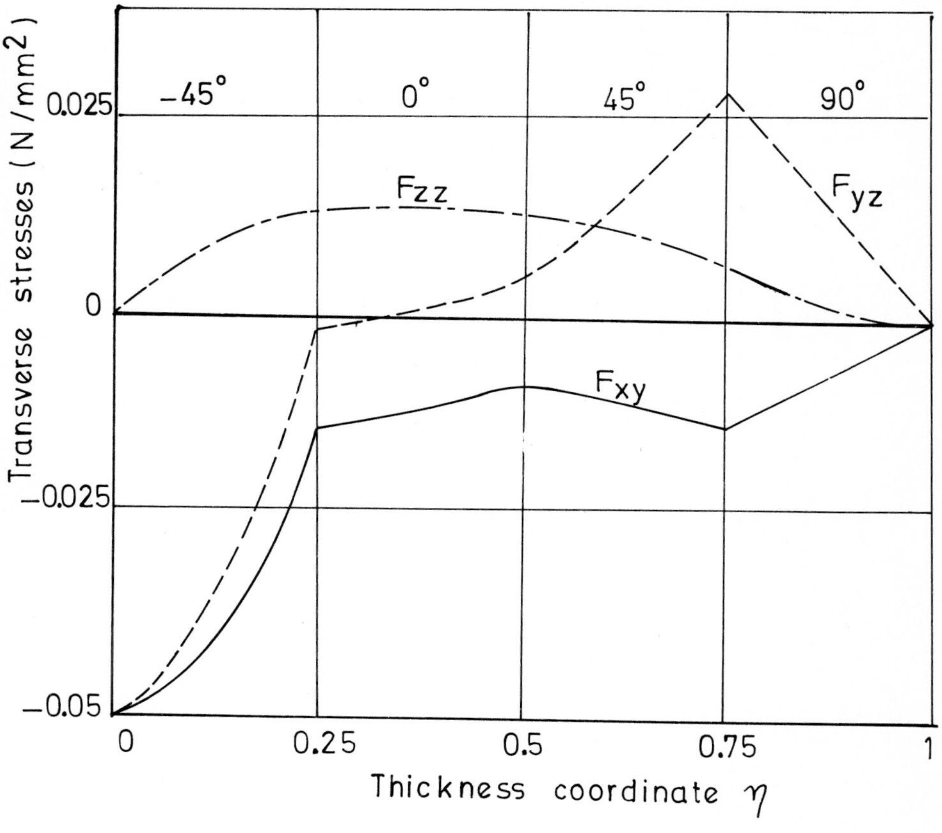

Figure 5 : Eight layers bent and twisted plate. Transverse edge stresses in the upper part, at $\xi = .7$; Mx = 10. ; Mxy = 1. (Nx mm)

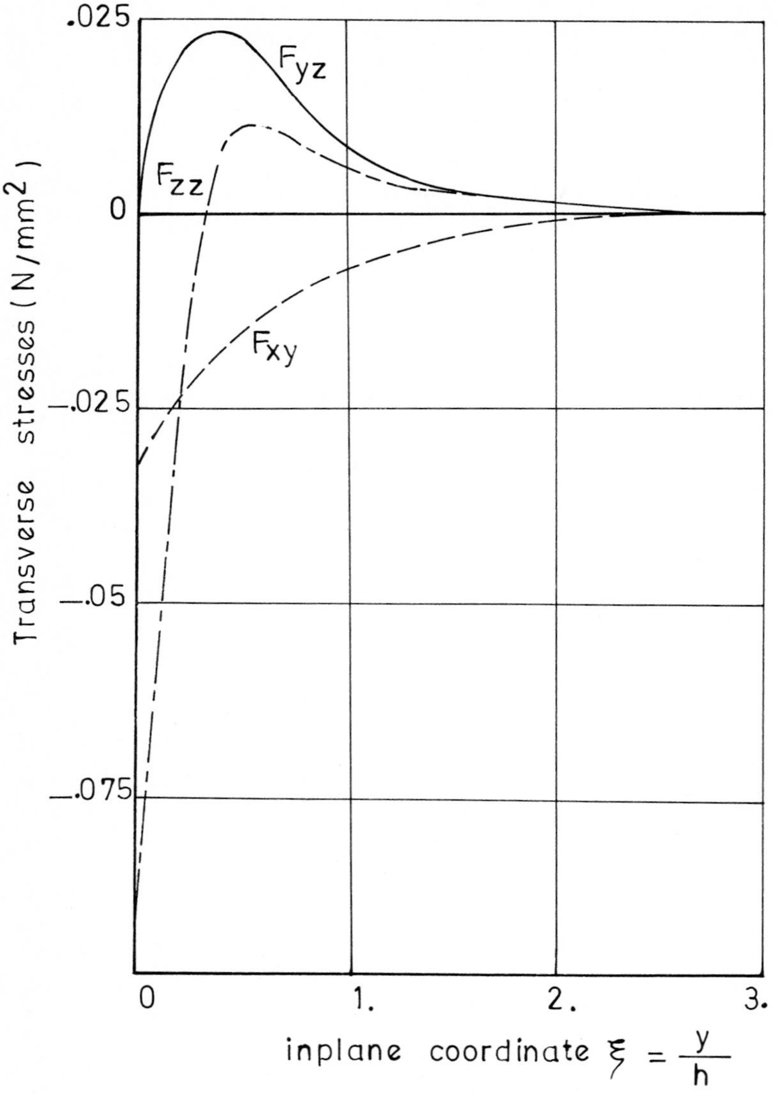

Figure 6 : Eight layers bent and twisted plate transverse edge stresses in the middle of + 45° layer Mx = 10.Nxmm M xy = 1.Nx mm

BOUNDARY LAYERS STRESSES IN ELASTIC COMPOSITES

Hélène DUMONTET (*)

Abstract : We consider an elastic stratified material with a periodic structure and we calculate the local stresses near a free boundary. Assuming that the microscopic displacements and stresses are periodic in the direction of the stratification, an homogenization method gives an approximation of the micro-stresses within the material. Since this approximation is not valid in the neighbourhood of the boundary, we define new micro-stresses as the sum of the microscopic stresses of the classical homogenization and boundary layers stresses which are periodic parallel to the boundary. These additional stresses satisfy a well-posed problem and decrease exponentially with the orthogonal boundary variable. We present some numerical results, which show the improvement in the stress calculation specially near the boundaries. This method gives also good results with other composite materials and with other types of boundary conditions.

1. - Introduction

The homogenization method, which consists in substituting a non homogeneous material for an homogeneous material with equivalent mechanic properties, has been a subject of intensive studies, (see |3|,|10|,|13|,|15|,|18| for example). We remind that the homogenization method, applied to the composite materials with periodic structures, occurs as an asymptotic problem with a small parameter ε, which should be converge to zero and which characterizes the dimension of the basic period Y. There are various methods to approach this problem. In this paper, we use the asymptotic expansion technique with two scales x and y, x is a spatial macroscopic variable and $y = x/\varepsilon$ a microscopic variable. In section 2, we present shortly this method. It consists in expanding the displacement and the stresses, according to the powers of ε, in Y-periodic terms and it gives the global displacement and stresses within the material. Since the micro-stresses are certainly responsible for the initiation of local damage in the forms of delamination fracture in laminated composites, we need a more precise analysis of the local stress field. After solving the homogenized problem, the asymptotic expansion technique gives an approximation of the local stresses within the material. Unfortunately, this approximation is not valid in the neighbourhood of the boundary. Indeed, the local stresses, obtained by this approximation do not satisfy the boundary conditions of Neumann, also they are Y-periodic in y and this hypothesis must be discussed near a boundary.
We propose to give a better approximation of the local stresses for an elastic material composed of homogeneous isotropic layers in the neighbourhood of a free boundary ; but this method is equally valid for other situations. In section 3, we present a local study ; it uses the results obtained by (|3|, |4|,|13) for the homogenization of the diffusion equation near a boundary

(*) Laboratoire de Mécanique Théorique, associé au C.N.R.S. (La 229), Université Pierre et Marie Curie - Tour 66 - 4, Place Jussieu 75230 Paris - france.

of Dirichlet. To the classical terms of the expansion of the displacements and stresses, we add boundary layers terms which are periodic parallel to the boundary. The problem, verified by the boundary layers terms, is firstly studied as a problem where the unknows are the stresses. The introduction of LEKHNITSKII's stress potentials allows us to divide this problem into two independant scalar problems, posed on a semi-infinite strip, in the direction which is perpendicular to the free edge. A mathematical study of these problems is presented in |8|. It consists in applying a lemma, due to TARTAR |13|, to these two elliptic problems of Dirichlet, one of them being to the 4th order. In this paper, we shall only remind that this lemma allows to prove that these problems are well-posed and that the boundary layers stresses decrease exponentially as a function of the variable which is perpendicular to the free edge. Secondly, the boundary layers problem is considered as a problem where the unknows are the displacements. This formulation leads to pose a second order elliptic system on the semi-infinite strip, with boundary conditions of Neumann on its free boundary part. We can find a mathematical study of this system in |8| which uses also the lemma of TARTAR. We confine ourselves here to this latter formulation to effect numerical calculations presented in section 4. The micro-stresses, the boundary layers stresses and the stresses prevailing in the heterogeneous material are computed near a free boundary, supposed to be an inclined plane at angle $\pi/4$ with the direction of the stratification. It is obvious on the results that the boundary layers stresses are decreasing exponentially and that the sum of the boundary layers stresses and the classical micro-stresses are a better approximation of the real stresses near a free boundary, than are micro-stresses only. We show also clearly the effects of the boundary layers displacements.

I am indebted to E. SANCHEZ-PALENCIA for the fruitful conversations that we had and for his careful reading of this study.

2. - Setting of the problem

Consider in $\mathbb{R}^3$ a bounded domain Ω, which smooth boundary $\partial\Omega$. The domain Ω is occupied by an elastic body, under the classical hypothesis of linear small deformation |11|. The body is subjected to forces of density $f = (f_1, f_2, f_3)$ and is fixed, for example, on a portion Γ_1 of its boundary ; we assume that the remainder of the boundary Γ_2 is free.
Let us set $x = (x_1, x_2, x_3)$ the Cartesian coordinates of a point of Ω referring to the system $R = (0, \vec{e}_1, \vec{e}_2, \vec{e}_3)$.
The composite is a laminated material composed of a finite number of elastic isotropic homogeneous layers with periodic structures in $\vec{e}_3$ direction. We denote ε - Z the period of the material, ε is a small parameter which should be converge to zero and $Z =]0, Z^*[$ is the basic period extended by homothety. The elastic behaviour of the material is characterized by the functions :

$$\mathcal{A}(z) = \{a_{ijkh}(z)\} \quad ,$$

defined on Z and extended to the whole space by Z periodicity.
The local elastic coefficients are given by :

$$\mathcal{A}^{\varepsilon}(x) = \mathcal{A}\left(\frac{x_3}{\varepsilon}\right) = \{a_{ijkh}\left(\frac{x_3}{\varepsilon}\right)\}$$

We assume that they satisfy the following properties :

$$\left.\begin{array}{l} \bullet \quad a_{ijkh} \quad \text{are bounded in } z \\ \bullet \quad a_{ijkh} = a_{khij} = a_{jikh} \\ \bullet \quad a_{ijkh}\, e_{ij}\, e_{kh} \leq c_\circ\, e_{ij}\, e_{ij} \quad , \quad c_\circ > 0 \quad , \quad (\forall e_{ij} = e_{ji}) \end{array}\right\} \qquad (2.1)$$

We know, (|3|,|10|,|15|,|18| ...), that the elastic homogenized behaviour, being the result of the convergence to zero of the coefficient ε, is obtained by the following expansion of the displacement $u^{\varepsilon}(x)$:

$$\left.\begin{array}{l} u^{\varepsilon}(x) = u^{\circ}(x) + \varepsilon\, u^{1}(x,z) + \varepsilon^{2}\, u^{2}(x,z) + \dots \,, \quad z = \dfrac{x_3}{\varepsilon} \\ \text{where the functions } z \longrightarrow u^{i}(x,z) \ (i \geq 1) \text{ are } Z \text{ periodic} \end{array}\right\} \quad (2.2)$$

$u^{\circ}(x)$ is the solution of the following homogenized problem :

$$\left.\begin{array}{l} \mathrm{Div}_x\, |\langle \sigma^{\circ}(x,z)\rangle_Z| = f \,, \quad \text{in } \Omega \\ \langle \sigma^{\circ}(x,z)\rangle_Z = \mathbb{Q}\, \varepsilon(u^{\circ}(x)) \,, \quad \text{in } \Omega \\ u^{\circ}(x) = 0 \,, \quad \text{on } \Gamma_1 \end{array}\right\} \quad (2.3)$$

$$\langle \sigma^{\circ}(x,z)\rangle_Z \cdot n = 0 \,, \quad \text{on } \Gamma_2 \qquad (2.4)$$

where Div_x represents the operator of the divergence : $(\mathrm{Div}_x\, A)_i = \dfrac{\partial A_{ij}}{\partial x_j}$,

ε is the strain tensor : $\varepsilon_{ij}(v) = \dfrac{1}{2}\left[\dfrac{\partial v_i}{\partial x_j} + \dfrac{\partial v_j}{\partial x_i}\right]$, n is the outside normal of Γ_2, where $\langle\ \ \rangle_Z$ denotes the average on Z :

$\langle f \rangle_Z = \dfrac{1}{Z^*}\displaystyle\int_0^{Z^*} f(z)\, dz$ and where the coefficients : $\mathbb{Q} = \{q_{ijkh}\}$ are given by :

$$\left.\begin{array}{l} q_{ijkh} = \langle \hat{b}_{ijkh}(z) \rangle_Z \\ \hat{b}_{ijkh}(z) = a_{ijkh}(z) - a_{ijpq}(z)\, e_{pq}(\chi^{kh}(z)) \end{array}\right\} \quad (2.5)$$

with the Z-periodic functions $\chi^{kh}(z)$, satisfying the equation :

$$\frac{d}{dz}\left[a_{i3pq}(z)\, e_{pq}\,(\chi^{kh}(z))\right] = -\frac{d}{dz}\left[a_{i3kh}(z)\right] \,, \quad \text{in } Z. \qquad (2.6)$$

where the not zero components of e are :

$$e_{\alpha 3}(v) = \frac{1}{2}\,\frac{dv_{\alpha}}{dz} \quad (\alpha = 1,2) \quad , \; e_{33}(v) = \frac{dv_3}{dz} \; .$$

After solving the homogenized problem [(2.3)(2.4)], the local stresses are obtained as follows :

$$\sigma^{\circ}_{ij}(x,z) = \hat{b}_{ijkh}(z)\; \varepsilon_{kh}(u^{\circ}(x)) \qquad (2.7)$$

Unfortunately, this approximation is not valid in the neighbourhood of the boundary (see |1|,|13|,|15|). It has clearly appeared in the study of an industrial laminated thrust, produced by SNIAS |9|. Indeed, on one hand, the local stresses (2.7) do not satisfy the boundary conditions on Γ_2, but satisfy only :

$$\sigma^{\circ}_{ij}(x,z) \cdot n_j \neq 0 \quad , \quad \text{on } \Gamma_2$$

and, on the other hand, near a boundary, there is no reason why the stresses $\sigma^{\circ}(x,z)$, in each period, are similar to those of the period beside.
We suggest to correct the expansion of the displacement (2.2) and the associated expansion of the stresses, in the neighbourhood of the free boundary Γ_2. Therefore, we add boundary layers terms for which the Z-periodicity hypothesis is replaced by an hypothesis of periodicity parallel to the boun-

dary. And we assume that these terms are defined on the free boundary so that the sum of the microscopic stresses and boundary layers stresses satisfy the free boundary condition at each point.

3. - Boundary layers problem

We specify that the free boundary Γ_2 is a plane (it is not a restriction ; providing that Γ_2 is smooth enough near a point, one can identify it with the tangent plane) supposed to be generated by the vectors :

$$\vec{e}_1 \quad \text{and} \quad \vec{e}\,'_2 = \cos\alpha\, \vec{e}_2 + \sin\alpha\, \vec{e}_3 \quad \text{with} \quad 0 < \alpha \leq \pi/2$$

Let us set $\vec{e}\,'_3$ the unit vector defined by : $\vec{e}\,'_3 = -\sin\alpha\, \vec{e}_2 + \cos\alpha\, \vec{e}_3$ and we denote R' the coordinates system : $R' = (0, \vec{e}_1, \vec{e}\,'_2, \vec{e}\,'_3)$, where 0 is the point of Γ_2 which is the object of this local study.
Let us set : $x' = (x'_1, x'_2, x'_3)$ the coordinates of a point of Ω, referring to the system R'. One introduces the following microscopic variables :

$$y_2 = \frac{x'_2}{\varepsilon} \quad \text{and} \quad y_3 = \frac{x'_3}{\varepsilon} \tag{3.1}$$

they satisfy the relation :

$$z = \sin\alpha\, y_2 + \cos\alpha\, y_3 \tag{3.2}$$

Let us set : $Y_2 =]0, Y_2^*[$ with $Y_2^* = \dfrac{z^*}{\sin\alpha}$

and $G = Y_2 \times]0, +\infty[$ the semi-infinite strip represented on the figure 1.

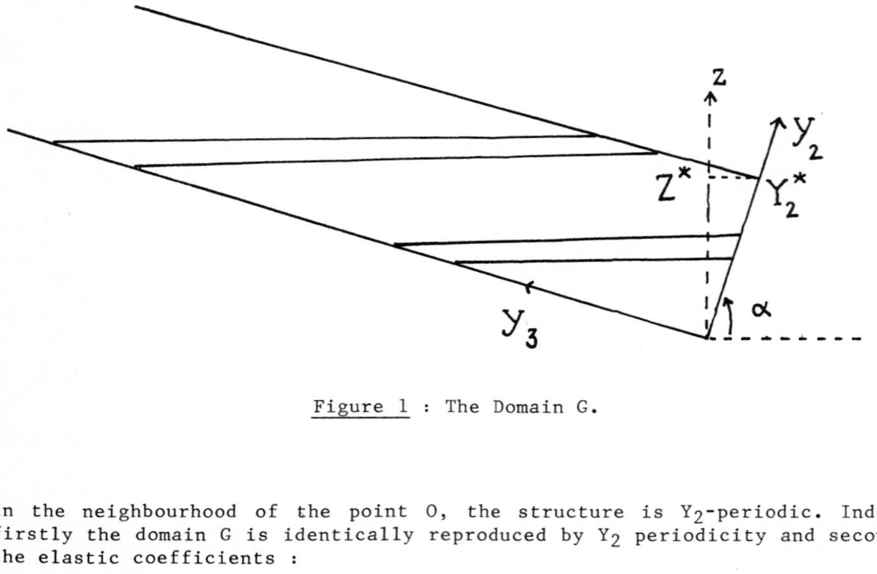

Figure 1 : The Domain G.

In the neighbourhood of the point 0, the structure is Y_2-periodic. Indeed, firstly the domain G is identically reproduced by Y_2 periodicity and secondly the elastic coefficients :

$$\tilde{\mathcal{Q}}(y_2, y_3) = \mathcal{Q}(z = \sin\alpha\, y_2 + \cos\alpha\, y_3)$$

are Y_2-periodic in y_2 at y_3 fixed.

So that, it is natural to try to find the displacement $u^\varepsilon(x')$ as follows :

$$\left.\begin{array}{l} u^\varepsilon(x') = u^\circ(x') + \varepsilon\,[u^1(x',z) + u^{1BL}(x',y_2,y_3)] + \dots , \\ \text{with } (y_2,y_3) \text{ and } z \text{ given by (3.1) and (3.2)} , \end{array}\right\} \quad (3.3)$$

where the boundary layers terms : $u^{iBL}(x',y_2,y_3)$, $(i \geq 1)$ are :

$$\left.\begin{array}{l} \bullet \text{ defined for } x' \in \Gamma_2 (\text{ie } x' = (x'_1,x'_2,0)) \text{ and } (y_2,y_3) \in G \\ \bullet\ Y_2\text{-periodic in } y_2 \end{array}\right\} \quad (3.4)$$

and where $u^\circ(x')$, $u^i(x',z)$ $(i \geq 1)$ are the classical terms of the expansion (2.2), expressed in the system R'.

In the same way, we shall loke for the stresses in the following form :

$$\left.\begin{array}{l} \sigma^\varepsilon(x') = [\sigma^\circ(x',z) + \sigma^{oBL}(x',y_2,y_3)] + \varepsilon \dots , \\ \text{with } (y_2,y_3) \text{ and } z \text{ given by (3.1) and (3.2)} \end{array}\right\} \quad (3.5)$$

where the boundary layers terms : $\sigma^{iBL}(x',y_2,y_3)$, $(i \geq 0)$ are :

$$\left.\begin{array}{l} \bullet \text{ defined for } x' \in \Gamma_2 \text{ and } (y_2,y_3) \in G \\ \bullet\ Y_2\text{-periodic in } y_2 \end{array}\right\} \quad (3.6)$$

and satisfy :

$$\left.\begin{array}{l} \bullet\ \sigma^{iBL}(x',y_2,y_3) \text{ converge to 0 as } y_3 \longrightarrow +\infty \\ \bullet\ \sigma^{iBL}_{p3}(x'_1,x'_2,0,y_2,0) = -\sigma^{i}_{p3}(x'_1,x'_2,0,y_2,0) \\ \qquad \forall\, y_2 \in Y_2 , \ (\forall\, p=1,2,3) \end{array}\right\} \quad (3.7)$$

and where $\sigma^i(x',z)$ $(i \geq 0)$ are the classical terms of the expansion of σ^ε associated with (2.2), expressed in system R'.
Once for all, we denote $b_{ijkh}(y_2,y_3)$ the coefficients defined below :

$$\sigma^\circ_{ij}(x',y_2,y_3) = b_{ijkh}(y_2,y_3) \quad \varepsilon_{kh}(u^\circ(x')) \quad (3.8)$$

We replace the asymptotic expansions (3.3) and (3.5) into :

$$\left.\begin{array}{l} \bullet \text{ equilibrium equations : } \mathrm{Div}_{x'}\ [\sigma^\varepsilon(x')] = f \\ \bullet \text{ constitutive relations : } \sigma^\varepsilon(x') = a^\varepsilon(x') \ \varepsilon\,(u^\varepsilon(x')) \end{array}\right\} \quad (3.9)$$

and we identify powers of ε, we find in this way :

$$\left.\begin{array}{l} \dfrac{\partial}{\partial y_2}\left[\sigma^{oBL}_{i2}\right] + \dfrac{\partial}{\partial y_3}\left[\sigma^{oBL}_{i3}\right] = 0 \quad (i=1,2,3) \\ \sigma^\circ_{ij}(x',y_2,y_3) = a_{ijkh}(y_2,y_3)\ e_{kh}(u^{1BL}(x',y_2,y_3)) \end{array}\right\} \quad (3.10)$$

where e denotes the strain tensor in (y_2,y_3) : $e_{ij}(v) = \frac{1}{2}\left[\dfrac{\partial v_i}{\partial y_j} + \dfrac{\partial v_j}{\partial y_i}\right]$

The boundary layers problem is defined by the equations (3.10),(3.4),(3.6),(3.7). Let us set, once for all, P^{BL}, this problem.

Remark 3.1 : The homogenized boundary conditions (2.4) is a necessary and sufficient condition for the existence of the boundary layer.

3.1. - Static formulation of the boundary layers problem

We choose to detail the formulation of the boundary layers problem, where the stresses are considered as unknows. Mathematically, this formulation is interesting, because it leads to study the infinite behaviour of Y_2-periodic solutions of 4th order elliptic equations, which extend to higher order equations the results obtained by LIONS |13| for the diffusion equation. Numerically, this formulation is equally interesting ; it amounts to solving scalar problems with boundary conditions of Dirichlet and their implements are easy. We introduce the LEKHNITSKII's stress potentials |12| :

$$\Psi(x',y_2,y_3) \text{ and } \Phi(x',y_2,y_3)$$

defined for $x' \in \Gamma_2$ and $(y_2,y_3) \in G$ such that :

$$\left.\begin{aligned} \frac{\partial\Psi}{\partial y_2}(x',y_2,y_3) &= -\sigma_{13}^{oBL}(x',y_2,y_3) \\ \frac{\partial\Psi}{\partial y_3}(x',y_2,y_3) &= \sigma_{12}^{oBL}(x',y_2,y_3) \end{aligned}\right\} \quad (3.11)$$

$$\left.\begin{aligned} \frac{\partial^2\Phi}{\partial y_2\partial y_2}(x',y_2,y_3) &= \sigma_{33}^{o}(x',y_2,y_3) \\ \frac{\partial^2\Phi}{\partial y_3\partial y_3}(x',y_2,y_3) &= \sigma_{22}^{o}(x',y_2,y_3) \\ \frac{\partial^2\Phi}{\partial y_2\partial y_3}(x',y_2,y_3) &= -\sigma_{23}^{o}(x',y_2,y_3) \end{aligned}\right\} \quad (3.12)$$

In this way, the static boundary layers problem separates into two independant problems of Dirichlet for the functions Ψ and Φ. We are going to study successively these problems.

. We can express the function $\Psi(x',y_2,y_3)$, which is defined up to the addition of an arbitrary function $\tilde{\Psi}(x')$, as follows :

$$\left.\begin{aligned} &\Psi(x'_1,x'_2,0,y_2,y_3) = \mathcal{C}^{rs}(y_2,y_3)\,\varepsilon_{rs}(u^o)|_{(x'_1,x'_2,0)} + \tilde{\Psi}(x'_1,x'_2,0) \\ &\text{with summation on the repeated indices } (r,s) \end{aligned}\right\} \quad (3.13)$$

where $u^o(x')$ is the solution of the homogenized problem (2.3) and (2.4) and where the functions $\mathcal{C}^{rs}(y_2,y_3)$ satisfy the following problem posed on G :

$$\frac{\partial}{\partial y_2}\left[\frac{1+\nu}{E}\frac{\partial\mathcal{C}^{rs}}{\partial y_2}\right] + \frac{\partial}{\partial y_3}\left[\frac{1+\nu}{E}\frac{\partial\mathcal{C}^{rs}}{\partial y_3}\right] = 0 \quad (3.14)$$

$$\mathcal{C}^{rs}(y_2,0) = g^{rs}(y_2) = \int_0^{y_2} b_{13rs}(t,0)dt - \frac{y_2}{Y_2^*}\int_0^{Y_2} b_{13rs}(y_2,0)dy_2 \quad (3.15)$$

$$\left(\frac{\partial\mathcal{C}^{rs}}{\partial y_2}, \frac{\partial\mathcal{C}^{rs}}{\partial y_3}\right) \text{ converge to 0 as } y_3 \longrightarrow +\infty \quad (3.16)$$

$\mathcal{C}^{rs}(y_2,y_3)$ are Y_2-periodic in y_2. (3.17) with $E(y_2,y_3)$ the Young's modulus and $\nu(y_2,y_3)$ the Poisson's ratio of the material.

The relation : $\dfrac{\partial^2 u_1^{1BL}}{\partial y_2 \partial y_3} = \dfrac{\partial^2 u_1^{1BL}}{\partial y_3 \partial y_2}$ and the formulas [(3.10).(3.11).(3.13)] implie (3.14). The equations (3.15) and (3.16) are obtained from the boundary conditions (3.7). And (3.15) follows from (3.6),(3.11),(3.13) and from the homogenized boundary condition (2.4).

By using mathematical results, one can show that the function $\tau^{rs}(y_2,y_3)$, under the hypothesis (2.1), satisfy a well-posed problem. It means that there exists a unique solution to the problem (3.14) ... (3.17). We can also prove that $\tau^{rs}(y_2,y_3)$ satisfy the following property : there exist (K_o,K_1) positive constants and ξ a real > 0 such that :

$$\left|\frac{\partial \tau^{rs}}{\partial y_2}\right| \leq K_o\, e^{-\xi y_3} \quad \text{and} \quad \left|\frac{\partial \tau^{rs}}{\partial y_3}\right| \leq K_1\, e^{-\xi y_3} \quad (\forall\, y_2 \in Y_2)$$

Consequently, taking account of (3.11) and (3.13), it means that the stresses σ_{12}^{oBL} and σ_{13}^{oBL}, solution of P^{BL}, exist, are unique and decrease exponentially as $y_3 \longrightarrow +\infty$.

For the readers, which are familiar with a mathematical analysis, we specify that this result of regularity is obtained by following the method used by LIONS |13| for the homogenization of the diffusion equation in the neighbourhood of a Dirichlet boundary. This method consists in applying a lemma due to TARTAR. One shall find the precise terms of this result in |8|.

• Then, consider the function $\Phi(x',y_2,y_3)$, defined up to the addition of arbitrary functions of x' : $\tilde{\Phi}_1(x')$, $\tilde{\Phi}_2(x')$, $\tilde{\Phi}_3(x')$, such that :

$$\Phi(x'_1,x'_2,0,y_2,y_3) = \varphi^{rs}(y_2,y_3)\, \varepsilon_{rs}(u^\circ)|_{(x'_1,x'_2,0)} + \tilde{\Phi}_1(x'_1,x'_2,0)y_2 + \tilde{\Phi}_2(x'_1,x'_2,0)y_3 + \tilde{\Phi}_3(x'_1,x'_2,0) \tag{3.18}$$

with summation on the repeated indices (r,s).

where $u^\circ(x')$ is the solution of the homogenized problem (2.3) and (2.4) and where the functions $\varphi^{rs}(y_2,y_3)$ satisfy the following problem on G :

$$A[\varphi^{rs}(y_2,y_3)] = 0 \tag{3.19}$$

$$\left.\begin{aligned} &\varphi^{rs}(y_2,0) = h^{rs}(y_2) \\ &\frac{\partial \varphi^{rs}}{\partial y_3}(y_2,0) = k^{rs}(y_2) \end{aligned}\right\} \tag{3.20}$$

$$\left(\frac{\partial^2 \varphi^{rs}}{\partial y_i \partial y_j}\right), \quad (i,j = 2,3) \text{ converge to 0 as } y_3 \longrightarrow +\infty \tag{3.21}$$

$$\varphi^{rs}(y_2,y_3) \text{ are } Y_2\text{-periodic in } y_2 \tag{3.22}$$

where A denotes the operator :

$$A = \frac{\partial^2}{\partial y_2 \partial y_2}\left[\frac{1-\nu^2}{E}\frac{\partial^2}{\partial y_2 \partial y_2}\right] + \frac{\partial^2}{\partial y_3 \partial y_3}\left[\frac{1-\nu^2}{E}\frac{\partial^2}{\partial y_3 \partial y_3}\right]$$

$$- \frac{\partial^2}{\partial y_2 \partial y_2}\left[\frac{\nu(1+\nu)}{E}\frac{\partial^2}{\partial y_3 \partial y_3}\right] - \frac{\partial^2}{\partial y_3 \partial y_3}\left[\frac{\nu(1+\nu)}{E}\frac{\partial^2}{\partial y_2 \partial y_2}\right] + 2\frac{\partial^2}{\partial y_2 \partial y_3}\left[\frac{1+\nu}{E}\frac{\partial^2}{\partial y_2 \partial y_3}\right]$$

and where $h^{rs}(y_2)$ and $k^{rs}(y_2)$ are the following functions defined on Y_2 :

$$\begin{cases} h^{rs}(y_2) = - \int_0^{y_2} \left[\int_0^{t} b_{33rs}\ (t_1,0)\ dt_1 \right] dt + \frac{y_2{}^2}{2Y_2^*} \int_0^{Y_2^*} b_{33rs}\ (t,0)\ dt \\ \qquad + \int_0^{Y_2^*} \left| \int_0^{t} b_{33rs}\ (t_1,0)\ dt_1 \right| dt - \frac{Y_2{}^*}{2} \int_0^{Y_2^*} b_{33rs}\ (t,0)\ dt \end{cases}$$

$$k^{rs}(y_2) = \int_0^{y_2} b_{23rs}\ (t,0)\ dt - \frac{y_2}{Y_2^*} \int_0^{Y_2^*} b_{23rs}\ (t,0)\ dt .$$

The equation (3.19) follows from the compatibility equations and formulas (3.10), (3.12), (3.18). By integrating the boundary conditions (3.7), one obtains (3.20) and (3.21). Finally [(3.6), (3.12), (3.13)] and the boundary homogenized condition (2.4) lead to (3.22).

Remark 3.2 : One can verify that the displacements u_1^{1BL} and (u_2^{1BL}, u_3^{1BL}), associated to $\Psi(x',y_2,y_3)$ and $\Phi(x',y_2,y_3)$ by the constitutive equations [(3.10), (3.11), (3.12)], are Y_2-periodic in y_2.

As before, we can summarize the mathematical study of the problem (3.19)...(3.21) as follows : there exist a unique solution φ^{rs}, for (r,s) fixed, of the problem (3.19)...(3.21) and this solution satisfy : $\forall y_2 \in Y_2$,

$$\left| \frac{\partial^2 \varphi^{rs}}{\partial y_i \partial y_j} \right| \leq K\ e^{-\xi y_3} \quad (i,j = 2,3) \quad \text{where K is a positive constant,}$$

ξ a real > 0.

Therefore, taking account of (3.12) and (3.18), it amounts to saying that the stresses $\sigma_{22}^{\circ BL}$, $\sigma_{23}^{\circ BL}$ and $\sigma_{33}^{\circ BL}$ exist, are unique and decrease exponentially as $y_3 \longrightarrow +\infty$. Here, we do not present the exact theorem for the existence and the regularity of the functions $\varphi^{rs}(y_2,y_3)$. The readers, which are interested, can refer to |8|, where one can find the precise terms of this theorem and the main steps of its proof. This result represents an extension to 4th order problem, of the method used by LIONS |13| for the diffusion equation.

Remark 3.3 : the calculation of the coefficients $b_{ijkh}(y_2,y_3)$, defined by (3.8), [by using the method proposed in |7|, we do not need to solve the problems satisfied by the functions $\chi^{kh}(z)$((2.5),(2.6))] shows that

$$\zeta^{rs} \equiv 0 \quad \text{for } (r,s) \neq (1,3) \text{ and } (r,s) \neq (1,2)$$

$$\text{and } \varphi^{rs} \equiv 0 \quad \text{for } (r,s) = (1,3) \text{ and } (r,s) = (1,2) .$$

In practice, the calculation of the boundary layers stresses need, following this formulation, firstly the resolution of the homogenized problem (2.3) and (2.4), which gives the homogenized displacement u° and secondly the resolutions of the problems (3.14) ... (3.17) and (3.19) ... (3.22) satisfied by ζ^{rs} and φ^{rs}, which are similar to problems of Laplacian's and Bilaplacian's type with Dirichlet's conditions.

3.2. - Kinematic formulation of the boundary layers problem

In this section, we present an other formulation of the boundary layers problem, where this time the unknow variables are the boundary layers displacement u^{1BL}.
One can express the boundary layers displacement $u^{1BL}(x',y_2,y_3)$, solution of P^{BL} and defined up to the addition of an arbitrary vector $\tilde{u}^{1BL}(x')$, as follows :

$$u^{1BL}(x'_1,x'_2,0,y_2,y_3) = \eta^{rs}(y_2,y_3)\ \varepsilon_{rs}(u^\circ)|_{(x'_1,x'_2,0)} + \tilde{u}^{1BL}(x'_1,x'_2,0) \tag{3.23}$$

with summation on the repeated indices (r,s) ,

where $u^\circ(x')$ denotes the solution of the homogenized problem (2.3) and (2.4) and where the functions $\eta^{rs}(y_2,y_3)$ satisfy the following problem posed on G :

$$\left\{\begin{array}{ll} \dfrac{\partial}{\partial y_2}\left[a_{i2kh}\, e_{kh}(\eta^{rs})\right] + \dfrac{\partial}{\partial y_3}\left[a_{i3kh}\, e_{kh}(\eta^{rs})\right] = 0 & (3.24) \\ a_{i3kh}(y_2,0)\, e_{kh}(\eta^{rs})|_{(y_2,0)} = -b_{i3rs}(y_2,0) + \dfrac{1}{Y_2^*}\displaystyle\int_0^{Y_2^*} b_{i3rs}(t,0)dt & (3.25) \\ \left(\dfrac{\partial \eta_i^{rs}}{\partial y_2}, \dfrac{\partial \eta_i^{rs}}{\partial y_3}\right) \ (i = 1,2,3) \text{ converge to } 0 \text{ as } y_3 \longrightarrow +\infty & (3.26) \\ \eta^{rs}(y_2,y_3) \text{ are } Y_2\text{-periodic in } y_2 & (3.27) \end{array}\right.$$

Mathematical results show that the functions $\eta^{rs}(y_2,y_3)$, solution of (3.24)...(3.27), are unique up to the addition of an arbitrary constant and satisfy : there exist (K_1,K_2,K_3) positive constants and (ξ_1,ξ_2) positive reals such that $\forall\ y_2 \in Y_2$:

$$\left[\frac{\partial \eta^{rs}}{\partial y_2}\right] \leq K_1\, e^{-\xi_1 y_3} \qquad \left[\frac{\partial \eta^{rs}}{\partial y_3}\right] \leq K_2\, e^{-\xi_2 y_3}$$

and $[\eta^{rs}] \leq K_3\, e^{-\xi_2 y_3}$. Consequently, it proves that the boundary layers displacement u^{1BL}, defined to the addition of an artibrary constant in (y_2,y_3) and the boundary layers strain $e(u^{1BL})$, which is unique, decrease exponentially as $y_3 \longrightarrow +\infty$. We can find more details in |8|.

Remark 3.4 : The problem (3.24) ... (3.27) is divided into a thermic problem for η_1^{rs} and a plane elastic problem for $(\eta_2^{rs},\eta_3^{rs})$. One can show :

$$\eta_1^{rs} \equiv 0 \ , \text{ for } (r,s) \neq (1,3) \text{ and } (r,s) \neq (1,2) \text{ and}$$

$$\eta_2^{rs} = \eta_3^{rs} \equiv 0 \ , \text{ for } (r,s) = (1,3) \text{ and } (r,s) = (1,2).$$

In practice, the calculation of the boundary layers stresses by this approach, leads to resolve, firstly the homogenized problem (2.3) and (2.4), which gives $u^\circ(x')$ and secondly the problem (3.24) ... (3.27), which is an elastic problem with boundary Neumann conditions.

3.3. - Approximation of local stresses near the free boundary

We obtain a good approximation of the micro-stresses in the neighbourhood of the free boundary Γ_2, by setting :

$$\sigma^{o*}(x'_1,x'_2,0,y_2,y_3) = \sigma^{o}(x'_1,x'_2,0,y_2,y_3) + \sigma^{oBL}(x'_1,x'_2,0,y_2,y_3) \quad (3.28)$$

From the formulas (3.8), (3.11), (3.12),(3.13), (3.18) and (3.23) and from the remarks 3.3 and 3.4, we can then write the local stresses σ^{o*} as functions of the homogenized displacement $u^{\circ}(x')$ in the form :

$$\sigma^{o*}(x'_1,x'_2,0,y_2,y_3) = c_{ijkh}(y_2,y_3)\, \varepsilon_{kh}(u^{\circ})_{|(x'_1,x'_2,0)} \quad (3.29)$$

where the not zero coefficients c_{ijkh} are given by :

- $(r,s) = (1,2)$ and $(r,s) = (1,3)$.

$$\begin{cases} c_{12rs} = b_{12rs} + \dfrac{\partial \tau^{rs}}{\partial y_3} = b_{12rs} + \tilde{a}_{12kh}\, e_{kh}\,(\eta^{rs}) \\ c_{13rs} = b_{13rs} - \dfrac{\partial \tau^{rs}}{\partial y_2} = b_{13rs} + \tilde{a}_{13kh}\, e_{kh}\,(\eta^{rs}) \end{cases}$$

- $(r,s) \neq (1,2)$ and $(r,s) \neq (1,3)$.

$$\begin{cases} c_{22rs} = b_{22rs} + \dfrac{\partial^2 \varphi^{rs}}{\partial y_3 \partial y_3} = b_{22rs} + \tilde{a}_{22kh}\, e_{kh}(\eta^{rs}) \\ c_{33rs} = b_{33rs} + \dfrac{\partial^2 \varphi^{rs}}{\partial y_2 \partial y_2} = b_{33rs} + \tilde{a}_{33kh}\, e_{kh}(\eta^{rs}) \\ c_{23rs} = b_{23rs} - \dfrac{\partial^2 \varphi^{rs}}{\partial y_2 \partial y_3} = b_{23rs} + \tilde{a}_{23kh}\, e_{kh}\,(\eta^{rs}) \\ c_{11rs} = b_{11rs} + (\frac{\nu}{E}) \left[c_{22rs} + c_{33rs} - b_{22rs} - b_{33rs} \right] \\ \qquad = b_{11rs} + \tilde{a}_{11kh}\, e_{kh}\,(\eta^{rs}) \end{cases}$$

with $\tau^{rs}(y_2,y_3)$ the solutions of (3.14) ... (3.17), $\varphi^{rs}(y_2,y_3)$ the solution of (3.19) ... (3.22) and $\eta^{rs}(y_2,y_3)$ the solutions of (3.24) ... (3.27).

Remark 3.5 : In this study, the components σ_{11}^{oBL} of the boundary layers stresses is never appeared. In fact, we have $\sigma_{11}^{oBL} = \frac{\nu}{E}\left[\sigma_{22}^{oBL} + \sigma_{33}^{oBL}\right]$ because $e_{11}(u^{1BL}) = 0$. It explains the expressions of the coefficients $c_{11rs}(y_2,y_3)$.

4. - Numerical Results

We consider a simple problem of structure and we are going to compute, near the free boundary, the micro-stresses σ^{o} and the boundary layers stresses σ^{oBL}. Then we shall compare their sum with the real stresses σ^{ε}, prevailing in the heterogeneous material near the boundary.
In order to simplify the computations, we assume that the material occupies a parallelepipedic domain in the $(\vec{e}_1,\vec{e'}_2,\vec{e'}_3)$ axis and that it is composed

of a stratification of 18 homogeneous layers in the direction $\vec{e}_3$ with $\vec{e}_3 = \frac{\sqrt{2}}{2}(\vec{e}'_2 + \vec{e}'_3)$ (bringing us back to the case $\alpha = \frac{\pi}{4}$). Each ply has a equal thickness and is supposed isotropic ; its mechanic properties, according it is composed of the material 1 or the material 2, are :

$$\begin{cases} E_1 = 0.84\ 10^{11}\ \text{Pa} & \nu_1 = 0.22 & \text{Material 1.} \\ E_2 = 0.4\ \ 10^{10}\ \text{Pa} & \nu_2 = 0.34 & \text{Material 2.} \end{cases}$$

We suppose that the planes $x'_3 = 0$ and $x'_3 = a > 0$ are free and the planes $x'_2 = b > 0$ and $x'_2 = -b$ subjected to uniform traction of density :

$$F_1 = F_3 = 0 \quad , \quad F_2 = 0.4\ 10^9$$

We assume that the displacement of a point of the structure and a component of the displacement of another point are fixed to zero, thus this problem admits a unique solution. And then, we suppose that the strains are planes parallely to $x'_1 = 0$.
These hypothesis are posed with the main of simplifying the calculations. Computations of an industrial structure, using this method, present no difficulty. We propose to apply the method to the thrust, described in |8|. Indeed, (see Remark 3.4), we never need tridimensional computations.

• First step : The first step of the computations consists in resolving the real equilibrium problem, where the material is supposed heterogeneous. It is the problem (3.9) with the preceding boundary conditions. One effects this computation with the object of use to reference. It is this resolution very expensive (which can present great difficulties for the industrial composite structures) that our method purposes to remove. We use a finite elements method with a P_1-Lagrange approximation for all computations. The mesh of the structure is composed of 2300 triangles (see figure 2). In figure 3, we have represented the deformed material.

• Second step : The second step of the computations consists in resolving the homogenized problem, which is the structure problem defined in the section 2 with the boundary conditions before precised. The material is considered homogeneous in the sense of the homogenization theory (see (2.3),(2.5), (2.6)). It is easy to solve this problem in the coordinates system R and after to effect a change of coordinates system. It leads to an anisotropic problem, which under the hypothesis of the loading, can be solve by simple algebraic calculations. We obtain a medium displacement $u^\circ(x)$, which is linear in x'_2 and x'_3 and we calculate the local stresses $\sigma^\circ(x', y_2, y_3)$ by means the formula (3.8). Therefore, the micro-stresses $\sigma^\circ(x', y_2, y_3)$ are constant in each material.

• Last step : The purpose of this step is to resolve the boundary layers problem P^{BL}. Since we use MODULEF-INRIA |13| Codes, which are particulary well adapted to the second order elliptic equations, we have chosen to solve this problem by using the kinematic formulation (see section 3.2). But, there is no difficulty to use the static formulation. On the contrary, the static formulation leads to boundary conditions of Dirichlet ; it is an advantage because numerically they are exactly satisfied and it is not the case of the boundary conditions of Neumann, which we are going to consider.
Under the simplifying hypothesis, we can remark that P^{BL} reduces to a plane elastic problem with two different loads, which is posed on the domain G truncated in y_3 direction far enough to $y_3 = 0$. The domain G is discretised with 504 triangles mesh (see figure 4). The boundary conditions prescribed on $y_2 = 0$ and $y_2 = 1$ are the conditions of periodicity, which are imposed

in each node. On $y_3 = 0$, one prescribes constant loads F^{rs} in each material : $F_i^{rs}(y_2) = -b_{i3rs}(y_2,0) + \frac{1}{Y_2^*}\int_0^{Y_2^*} b_{i3rs}(t,0)\,dt$ (see (3.25)) for $(r,s = 2,2)$ $(r,s = 2,3)$ and $(i = 2,3)$. And on $y_3 = 4$, we impose, for example, $\eta^{rs} \equiv 0$. The expression of the coefficients b_{ijkh} leads to show that $\eta^{22} = \eta^{33}$ and $\eta^{32} = \eta^{23}$.

The deformation of the domain G, presented in the figure 5, show that the displacement η^{rs} and therefore the boundary layers displacement $u^{1BL}(x',y_2,y_3)$ restitutes a strain on the free boundary, which appears in the computations of the real structure (see figure 3). The classical homogenization theory leads to a rectilinear free boundary and in this sense, we see another reason why we have introduced the boundary layers terms. The displacements η^{22} are presented on the figures 6 and 7 as functions of y_3 (which measure the distance of the boundary) for two fixed values of y_2 [The variations of η^{23} are similar]. These figures show clearly the exponentially decrease of η^{rs} when $y_3 \longrightarrow +\infty$, as it is announced in section 3.2. Once these problems are resolved, we can obtain the boundary layers stresses by setting :

$$\sigma_{ij}^{oBL}(x',y_2,y_3) = s_{ij}^{rs}(y_2,y_3)\,\varepsilon_{rs}(u^o)|(x'_1,x'_2,0)$$

with $s_{ij}^{rs}(y_2,y_3) = a_{ijkh}(y_2,y_3)\,e_{kh}(\eta^{rs})$.

We present, in the figure 8, the principal stresses s_{ij}^{rs} with $(r,s) = (2,2)$ in the neibourhood of the free boundary. In the figures 9 and 10, we have represented the boundary layers stresses as functions of y_3 for two fixed values of y_2 in the neighbourhood of the point 0. It appears, as planned in the section 3.1, that these stresses decrease exponentially. On the figures (10,11,12) and (13,14,15), we have superposed the different components of the stresses : σ^o, $\sigma^o + \sigma^{oBL}$ and σ^ε as functions of y_3 for two fixed values of y_2 in the neighbourhood of the point 0. Far enough to the free boundary, it appears that the stresses σ^o, $(\sigma^o + \sigma^{oBL})$ are a good approximation of σ^ε. But, near the free boundary, one can see that σ^o do not coincide with σ^ε, since the sum $(\sigma^o + \sigma^{oBL})$ remains a good approximation of σ^ε. In addition, one can remark that the free boundary condition is correctly satisfied (in an approximative sense) by the sum $(\sigma^o + \sigma^{oBL})$ when it is not the case of only stresses σ^o.

Remark 4.1 : The question of the possible presence of stress singularities in the intersection of the free boundary and an interface between any two elastic materials is very discussed, (see |4|,|6|,|16|,|17|,|19|,|20|). In some cases, it seems that the stresses are singular. By constructing an adapted mesh, our method, described the stress field near the boundary, should be allow the numerical studies of the singularities presence.

These results are obtained on CDC cyber 750 computer and Mini6-Bull computer.

Conclusion :

This method, which gives an approximation of the local stresses in the neighbourhood of a free boundary, should be consider as a complement of the classical homogenization method and of the local stresses calculation. The introduction of the boundary layers terms in the asymptotic expansion of the displacement and stresses is fully justified by the numerical results. The numerical implement of these terms is very easy and inexpensive, even in the case of structures requiring tridimensional computations ; then the point of view of the homogenization theory is preserved. Finally, let us point out that this method can be applied to other materials, such as the stratified materials with anisotropic layers, the materials reinforced by unidirectional fibers and other boundaries.

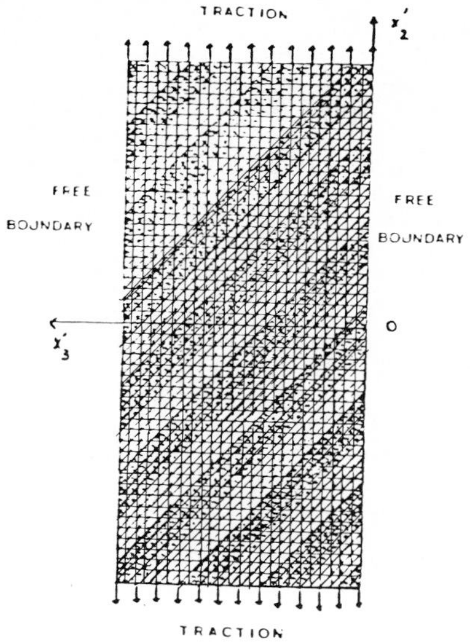

Figure 2 :

The triangulation of the domain Ω.
Number of triangles : 2304.
The triangles of the material 2 are distinguished by a point.

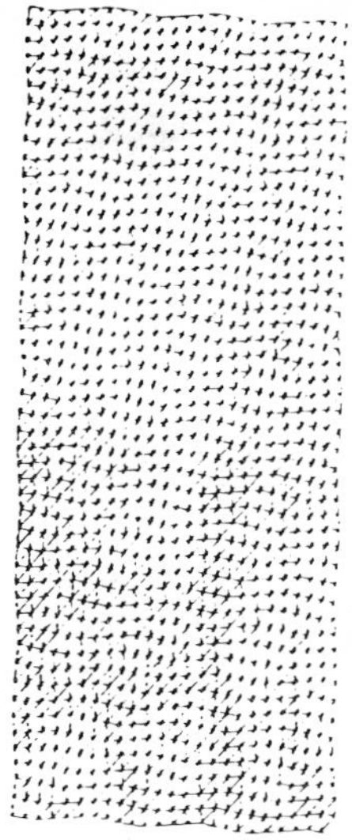

Figure 3 :

The deformed mesh of the domain Ω.

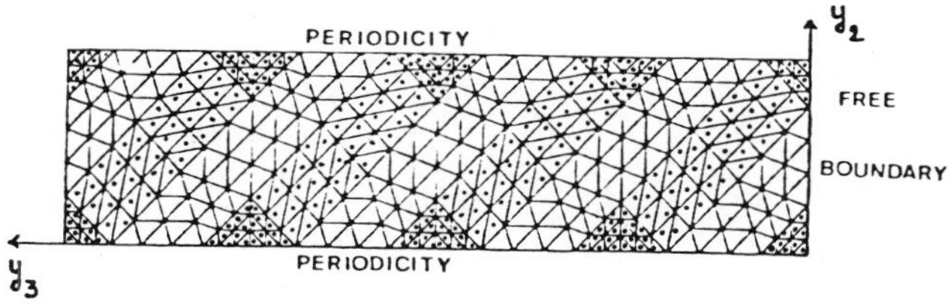

Figure 4 :

The triangulation of the domain G.
Number of triangles : 504.
The triangles of the material 2 are distinguished by a point.

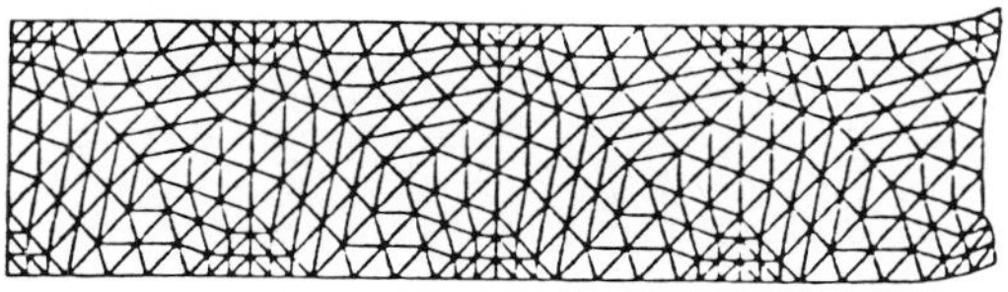

Figure 5 :

The deformed mesh of the domain G.

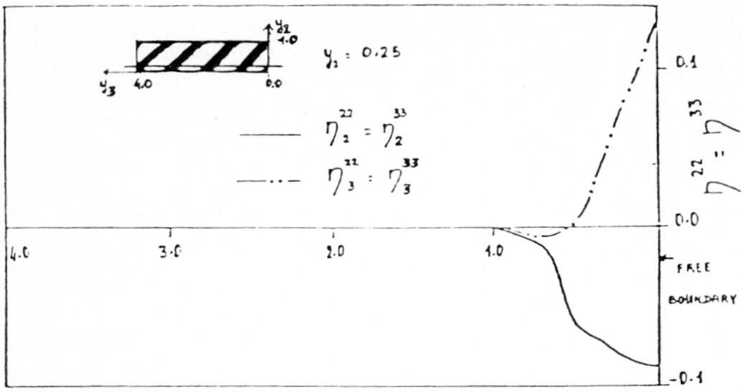

Figure 6

The elementary boundary layers displacement η^{22} ploted against y_3 (the distance of the free boundary) at y_2 = 0.25 fixed.

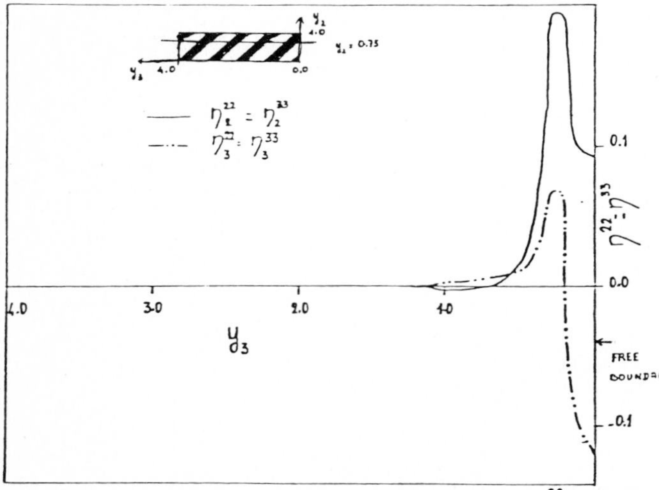

Figure 7

The elementary boundary layers displacement η^{22} ploted against y_3 at y_2 = 0.75 fixed.

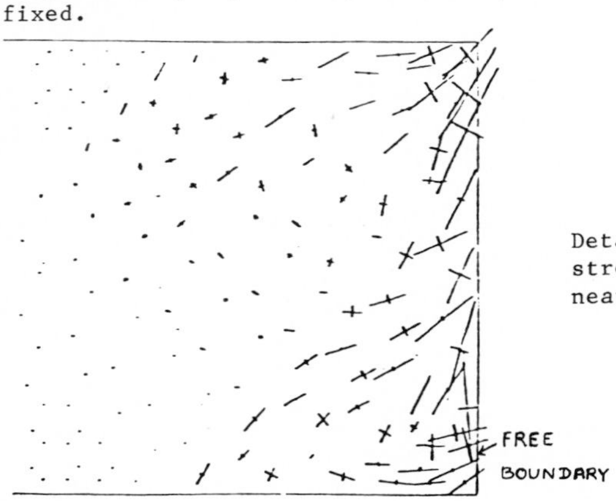

Figure 8

Detail of the principal stresses $s^{22} = \tilde{a}\, e(\eta^{22})$ near the free boundary.

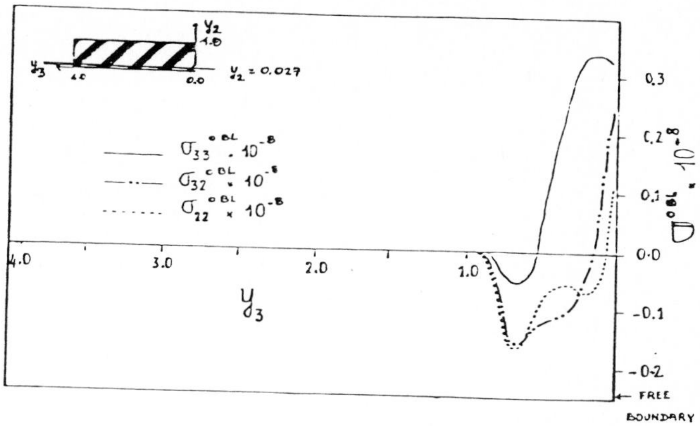

Figure 9 :

The boundary layers stresses σ^{oBL} ploted against y_3 at $y_2 = 0.319$ fixed.

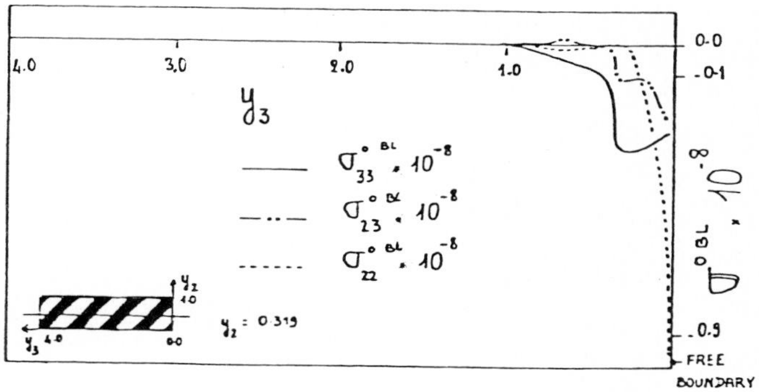

Figure 10 :

The boundary layers stresses σ^{oBL} ploted against y_3 at $y_2 = 0.319$ fixed.

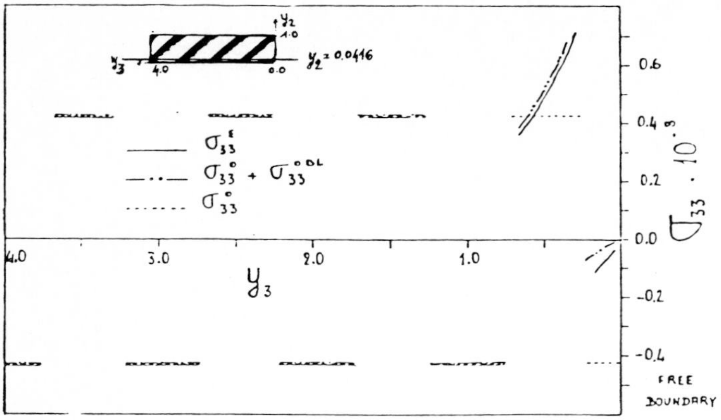

Figure 11 : The (3,3) components of the real stresses σ^{ε}, of the micro-stresses σ^{o} and of the boundary layers stresses σ^{oBL} added to σ^{o} ploted against y_3, at $y_2 = 0.0416$ fixed.

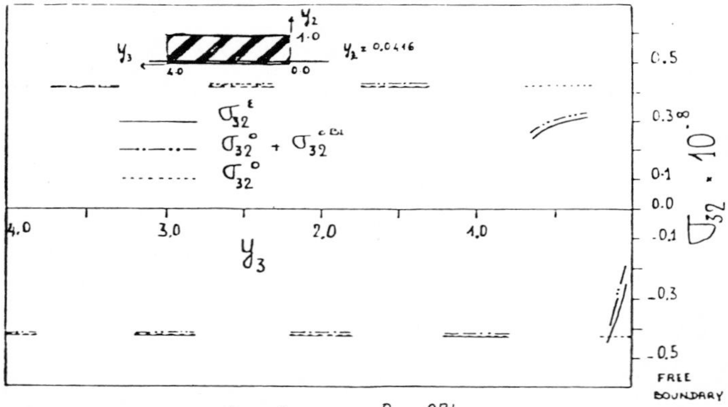

Figure 12 : The (2,3) components of σ^{ε}, σ^{o} and $(\sigma^{o} + \sigma^{oBL})$ ploted against y_3, at $Y_2 = 0.0416$ fixed.

Figure 13 : The (2,2) components of σ^{ε}, σ^{o} and $(\sigma^{o} + \sigma^{oBL})$ ploted against y_3, at y_2 : 0.0416 fixed.

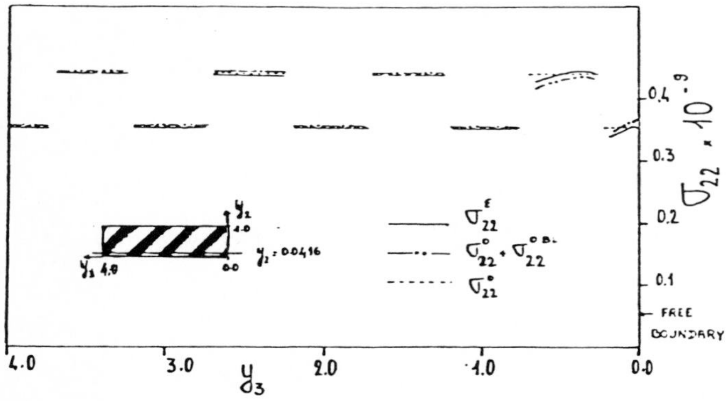

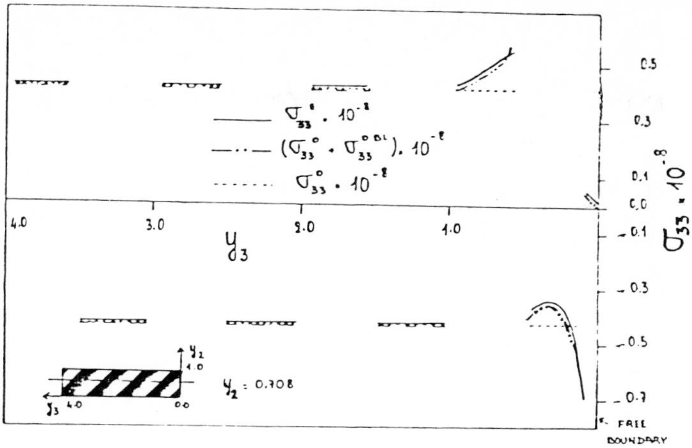

Figure 14 : The (3,3) components of the real stresses σ^{ε}, of the micro-stresses σ^{o} and the boundary layers stresses $\sigma^{o\,BL}$ added to σ^{o}, ploted against y_3 at $y_2 = 0.708$ fixed.

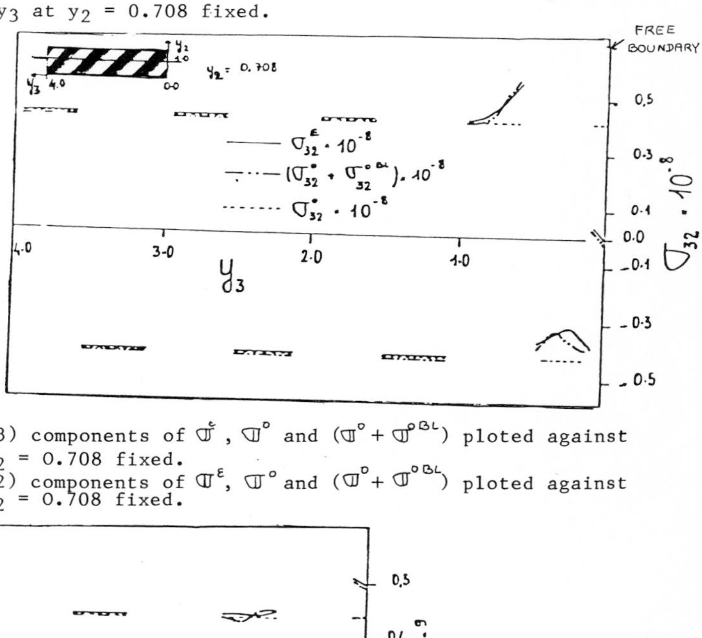

Figure 15 : the (2,3) components of σ^{ε} , σ^{o} and $(\sigma^{o} + \sigma^{o\,BL})$ ploted against y_3 at $y_2 = 0.708$ fixed.

Figure 16 : The (2,2) components of σ^{ε}, σ^{o} and $(\sigma^{o} + \sigma^{o\,BL})$ ploted against y_3 at $y_2 = 0.708$ fixed.

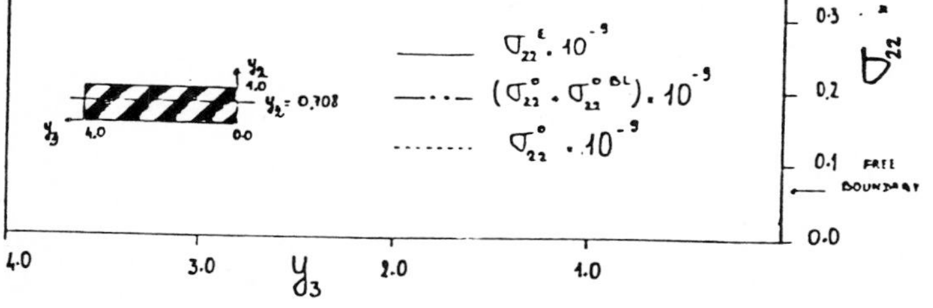

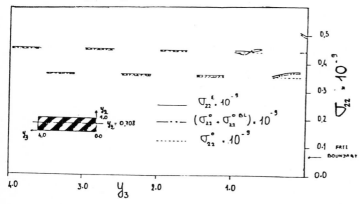

REFERENCES

|1| C. AURIAL, G. BOUBAL, P. LADEVEZE : Sur une méthode de calculs des effets locaux, Comptes Rendus des Troisièmes Journées Nationales sur les Composites, JNC3, PARIS, Sept. 82, p 279,288.

|2| R.S. BARSOUM and C.E. FREESE : An Iterative Approach for the Evaluation of Delamination Stresses in Laminated Composites, Int. Jour. Num. Meth. Engng, 20(1984), p 1415,1431.

|3| A. BENSOUSSAN, J.L. LIONS and G. PAPANICOLAOU : Asymptotic Analysis for Periodic Structures, North Holland, Amsterdam, 1978.

|4| A. BENSOUSSAN, J.L. LIONS and G. PAPANICOLAOU : Boundary Layer Analysis in Homogenization of Diffusion Equation with Dirichlet conditions, K. ITO ed, J. WILEY and Sons, New-York, 1978.

|5| H. BREZIZ : Analyse Fonctionnelle, Théorie et Applications, Masson ed, Paris, 1983.

|6| J.L. DAVET, P. DESTRUYNDER and T. NEVERS : Asymptotic and Boundary Layer Methods in the Delamination for Composite Materials, in Euromech 84, the Inclusion of Local Effects in the Analysis of Structures, Sept 84, Cachan.

|7| H. DUMONTET : Thèse de Doctorat de 3ème cycle, Université P et M Curie, 1983.

|8| H. DUMONTET : Study of a Boundary Layers Problem in Elastic Composites, to appear in RAIRO.

|9| H. DUMONTET and D. LEGUILLON : Etude d'une butée lamifiée soumise à des efforts de Compression, Rapport S.N.I.A.S., 1984.

|10| G. DUVAUT : Analyse Fonctionnelle et Mécanique des Milieux Continus, In Theoretical and Applied Mechanic, W. Koiter ed, North-Holland, 1976, p 119-132.

|11| P. GERMAIN : Cours de Mécanique des Milieux Continus, Masson ed, Paris 1973.

|12| S.G. LEKHNITSKII : Theory of Elasticity of an Anisotropic Elastic Body, Holden-day ed, San Francisco, 1963.

|13| J.L. LIONS : Some Methods in the Mathematical Analysis of Systems and their Control, Gordon and Breach, New-York, 1981.

|14| A. PERRONET : Presentation du Club Modulef, INRIA, 1977.

|15| E. SANCHEZ-PALENCIA : Non Homogeneous Media and Vibration Theory, Lecture Notes in Physic, 127, Springer-Verlag ed, Heidelberg, 1980.

|16| E. SANCHEZ-PALENCIA : Influence de l'Anisotropie sur l'Apparition de Singularités de bord dans les problèmes aux limites relatifs aux matériaux Composites, Compt. Rend. Acad. Sci. Paris, ser I, 1984.

|17| E. SANCHEZ-PALENCIA : On the edge Singularities in Composite Media. Influence of the Anisotropy, to appear in the Proceedings of the Stephan Banach Center, Warsaw, Semester on PDE 1984, Prof Bojarshied.

|18| P. SUQUET : Thèse de Doctorat ès Sciences, Homogénéisation et Plasticité Université P et M Curie, 1982.

|19| S.S. WANG and I. CHOI : Boundary Layer Effects in Composite Laminates, I Free Edge Stress Singularities, II Free Edge Solutions and Basic Charascteristics, Jour. Appl. Mech., 49, 1982, I p 541-548, II p 549-561.

|20| R.I. ZWIERS, T.C. TING and R.L. SPILKER : On the logarithmic Singularity of Free Edge Stress in Laminated Composites under Uniform Extension, Jour. Appl. Mech., 49, 1982, p 561-569.

CHAPTER 3 :

LOCAL EFFECTS IN DYNAMICS

HIGH STRESS INTENSITIES IN FOCUSSING ZONES OF WAVES

J.BALLMANN, H.J.RAATSCHEN and M.STAAT
Lehr- und Forschungsgebiet Mechanik, RWTH Aachen,
Templergraben 64, 5100 Aachen (FRG)

ABSTRACT

The propagation of mechanical waves in plates of isotropic elastic material is investigated. After a short introduction to the understanding of focussing of stress waves in a plate with a curved boundary the method of characteristics is applied to a plate of hyperelastic material. Using this method the propagation of acceleration waves is discussed. Based on this a numerical difference scheme is developed for solving initial-boundary-value problems and applied to two examples: propagation of a point disturbance in a homogeneously finitely strained non-linear elastic plate and geometrical focussing in a linear elastic plate.

INTRODUCTION

Modern investigations on mechanical wave propagation show increasing interest in local stress concentration under transient loading (ref.13).

Experimental investigations of specimens of revolution showed internal cracks due to stress wave focussing (ref.14). The phenomenon of stress concentration caused by focussing can be demonstrated by the ray method in the sequence of figures 1a-c. A plane longitudinal wave travelling along rays is reflected into a longitudinal and a transversal wave due to the free boundary conditions (see Fig. 1a). The envelopes of the reflected rays form two caustics (an incident transversal wave forms a second pair of caustics (see Fig. 1b)), which are the traces of singular points of the wave fronts (see Fig. 1c). Loading plates of transparent materials (PMMA e.g.) in shock tubes the wave fronts can be made visible by shadow photographs (see Fig. 1d and ref.11).

From arguments of geometrical acoustics and from experimental evidence high stresses are expected near the cusps of the caustics. The ray method neglects field effects,however, and gives solutions at the wave front and this only if the solution ahead of the wave is known. But generally this solution is also unknown.

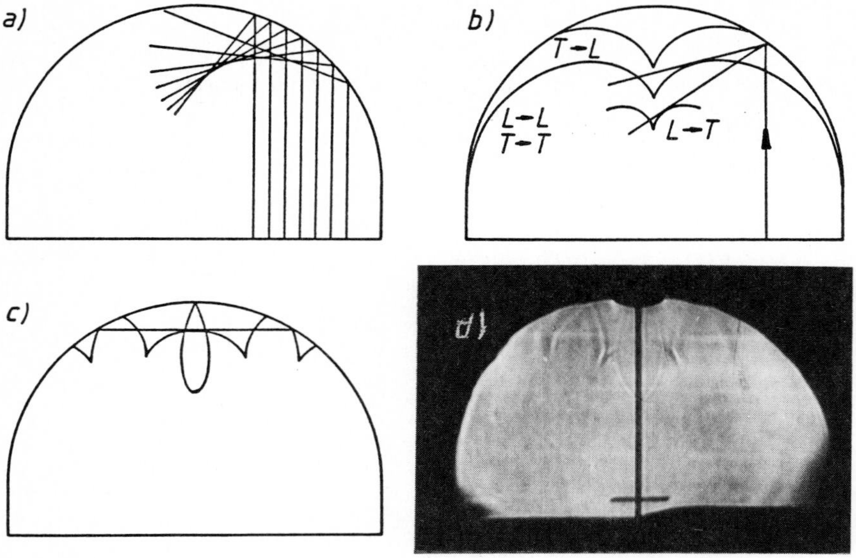

Fig.1. a) Rays of an incident longitudinal wave; b) Caustics; c) Wave fronts at successive times; d) Shadow photograph of the wave fronts (ref.11)

For the numerical solution of elliptical PDEs the application of spatial discretisation procedures especially the finite element method (FEM) is generally accepted. But the great success could not be repeated in the application to hyperbolic PDEs. Even in problems with only one space dimension these methods produce suprious oscillations thus smearing sharp wave fronts (ref.3,17). This is because a local disturbance immediately affects the whole domain of calculation. But physically the wave speed is finite. Moreover, these methods do not treat wave surfaces as discontinuities. Explicit difference schemes e.g. face severe problems with stability because they employ expansions neglecting these discontinuities. Commonly dispersion is controlled by lumping and stability is enforced by using artificial viscosity. A more promising way is the development of finite element methods based on characteristic variables (ref.15).

The method of characteristics was developed for hyperbolic PDEs and became a well established tool for modelling non-linear wave propagation and shock waves in nonsteady gasdynamics (refs.1,10, 12,20). Surfaces that may support discontinuities of some derivatives of the dependent variables are calculated. By integrating in characteristic surfaces these jumps are not smeared. Explicit difference schemes based on the method of characteristics are stable and reproduce details of the solution with only small dispersion.

Numerical solutions of the complete field equations for the above problem of a semicircular plate were obtained by the authors using the method of characteristics (ref.2).They also applied it to other linear and non-linear problems of elastic plates. Before discussing some results it is felt necessary to explain the method in more detail since it is not yet commonly used in elastodynamics.

The characteristic directions and the so-called compatibility equations for a non-linear elastic plate are derived which are the basis of the numerical scheme. One can find the related equations for linear elastic plates in a similar way or by simplifications of the difference equations.

BASIC EQUATIONS

The position of a material point is given by the Cartesian coordinates x^i and $\bar{x}^i$ $(i=1,2,3)$ in the unstressed reference configuration and in the actual configuration, respectively. Let both sets be connected by a smooth bijective one parameter family of mappings $\bar{x}=\bar{x}(x;\tau)$ with time τ as parameter. The deformation gradient

$$\mathbf{F} = \frac{\partial \bar{x}}{\partial x} \qquad (1)$$

is the linear approximation of the mapping. With the displacement $\bar{\mathbf{u}} = \bar{\mathbf{x}} - \mathbf{x}$ the particle velocity $\bar{\mathbf{v}}$ and the gradient of displacement are defined as usual (unit tensor $\mathbf{1}$)

$$\bar{\mathbf{v}} = \frac{\partial \bar{\mathbf{u}}}{\partial \tau} \qquad (2)$$

$$\mathbf{H} = \frac{\partial \bar{\mathbf{u}}}{\partial \mathbf{x}} = \mathbf{F} - \mathbf{1} \quad . \qquad (3)$$

As strain measures the right CAUCHY-GREEN tensor $\mathbf{C}$ and the GREEN strain tensor $\mathbf{G}$ are convenient for imcompressible and compressible materials, respectively

$$\mathbf{C} = \mathbf{F}^T \mathbf{F} \qquad (4)$$

$$\mathbf{G} = \frac{1}{2} (\mathbf{C} - \mathbf{1}) = \frac{1}{2} (\mathbf{H}^T + \mathbf{H} + \mathbf{H}^T \mathbf{H}) \quad . \qquad (5)$$

Our discussion is restricted to materials which are hyperelastic, i.e. which have a stored energy density U(C) or U(G). With the symmetric KIRCHHOFF stress $\boldsymbol{\sigma}$ its variation is

$$\delta U = \frac{1}{2\varrho}\,\boldsymbol{\sigma} : \delta\mathbf{C} \qquad , \qquad \delta U = \frac{1}{\varrho}\,\boldsymbol{\sigma} : \delta\mathbf{G} \qquad (6)$$

The mass density ϱ is taken in the reference configuration. The gradient of eq.(6) with respect to the strain measure yields the purely mechanical constitutive equations

$$\boldsymbol{\sigma}_C = \boldsymbol{\sigma}(\mathbf{C}) \qquad , \qquad \boldsymbol{\sigma}_G = \boldsymbol{\sigma}(\mathbf{G}) \quad . \qquad (7)$$

The balance of momentum is given by

$$\varrho\,\frac{\partial^2 \bar{\mathbf{u}}}{\partial \tau^2} - \operatorname{div}\,(\mathbf{F}\,\boldsymbol{\sigma}) - \varrho\,\bar{\mathbf{k}} = \mathbf{0} \quad .$$

Using eqs.(1)-(7) the balance of momentum can be rewritten into first order PDEs

$$\varrho\,\frac{\partial \bar{\mathbf{v}}}{\partial \tau} - \frac{\partial}{\partial \mathbf{F}}\,(\mathbf{F}\,\boldsymbol{\sigma}) \,\vdots\, \operatorname{grad}\,\mathbf{H}^T + \varrho\,\bar{\mathbf{k}} = \mathbf{0}$$

and finally, with the fourth order elasticities $\mathbf{A}$,

$$\delta\mathbf{F} : \mathbf{A} : \delta\mathbf{F} = \varrho\,\delta^2 U \qquad (8)$$

it reads

$$\varrho\,\frac{\partial \bar{\mathbf{v}}}{\partial \tau} - \mathbf{A} \,\vdots\, \operatorname{grad}\,\mathbf{H}^T + \varrho\,\bar{\mathbf{k}} = \mathbf{0} \quad . \qquad (9)$$

Combining eqs.(2) and (3) yields the integrability equation for $\bar{\mathbf{u}}$

$$\frac{\partial \mathbf{H}}{\partial \tau} - \operatorname{grad}\,\bar{\mathbf{v}} = \mathbf{0} \quad . \qquad (10)$$

The material differential operators div and grad are used in the usual manner. Multiple dots denote multiple transvection.

The quasi-linear system of first order PDEs (9),(10) is hyperbolic for the materials under consideration. Therefore certain derivatives of $\bar{\mathbf{v}}$ and $\mathbf{H}$ may be discontinuous and may propagate as so-called acceleration waves (ref.22). The following discussion is confined to plane stress problems and for compressible materials also to plane strain problems.

THE METHOD OF CHARACTERISTICS

The object of the method of characteristics is twofold. First the characteristic condition determines the directions $\mathbf{n}^*$ in which the first order PDEs allow jumps in certain derivatives of the dependent variables. Next these undefined derivatives can be eliminated by forming a linear combination of the original PDEs. The resulting so-called compatibility equation contains only continuous derivatives in a characteristic surface with normal $\mathbf{n}^*$. A star * is used in the text to denote quantities in space and time.

With the independent variables $x^0=c\tau, x^1, x^2$ (c being an arbitrary constant velocity) and the covariant and contravariant base vectors, respectively, $\mathbf{e}_i$ and $\mathbf{e}^j$ $(i,j=0,1,2)$

$$e_i \; e^j = \delta_i{}^j \qquad (11)$$

a gradient in space and time is introduced by ($\alpha = 0,1,2$)

$$\nabla^* f = \frac{\partial f}{\partial x^\alpha} \circ \; e^\alpha \qquad (12)$$

The dyadic product is indicated by the symbol o. Thus the PDEs (9),(10) can be abbreviated ($r,s,\varrho,\sigma = 1,...,6$)

$$a_r^{*\sigma} \; \nabla^* z_\sigma - c_r = 0 \qquad (13)$$

with unknown functions $z = (\bar{v}^1, \bar{v}^2, H^1{}_1, H^2{}_1, H^1{}_2, H^2{}_2)$, the coefficient matrix of vectors a_r^{*s} and terms c_r containing the unknowns only in a non-differentiated form. The matrix of vectors a_r^{*s} is given by

$$a_r^{*s} = \begin{bmatrix} \varrho c e_o & 0 & e_1 o \bar{e}^1 o e_1 \vdots A & e_1 o \bar{e}^1 o e_2 \vdots A & e_1 o \bar{e}^2 o e_1 \vdots A & e_1 o \bar{e}^2 o e_2 \vdots A \\ 0 & \varrho c e_o & e_2 o \bar{e}^1 o e_1 \vdots A & e_2 o \bar{e}^1 o e_2 \vdots A & e_2 o \bar{e}^2 o e_1 \vdots A & e_2 o \bar{e}^2 o e_2 \vdots A \\ -e_1 & 0 & c e_o & 0 & 0 & 0 \\ 0 & -e_1 & 0 & c e_o & 0 & 0 \\ -e_2 & 0 & 0 & 0 & c e_o & 0 \\ 0 & -e_2 & 0 & 0 & 0 & c e_o \end{bmatrix}$$

The linear combination of eq.(13) with multipliers η^r reads

$$\eta^\varrho \; a_\varrho^{*\sigma} \; \nabla^* z_\sigma - \eta^\varrho \; c_\varrho = 0 \qquad (14)$$

The condition that all remaining derivatives lie in a so-called characteristic surface with normal n^* is equivalent to

$$\eta^r \; a_r^{*s} \; n^* = 0 \qquad (15)$$

This homogeneous system of linear equations has a nontrivial solution η if and only if the coefficient determinant vanishes, i.e. the characteristic equation holds

$$\det \left(a_r^{*s} \; n^* \right) = 0. \qquad (16)$$

Choosing the ansatz

$$n^* = -\frac{v}{c} \; e^o + n \qquad (17)$$

with a space-like unit normal $n = \cos\varphi \; e^1 + \sin\varphi \; e^2$ the characteristic condition eq.(16) becomes a one parameter form for the wave speed v

$$v = v \; (n) = v \; (\varphi) \quad .$$

For all the angles $o \leqslant \varphi < 2\pi$ the vectors n^* generate the normal cone while the corresponding characteristic surfaces envelop the MONGE-cone (see Fig. 2). The generators m^* of the MONGE-cone

$$m^* = c \; e_o + m \qquad , \quad m = \frac{\partial v}{\partial n} \qquad (18)$$

are called bicharacteristics and ensue from the conditions of orthogonality and enveloping

$$\mathbf{m}^* \mathbf{n}^* = 0 \quad , \quad \mathbf{m}^* (\mathbf{n}^* + \delta\mathbf{n}^*) = 0$$

Together with a unit space-like tangent $\mathbf{t}=\mathbf{e}^o \times \mathbf{n}$ the bicharacteristic $\mathbf{m}^*$ spans the characteristic surface element and so does any near-characteristic $\mathfrak{m}$ (ref.21). The possibly natural choice is

$$\mathfrak{m}^* = c\,\mathbf{e}_o + \mathfrak{m} \quad , \quad \mathfrak{m} = v\,\mathbf{n}$$
$$\mathfrak{m}^* = \mathbf{m}^* - m\,\mathbf{t} \quad , \quad \mathfrak{m}\,\mathbf{t} = 0 \quad . \qquad (19)$$

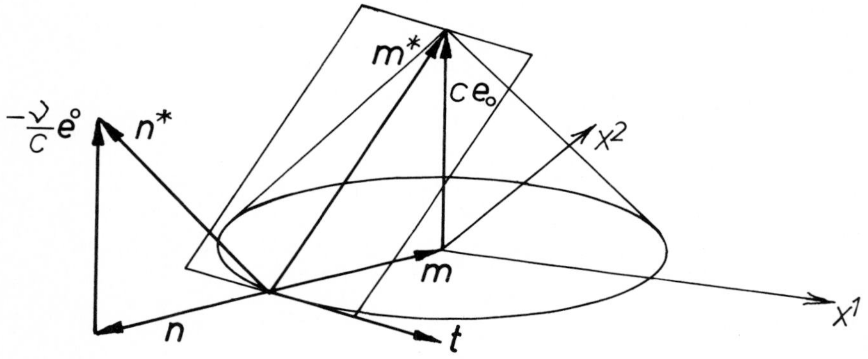

Fig.2. Geometry of characteristic surfaces

Discussion of the Characteristic Condition

After some calculations the characteristic condition reads

$$v^2 \det (\bar{\mathbf{Q}}(\mathbf{n}) - \varrho v^2 \bar{\mathbf{1}}) = 0 \qquad (20)$$

revealing the eigenvalue problem of the acoustic tensor $\bar{\mathbf{Q}}$ in the actual configuration

$$\bar{\mathbf{Q}} = (\bar{\mathbf{e}}_\lambda \circ \mathbf{n} : \mathbf{A} : \bar{\mathbf{e}}^\mu \circ \mathbf{n})\, \bar{\mathbf{e}}^\lambda \circ \bar{\mathbf{e}}_\mu \qquad (21)$$

The root $v_o=0$ originates from the special choice of the dependent variables. The related MONGE-cone degenerates to the pathline of a material point. The other roots of the characteristic equation are the eigenvalues of $\bar{\mathbf{Q}}$. They can be written with the principal invariants ($I=\mathrm{tr}\bar{\mathbf{Q}}$, $II=0.5(\mathrm{tr}\bar{\mathbf{Q}}^2-I^2)$) leading to

$$\varrho v_\varepsilon^{\pm} = \pm(-\,0.5\,I \pm \sqrt{0.25\,I^2+II}\,) \quad , \quad \varepsilon=L,T \qquad (22)$$

Positive and negative roots $v_\varepsilon(\mathbf{n})$ determine the forward and the backward MONGE-cones, respectively. The latter were computed for a highly non-linear material for some state of strain at point P (see Fig.3). For a non-homogeneous deformation these cones are the local linear approximations of the global MONGE-conoids while the bicharacteristics m^* are tangent to their generators which are not plane curves generally.

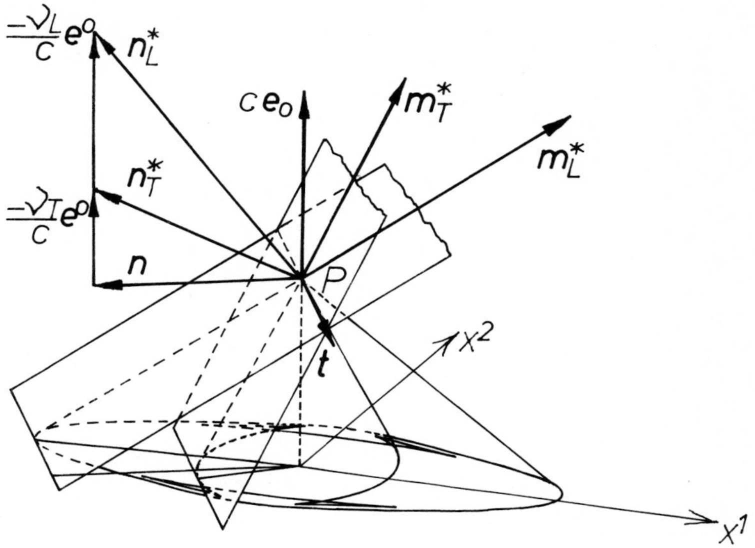

Fig.3. Pathline, quasi-longitudinal and quasi-transversal MONGE-cones

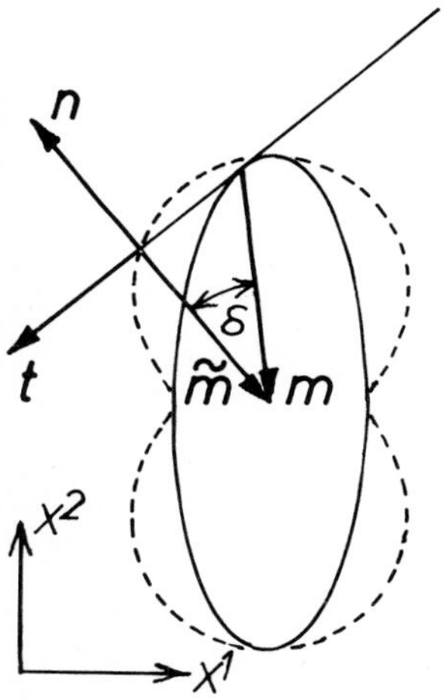

Fig.4. Characteristic loci and FRIEDRICHS diagram

Figure 4 shows a plane intersection of longitudinal MONGE-cones with the plane $c\tau$ = const. together with the solutions of the characteristic condition (20) in broken lines. In gasdynamics these curves are called characteristic loci and FRIEDRICHS diagram, respectively. A plane wave is the trace in space of the characteristic surface. The characteristic loci can either be seen as envelopes of all plane waves that passed through its center P at a time Δt in the past or can be interpreted as the wave fronts in space of a point disturbance at P at the same time (refs.16,18).

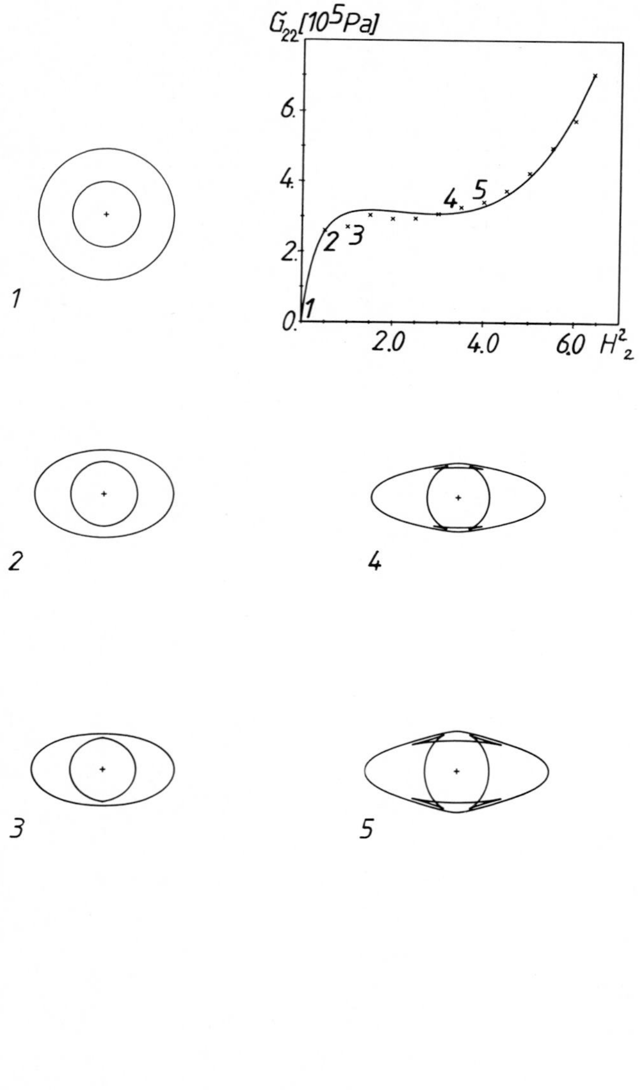

Fig.5. Simple elongation strain path of TRELOAR-material (ref.9 experimental values are denoted by x) and corresponding wave fronts

Plane waves with normal $\mathbf{n}$ travel at their normal speed $\hat{m}_\varepsilon(\mathbf{n})$ while the accompanying discontinuities move along their rays $\mathbf{m}_\varepsilon(\mathbf{n})$ with the angle δ between $\mathbf{m}_\varepsilon$ and the normal direction $\mathbf{n}, \hat{m}_\varepsilon$. Thus waves are generally quasi-longitudinal and quasi-transversal on the outer and inner MONGE-cone, respectively. For non-linear materials pure modes propagate along the axes of symmetry where $\delta=0$. Since these coincide with the principal axes of $\boldsymbol{\sigma}, \mathbf{C}$ and $\mathbf{G}$ for isotropic materials corresponding waves were termed principal waves (ref.22).

For some isotropic materials the variety of wave fronts is achievable as observed for different linear anisotropic materials (refs.4,16,19). The example in figure 5 shows an incompressible material with simple elongation strain path in direction $\mathbf{e}_2$. The KIRCHHOFF stress σ_{22} was calculated from data given for TRELOAR-material (ref.9). For the numbered deformations the calculated wave fronts are shown. Note that there are either 0,2 or 4 cusped triangles on the quasi-transversal front. It could be proved for linear anisotropic materials that the interior of the wave front cuspoidal triangles are stable lacunae that are gaps in the range of influence where the displacement, associated with an impulsive disturbance at the vertex of the forward MONGE-cone, vanishes identically (refs.5,18). In the linear elastic case the wave fronts emenating from a point are circles as in figure 5 for the undeformed state (N$^\circ$ 1).

<u>Compatibility Equations</u>

Eq.(15) provides a set of multipliers η^r_ε $(r=1,\ldots,6)$ for each solution v_ε of the characteristic condition (16),(20). For each set of η^r_ε eq.(15) can be formulated;

along the pathline $(\varepsilon=0)$:

$$\nabla^*\mathbf{H} : \mathbf{t} \circ \mathbf{m}^*_\varepsilon = \nabla^*\bar{\mathbf{v}}\ \mathbf{t} \qquad (23)$$

along the cones $(\varepsilon=L,T)$:

$$\varrho v_\varepsilon\ \bar{\mathbf{q}}_\varepsilon \circ \mathbf{m}^*_\varepsilon : \nabla^*\bar{\mathbf{v}} - \bar{\mathbf{q}}_\varepsilon \circ \mathbf{n} : \mathbf{A} \circ \mathbf{m}_\varepsilon \,\vdots\, \nabla^*\mathbf{H} = r_\varepsilon \qquad (24)$$

with $\quad r_\varepsilon = \bar{\mathbf{q}}_\varepsilon \circ \mathbf{n} : \mathbf{A}\ \mathbf{t} \circ \mathbf{t} : \nabla^*\bar{\mathbf{v}} - \quad v_\varepsilon\ \bar{\mathbf{q}}_\varepsilon \circ \mathbf{t} : \mathbf{A} \circ \mathbf{t} \,\vdots\, \nabla^*\mathbf{H} - \varrho v_\varepsilon \bar{\mathbf{q}}_\varepsilon \bar{\mathbf{k}}$

$$-\varrho v_\varepsilon (\mathbf{m}_\varepsilon \mathbf{t}) \bar{\mathbf{q}}_\varepsilon \circ \mathbf{t} : \nabla^*\bar{\mathbf{v}} + (\mathbf{m}_\varepsilon \mathbf{t}) \bar{\mathbf{q}}_\varepsilon \circ \mathbf{n} : \mathbf{A} \circ \mathbf{t} \,\vdots\, \nabla^*\mathbf{H}$$

Herein $\bar{\mathbf{q}}_\varepsilon$ denotes the normed eigenvector of $\bar{\mathbf{Q}}$.

Eqs.(23),(24) contain only interior derivatives in the direction of $\mathbf{m}^*_\varepsilon$ and $\mathbf{t}$ and lack any outward derivative in the direction of $\mathbf{n}^*$. Hence all derivatives lie on the characteristic surface and eq.(23) cannot lead out of it. As a consequence all functions and gradients on the characteristic surface are continuous, while they may be discontinuous across it. Eqs. (23),(24) are the so-called compatibility equations. Initial values on characteristic surfaces must be prescribed compatible with them.

Choosing $\mathfrak{m}_\varepsilon^*$ of eq.(19) instead of $\mathbf{m}_\varepsilon^*$ may be advantageous to simplify eq.(24). Note, there is no difference between $\mathfrak{m}_\varepsilon^*$ and $\mathbf{m}_\varepsilon^*$ on principal axes.

$$\varrho v_\varepsilon \, \bar{q}_\varepsilon \circ \mathfrak{m}_\varepsilon^* : \nabla^* \bar{v} - \bar{q}_\varepsilon \circ \mathbf{n} : \mathbf{A} \circ \mathfrak{m}_\varepsilon^* \vdots \nabla^* \mathbf{H} = \Psi_\varepsilon \qquad (25)$$

$$\Psi_\varepsilon = \bar{q}_\varepsilon \circ \mathbf{n} : \mathbf{A} \; \mathbf{t} \circ \mathbf{t} : \nabla^* \bar{v} - v_\varepsilon \bar{q}_\varepsilon \circ \mathbf{t} : \mathbf{A} \circ \mathbf{t} \vdots \nabla^* \mathbf{H} - \varrho v_\varepsilon \bar{q}_\varepsilon \bar{k} \quad .$$

A difference scheme for the numerical solution of elastic problems is developed from the physically reasoned eqs.(18),(23),(24) or (19),(23),(25) in the next section.

DIFFERENCE EQUATIONS

Given the initial data on the surface $\tau = t_o$ a solution at point P on the surface $\tau = t_o + \Delta t$ is obtained numerically by integrating the compatibility equations along characteristics passing through P. Having done this for all points of the solution surface $\tau = t_o + \Delta t$ it is used as new initial surface, and the process repeated until the complete range of influence specified by the initial data has been determined.

For convenience introduce a differential operator on characteristic ε (ε=O,L,T)

$$\left.\frac{Dy}{D\tau}\right|_\varepsilon = \nabla^* y \; \mathbf{m}_\varepsilon^* \qquad (26)$$

with y to be integrated. Time τ is used as parameter of integration. Integrating eq.(26) gives

$$y(t_o+\Delta t) - y(t_o) = \int_{to}^{t_o+\Delta t} (\nabla^* y \; \mathbf{m}_\varepsilon^*) \, D\tau$$

and with a TAYLOR-expansion of the integrand

$$y(t_o+\Delta t) - y(t_o) = 0.5\left(\left.(\nabla^* y \; \mathbf{m}_\varepsilon^*)\right|_{t_o+\Delta t} + \left.(\nabla^* y \; \mathfrak{m}_\varepsilon^*)\right|_{t_o}\right)\Delta t + o(\Delta t^3) \qquad (27)$$

Here $(\nabla^* y \; \mathfrak{m}_\varepsilon^*)|_{to}$ is preferred rather than $(\nabla^* y \; \mathbf{m}_\varepsilon^*)|_{to}$ to maintain the space-like normal n along the path of integration. The domain of dependence of $P(t_o+\Delta t)$ is found by applying eq.(27) to the position vector, y=x

$$\mathbf{x}(t_o+\Delta t) - \mathbf{x}(t_o) = 0.5 \left(\left.\mathbf{m}_\varepsilon\right|_{t_o+\Delta t} + \left.\mathfrak{m}_\varepsilon\right|_{t_o}\right) \Delta t + o(\Delta t^3) \; . \qquad (28)$$

Note, $\mathbf{m}_\varepsilon$ and $\mathfrak{m}_\varepsilon$ depend on the solution at P; ε=L,T.

Integration of the compatibility equations leads to the following difference equations

along the pathline:

$$H\Big|_{t_o+\Delta t} - H\Big|_{t_o} = 0.5(((\nabla^*\bar{v})1)\Big|_{t_o+\Delta t} + ((\nabla^*\bar{v})1\Big|_{t_o})\Delta t + o(\Delta t^3) \qquad (29)$$

along the cones:

$$\begin{aligned}&((\varrho v_\varepsilon\, \bar{q}_\varepsilon)\Big|_{t_o+\Delta t} + (\varrho v_\varepsilon\, \bar{q}_\varepsilon)\Big|_{t_o})(\bar{v}\Big|_{t_o+\Delta t} - \bar{v}\Big|_{t_o}) - \\ &-((\bar{q}_\varepsilon o\ n:A)\Big|_{t_o+\Delta t} + (\bar{q}_\varepsilon o\ n:A)\Big|_{t_o}):(H\Big|_{t_o+\Delta t} - H\Big|_{t_o}) = \\ &= (r_\varepsilon\Big|_{t_o+\Delta t} + r_\varepsilon\Big|_{t_o})\ \Delta t + o(\Delta t^3)\end{aligned} \qquad (30)$$

The TAYLOR-expansion of the coefficients and the right side of eqs.(29),(30) is admissible because all quantities are continuous as long as no shocks occur and have continuous first derivatives in the characteristic surface.

Difference equations (30) may be formulated along any characteristic. They employ the values of the six unknown functions and their space-like inner derivatives. Choosing the eigenvectors of σ at point P as a basis for a local scheme four of the eighteen unknowns become decoupled. Four equations are formulated along the pathline and both cones, respectively. The balance of momentum integrated along the pathline is used to complete the system by two (non-characteristic) equations (refs.6,7). The space-like inner derivatives are eliminated explicitely, leaving a set of six non-linear algebraic equations for the solution at P which is solved iteratively for the non-linear case. For an initial guess $\bar{v}$ is not needed because it is not employed in the constitutive law eq.(6) and thus does not enter the coefficients in eq.(30). Hence it is sufficient to integrate the compatibility equations along the pathline

$$H\Big|_{t_o+\Delta t} - H\Big|_{t_o} = ((\nabla^*\bar{v})1)\Big|_{t_o}\ \Delta t + o(\Delta t^2)\ . \qquad (31)$$

A suitable net is employed for the spatial discretisation. Then the required values and their derivatives on the initial surface are calculated from the local approximation by a second order surface in a least squares sence. The COURANT-FRIEDRICHS-LEWY stability condition (ref.8) is satisfied if the analytical domain of dependence is a subset of the numerical domain of dependence as represented by the points employed in the approximation. Numerical dispersion is reduced by choosing points that are close to the outer cone. Unfortunately the stability condition is but necessary and is sufficient for linear systems of PDEs only. In figure 6 some typical numerical schemes as used for the calculation are shown for inner points and for points on boundaries.

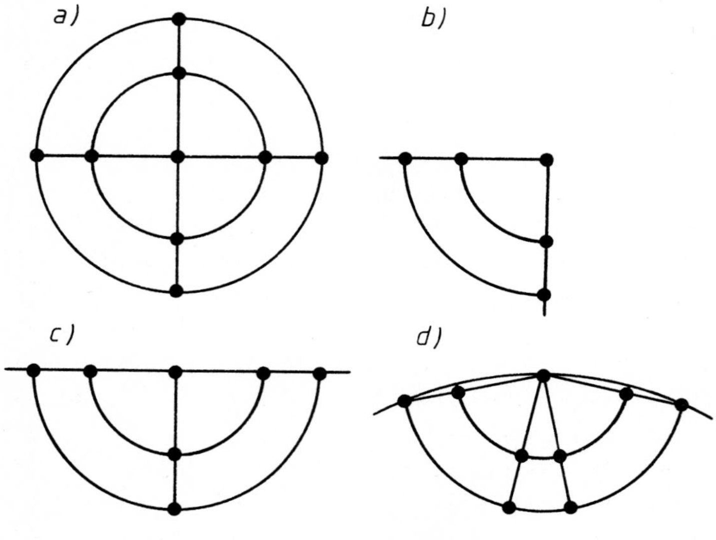

Fig.6 Numerical schemes: a) inner point, b) corner point, c) point on straight boundary, d) point on curved boundary

EXAMPLES

Point Disturbance in a Nonlinear-Elastic Plate

For testing the properties of the numerical scheme a point disturbance in an homogeneously deformed plate is investigated. The material of figure 5 in state N° 5 was used for the example, with the higher principal strain C_{22} so that the faster principal longitudinal wave propagates in direction $\mathbf{e}_2$. At time $\tau=0$ an initial disturbance $\bar{v}^2$ is introduced at the centre of the plate. It propagates along the quasi-transversal MONGE-conoid (see fig. 7a). The perspective view and the conture lines clearly show the influence of the lacunae (see fig. 5). The components $H^2{}_1$ and $H^2{}_2$ move also along the inner cone whereas $\bar{v}^1$ (see fig. 7b), $H^1{}_2$ and $H^1{}_1$ propagate along the quasi-longitudinal MONGE-conoid predominantly.

Apparently the numerical scheme maintains the structure of the local wave fronts in the global field. There is no sign of any precursors of the fronts.

Geometrical Focussing in a Linear Elastic Plate

The local stress concentration due to geometrical focussing is calculated for a rectangular, linear elastic plate with one semicircular boundary. The front side (opposite the curved boundary) is subjected to constant stress σ_{22} while all other boundaries are free of stresses. The incident longitudinal wave undergoes a phase shift when reflected at the boundary. Tensile stresses increase when the wave front approaches the geometrical focus. From geometrical accoustics we expect the highest stresses at the cusps of the caustics. The cusp of the caustic of the reflected longitudinal wave is located on the axis of symmetry at 0.5 of the radius.

The sequences in figures 8a,b show the principal stresses σ_I and σ_{II}. The first picture is taken when the whole domain is disturbed. In the following time steps the maximum values move towards the focus and increase. The highest values are reached in the third picture in the assumed area. Then the wave amplitudes decrease.

CONCLUSIONS

The numerical method of bicharacteristics is appropiate to the computation of transient wave motion. Also for the non-linear problem it models the physically anisotropic propagation of waves correctly. The described difference scheme can be applied also to waves in linear transversely isotropic elastic plates. The method proved suitable for strong discontinuities with focussing effects. Furthermore shocks can be included in the numerical scheme as sharp discontinuities as is well-known in gasdynamics.

ACKNOWLEDGEMENTS

We are indebted to the Deutsche Forschungsgemeinschaft which partly supported our work through the SFB 27. We are grateful to Prof. H. Grönig and A. Henckels of the Stoßwellenlabor, Aachen for supplying the shadow photograph.

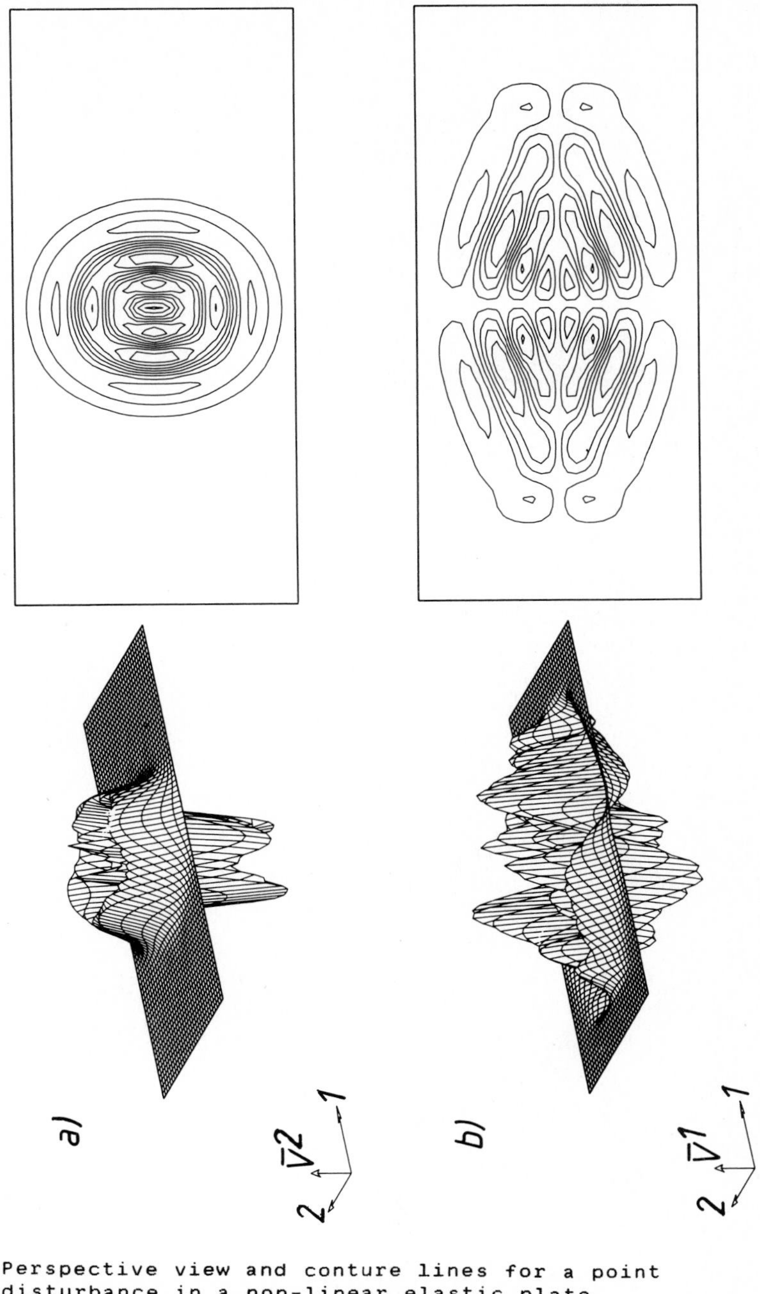

Fig.7. Perspective view and conture lines for a point disturbance in a non-linear elastic plate a) velocity $\bar{v}^2$, b) velocity $\bar{v}^1$

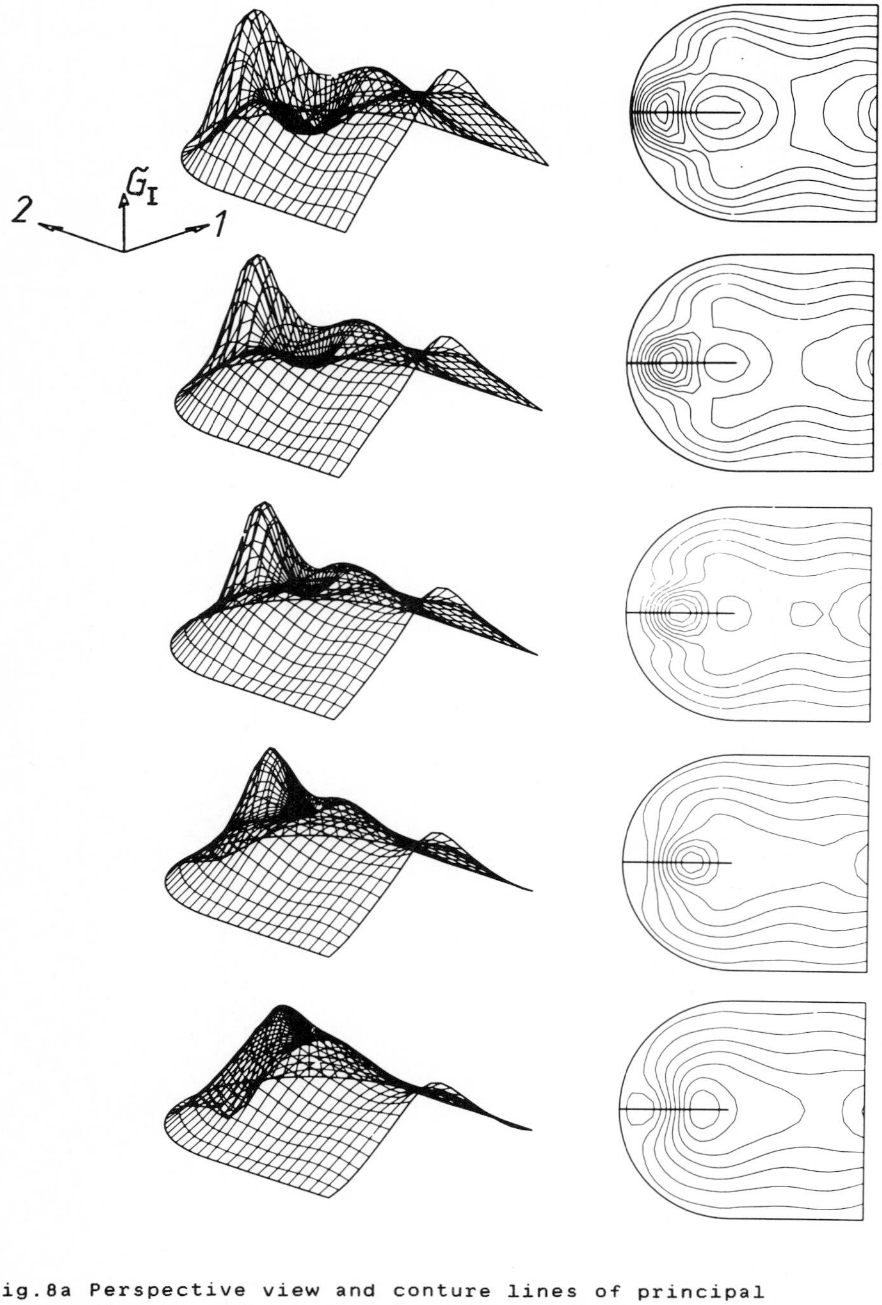

Fig.8a Perspective view and conture lines of principal stress σ_I for successive time steps

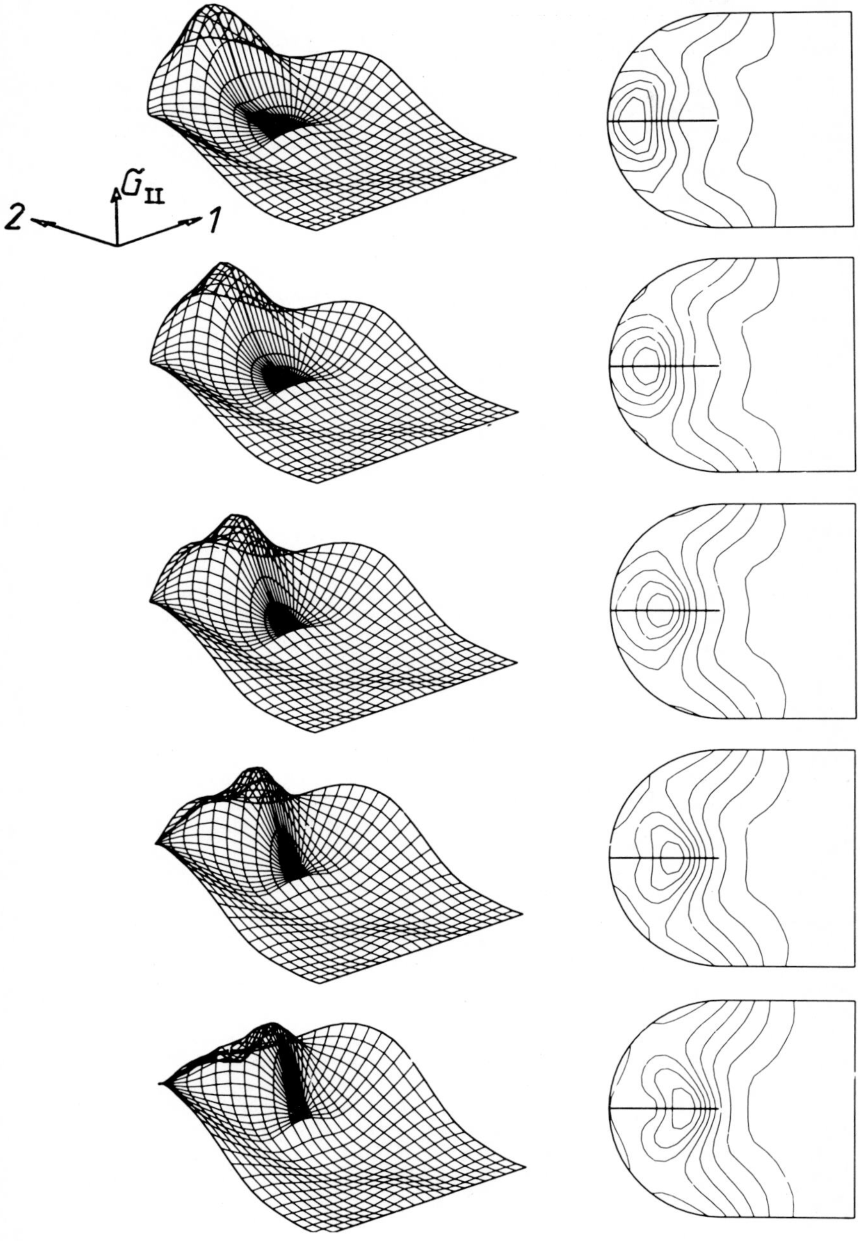

Fig.8b Perspective view and conture lines of principal stress σ_{II} for successive time steps

REFERENCES

1 W.Arrenbrecht and J.Ballmann, in P.L.Butzer and F.Fehėr (Eds.), E.B.Christoffel, Birkhäuser, Basel, 1981, pp.449-460

2 J.Ballmann, H.J.Raatschen and M.Staat, Fokussierung von Spannungswellen in Scheiben, SFB 27-Jahresberichte, Aachen, 1981/82

3 T.Belytschko and R.Mullen, in J.Miklowitz and J.D.Achenbach (Eds.), Modern Problems in Elastic Wave Propagation, Wiley, New York, 1978, 67-82

4 M.Braun, in U.Nigul and J.Engelbrecht (Eds.), Proceedings of the IUTAM Symposium on Nonlinear Deformation Waves 1982 Tallinn, Springer, Berlin, 1983,pp379-384

5 R.Burridge, Lacunas in Two-Dimensional Wave Propagation, Proc. Camb. Phil. Soc.,63 (1967) 819-825

6 D.S.Butler, The Numerical Solution of Hyperbolic Systems of Partial Differential Equations in Three Independent Variables, Proc.Roy.Soc.London, 255A (1960) 232-252

7 R.J.Clifton, A Difference Method for Plane Problems in Dynamic Elasticity, Quart.Appl.Math. 25 (1967) 97-116

8 R.Courant, K.O.Friedrichs and H.Lewy, Über die partiellen Differenzengleichungen der mathematischen Physik, Math.Ann., 100 (1928/29) 32-74

9 D.W.Haines and W.D.Wilson, Strain-Energy Density Function for Rubber-Like Materials,J.Mech.Phys.Solids,27 (1979), 331-343

10 C.Heinz, Theoretische Gasdynamik, Vorlesungen an der RWTH Aachen, 1974/75

11 A.Henckels and H.Grönig, in M.Pichal (Ed.), Proceedings of the IUTAM Symposium on Optical Methods in Dynamics of Fluids and Solids 1984 Prague, Springer, Berlin, to be published

12 A.Jeffrey and T.Taniuti, Non-Linear Wave Propagation, Academic Press, New York, 1964

13 W.Johnson, Impact Strength of Materials, Arnold, London, 1983

14 W.Johnson and A.G.Mamalis, in K.Kawata and J.Shiori (Eds.), Proceedings of the IUTAM Symposium on High Velocity Deformation of Solids 1977 Tokyo, Springer,Berlin,1978,pp228-246

15 R.Löhner, K.Morgan and O.C.Zienkiewicz, The Solution of Non-Linear Hyperbolic Equation Systems by the Finite Element Method, Int.J.Num.Meth.Fluids, 4 (1984) 1043-1063

16 M.J.P.Musgrave, Crystal Acoustics, Holden-Day, San Francisco, 1970

17 N.Nakagawa, Y.Fujiwara and R.Kawai, Proceedings of the 27th Japan Congress on Materials Research, Kyoto, 1984

18 R.G.Payton, Elastic Wave Propagation in Transversely Isotropic Media, Nijhoff, The Hague, 1983

19 R.G.Payton, Two Dimensional Wave Front Shape Induced in a Homogeneously Strained Elastic Body by a Point Perturbing Body Force, Arch.Rational Mech.Anal.,32 (1969) 311-330 and 35 (1969) 402-408

20 R.Sauer, Anfangswertprobleme bei partiellen Differentialgleichungen, Springer, Berlin, 1958

21 R.Sauer, Differenzenverfahren für hyperbolische Anfangswertprobleme bei mehr als zwei unabhängigen Veränderlichen mit Hilfe von Nebencharakteristiken, Numerische Mathematik, 5 (1963) 55-67

22 C.Truesdell, General and Exact Theory of Waves in Finite Elastic Strain, Arch. Rational Mech. Anal., 8 (1961) 263-296

THE LOCAL EFFECTS IN THE LINEAR DYNAMIC ANALYSIS OF STRUCTURES IN THE MEDIUM FREQUENCY RANGE

C.H. SOIZE[1]

[1]ONERA, 29, Avenue de la Division Leclerc, 92320 Châtillon (France)

ABSTRACT

The linear dynamic response of an elastic continuum subjected to a low-frequency localized force is generally global. As the frequency increases to the medium range, the vibratory energy is generally localized around excitation force. In this paper we present a general method for analyzing the dynamic behavior of 3-D elastic structure in this medium frequency range. An example is given and the results are compared with experimental data.

INTRODUCTION

Ideas on the local effects in the medium frequency range.

When analyzing the linear vibrations of an elastic, viscous anisotropic structure occupying a bounded domain in space, slightly damped, it is common to define the low frequency (LF) domain, such that the response to a point force, depends only on the first eigenmodes of vibrations for the associated undamped system. The spectral (or frequency) response then exhibits certain rays from the response of the first isolated eigenmodes of the structure. In this case, the vibratory energy propagates broadly through the structure, because the first eigenmodes are generally global, and the vibratory energy is not localized around the point where the excitation force is located. For instance, let us consider a slender circular shell cylinder with some inside transverse stiffeners and with a floor. The figure 4-a shows the spatial distribution of energy E of the radial acceleration in the low frequency band (300-400 Hz). The excitation is a concentrated force. We see in this figure that the energy propagates all through the structure. The figure 5-a shows the same system for an excitation in the medium frequency (MF) band (2000-2100 Hz). We see that the vibratory energy is localized around the excitation force. The energy does not propagate through the structure. The reason, in this case, is that the modal density is high enough in this frequency band. This example gives us a way of defining what we mean by the local effects in the linear dynamic analysis of structures in the medium frequency range.

To clarify the explanations, we first recall some elements of the linear dynamic analysis of systems with a finite number of degrees of freedom (DOF).

Linear oscillator of finite dimension (ref.4,6,28)

Let $U = (U_1,\dots,U_m) \in \mathbb{C}^m$. We note by the same U symbol the column matrix of the U_j components. The space $\mathbb{C}^m$ is equipped with the scalar product and the associated norm :

$$(U,V) = \langle U,\bar{V} \rangle = \sum_{j=1}^{m} U_j \bar{V}_j \quad ; \qquad ||U|| = (U,U)^{\frac{1}{2}} \tag{1}$$

Let $W = L^2(\mathbb{R},\mathbb{C}^m)$ be the vector space of all $\mathbb{C}^m$-valued function $t \to V(t)$ a.s defined on $\mathbb{R}$ such that $\forall\, j \in \{1,\dots,m\}$, $\int_{\mathbb{R}} |V_j(t)|^2 dt < +\infty$. For any $V \in W$ we denote by $\hat{V} \in W$ the Fourier transform (FT) of V. Let M, C and K be respectively the mass, damping and stiffness (m x m) matrices, which are symmetric, real and positive definite matrices. The linear vibrations of this m-oscillator is governed by :

$$M\,\ddot{U}(t) + C\,\dot{U}(t) + K\,U(t) = F(t) \tag{2}$$

where U, $\dot{U}$ and $\ddot{U}$ are respectively the nodal displacements, velocities and accelerations, and where $F = (F_1,\dots,F_m) \in W$ represents the nodal excitation forces.

Let $\omega \to T(\omega)$ be the frequency response function, defined on $\mathbb{R}$ and with values in the (m x m) complex matrices :

$$\forall\, \omega \in \mathbb{R}, \qquad T(\omega) = [-\omega^2 M + i\omega C + K]^{-1} \tag{3}$$

Let $\hat{U}$ be the FT of U defined by

$$\forall\, \omega \in \mathbb{R}, \qquad \hat{U}(\omega) = T(\omega)\,\hat{F}(\omega) \tag{4}$$

Then $\hat{U} \in W$ and $\hat{U}$ is usually called the forced vibrations. Indeed, if $\underset{\sim}{U}(t)$ is the solution of eq.(2) for $t > 0$, such that $\underset{\sim}{U}(0) = U_0$ and $\dot{\underset{\sim}{U}}(0) = U_1$, then we have $\lim ||U(t) - \underset{\sim}{U}(t)|| = 0$ for $t \to +\infty$, where U is such that its FT $\hat{U}$ is givenby eq.(4).

The generalized eigen-problem associated with the equation (2) is written :

$K\,\Phi_j = \omega_j^2 M\,\Phi_j$ with the normalization $\langle M\Phi_j,\Phi_j\rangle = m_j$. Let $f_j(t) = (F(t),\Phi_j)$ and $u_j(t) = (U(t),\Phi_j)$. Then, by writing $U(t) = \sum_{j=1}^{m} u_j(t)\,\Phi_j$, the equations (2) yields :

$$\mathcal{M}\,\ddot{u}(t) + \mathcal{C}\,\dot{u}(t) + \mathcal{K}\,u(t) = f(t) \tag{5}$$

where the (m x m) real matrices $\mathcal{M}$, $\mathcal{C}$ and $\mathcal{K}$ are such that :

$$\mathcal{M}_{jl} = \delta_{jl}\, m_j, \qquad \mathcal{C}_{jl} = \langle C\,\Phi_j, \Phi_l \rangle \qquad \mathcal{K}_{jl} = \delta_{jl}\, m_j\, \omega_j^2.$$

Let $H(\omega)$ be the (m x m) complex matrix such that :

$$\forall\, \omega \in \mathbb{R}, \qquad H(\omega) = [-\omega^2\mathcal{M} + i\omega\mathcal{C} + \mathcal{K}]^{-1}. \tag{6}$$

Then we obtain for any ω in $\mathbb{R}$:

$$T(\omega) = \sum_{j,l=1}^{m} H(\omega)_{jl}\, \Phi_j\, {}^t\Phi_l. \tag{7}$$

The difficulties of analyzing the structures in the MF domain

Theoretically, the linear vibrations of an elastic, viscous, anisotropic structure occupying a bounded domain in space can be analyzed easily if we know explicitly the spectrum. $\{\omega_j, j \in \mathbb{N}\}$ of eigenfrequencies for the associated undamped system, and the corresponding modal base $\{\phi_j, j \in \mathbb{N}\}$. In practice, for anisotropic elastic structures of any given a priori geometry, the modal base $\{\phi_j\}$ is not explicitly known and must be calculated numerically.

For the high frequency range, the excitation frequencies are high enough that specific methods such as asymptotic methods (ref.5,20,41) or the SEA method (ref.34) can be used. For the LF range, the excitation frequencies are low enough for the response to be of the modal type, i.e. only the first modes of the structure intervene in the response.

We are interested here in the MF, or intermediate frequency range. The first characteristic of this range is that the response brings a very large number of high-order vibration eigenmodes while the modal density can be set high a priori. On the other hand, for this range, we can obtain modal behaviors as in low frequency, locally or not, as well as global dynamic behaviors, grouping by eigenmode packages when the modal density is large enough. All of the behaviors may appear simultaneous at certain frequencies in the MF band considered. Therefore, for this MF range, all the principal elements of the structure, the "exact" geometry, the boundary conditions, play a fundamental part in the response of the mechanical system. Let us note first of all that the structure must be discretized finely into finite elements in this MF range. This leads us to reason using discretized systems with a large number of degrees of freedom. For instance, thirty or fourty thousands DOF may be necessary.

Let us investigate the possibilities of applying the LF method to an MF problem. The m-discretized system is governed by a linear oscillator of m-finite dimension, with a high value of m. In the low frequency range, linear vibrations of structures can classically be solved [see the references] :

M1 - By direct numerical time integration of equation (2),

M2 - By a numerical time integration of equation (5), with m = p, in the truncated modal basis $\{\phi_1,\dots,\phi_p\}$ with $p \ll m$. This is the modal synthesis in the time domain

M3 - By calculating the frequency response function defined by equation (3), which can be : (a) carried out using the truncated modal basis $\{\phi_1,\dots,\phi_p\}$. Then $T(\omega)$ is given by (7) replacing m by p. This is the modal synthesis in the frequency domain ; (b) or carried out directly calculating, for each value of ω considered, the $T(\omega)$ matrix given by (3).

If applied to medium frequencies, the above methods would have to be adapted as follows :

- For a broad-band MF excitation, the method M1 requires a long integration time with a very small integration time step, resulting in a large number of integration time steps. As the order of the matrices is very high (e.g. m = 30000), there are many numerical difficulties (numerical noises, error propagation, etc.) and the data processing gets very involved. The substructuring technique, on the other hand, is easy.
- Methods M2 and M3 a) call for calculating the vibration eigenmodes up to a high order (e.g. 1000 eigenmodes). As the discretized system has a great many DOF's, it seems difficult to determine the necessary truncated modal basis directly and with enough precision to separate the eigenmodes, considering the current state of numerical precision and knowledge of the algorithms used to solve a large generalized eigenvalue problem. The substructuring dynamic analysis is possible but not very easy. At the present time and for the MF problem, the modal approach does not seem really efficient and is very costly in CPU time and computer I/O.
- Method M3 b) involves finding the solution to the complex linear system associated with equ. (4), carried out for each ω. If the calculation is to be carried out for only a few values of ω, this method is then very effective in the MF range. Unfortunately, in the general case, the excitation is broad-band, and many values of ω must then be considered a priori, for dynamic identification.
- We are going proposing here a method of solution for the MF range, based on physical and numerical reasons. The MF vibratory states will be characterized by the energies in a relatively narrow frequency band.

The fact of considering certain narrow bands, makes it possible to build a direct efficiency numerical method based on the technique of multiple scales. Then this numerical approach is a combined time-frequency method using methods M1 et M3 b) simultaneously without getting their disadvantages. The substructuring technique is also very easy to use.

MATHEMATICAL FORMULATION OF THE MF ANALYSIS OF A LINEAR ELASTIC CONTINUUM MEDIUM

Notation and assumptions

We study the linear vibrations of an elastic anisotropic body that, in its reference configuration, occupies an open bounded domain Ω of $\mathbb{R}^3$, with boundary $\partial\Omega$ supposed C^0 and C^1 by parts.

We call $x = (x_1, x_2, x_3)$ a point of $\mathbb{R}^3$, ∂_t (resp. ∂_j) the partial derivative with respect to t (resp. x_j). The common convention of summation on repeated dummy indices is used. Let $u = (u_1, u_2, u_3)$ be the field of displacement of the elastic body. In the part Γ_u of $\partial\Omega$ with a positive measure, we set $u|_{\Gamma_u} = 0$, $\forall$ t. The variational formulation, (ref.9,14,33,40,43) of this elastodynamic problem involves a classical introduction of the following Complex Hilbert Spaces (CHS) $H = \{v \; ; \; v_j \in L^2_{dx}(\Omega, \mathbb{C})\}$ equipped with the scalar product (s.p.) and the associated norm :

$$(v,w)_H = \int_\Omega v_j(x)\, \overline{w_j(x)}\, dx \; ; \qquad ||v||_H = (v,v)_H^{\frac{1}{2}} \tag{8}$$

and the CHS V :

$$V = \{v \; ; \; v_j \in H^1_{dx}(\Omega,\mathbb{C}) \; ; \; v_j|_{\Gamma_u} = 0\} \tag{9}$$

equipped with the s.p. of H^1 :

$$(v,w)_{H^1} = (v,w)_H + \sum_{j=1}^{3} (\partial_j v, \partial_j w)_H \tag{10}$$

where $\partial_j v = (\partial_j v_1, \partial_j v_2, \partial_j v_3)$. We have $V \subset H = H' \subset V'$ where V' is the dual of V, H is identified to its dual H'. The space V is a dense subset of H and the injection $V \hookrightarrow H$ is compact.

In our elastodynamic problem, we must introduce another CHS, because the time parameter intervenes. Let X be a CHS equipped with the s.p. $(v,w)_X$ and the norm $||v||_X$. If $X = \mathbb{C}$, the s.p. is $v\bar{w}$ and the norm is $|v|$. If $X = H$ the s.p. is given by (8). Let $W^X = L^2(\mathbb{R},X)$ the CHS of the X-valued functions dt-almost everywhere (a.e.) defined on $\mathbb{R}$, equipped with the s.p. and the norm :

$$(u,v)_{W^X} = \int_{\mathbb{R}} (u(t), v(t))_X\, dt, \quad ||u||_X = (u,u)_X^{\frac{1}{2}} \tag{11}$$

For any u in W^X, the FT with respect to t is $\hat{u}(w) = \mathcal{F}u \in W^X$ such that for a.e. ω in $\mathbb{R}$:

$$\hat{u}(\omega) = \int_{\mathbb{R}} e^{-i\omega t}\, u(t)dt, \qquad u(t) = \frac{1}{2\pi}\int_{\mathbb{R}} e^{i\omega t}\, \hat{u}(\omega)d\omega \tag{12}$$

Using Plancherel's formula, we have $\forall$ u, v $\in W^X$:

$$(u,v)_{W^X} = \frac{1}{2\pi}(\hat{u},\hat{v})_{W^X}, \quad ||u||^2_{W^X} = \frac{1}{2\pi}||\hat{u}||^2_{W^X} \tag{13}$$

Let K be a compact subset of $\mathbb{R}$. For instance, we shall take $K = [\omega_1,\omega_2]$ a closed bounded interval of $\mathbb{R}$. We denote by W^X_K the subset of W^X such that $\forall$ u $\in W^X_K$, the FT $\hat{u}$ has the compact support K. Consequently, $\forall\, \omega \notin K$, $\hat{u}(\omega) = 0$. Therefore, we have : $W^X_K = \{u \in W^X \; ; \; \mathrm{supp}_\omega\, \hat{u} = K\}$, and W^X_K is a CHS equipped with the s.p. and the norm :

$$(u,v)_{W^X_K} = \int_{\mathbb{R}} (u(t),v(t))_X\, dt = \frac{1}{2\pi}\int_K (\hat{u}(\omega),\hat{v}(\omega))_X d\omega \tag{14}$$

$$||u||_{W^X_K} = (\int_{\mathbb{R}} ||u(t)||^2_X\, dt)^{\frac{1}{2}} = (\frac{1}{2\pi}\int_K ||\hat{u}(\omega)||^2_X\, d\omega)^{\frac{1}{2}} \tag{15}$$

when $X = H$, (resp. $X = \mathbb{C}$ we denote by W_K^H (resp. $W_K^{\mathbb{C}}$) the space W_K^X.

Mass, damping, stiffness and observation operators

Let M (resp. C), the linear mass (damping resp.) operator, be real continuous, symmetric and positive definite on H. We then have for the mass operator :

$$\forall\ v,w \in H, \quad (Mv,w)_H = \int_\Omega \rho(x)\ v_j(x)\ \overline{w_j(x)}\ dx \tag{16}$$

where ρ is the density of the elastic medium, which verifies the hypothesis $\forall$ $x \in \Omega$, $0 < \rho_1 \leqslant \rho(x) \leqslant \rho_2 < +\infty$.

Let K be the linear stiffness operator such that :

$$\forall\ v,w \in V,\ (Kv,w)_{V',V} = \int_\Omega a_{ijkh}(x)\ \varepsilon_{kh}(v)\ \varepsilon_{ij}(\bar{w})\ dx, \tag{17}$$

where $\varepsilon_{ij} = (\partial_j v_i + \partial_i v_j)/2$. On the strain tensor, the elasticity constants $a_{ijkh}(x)$ verify the usual properties of symmetry and positiveness. Under these conditions the linear stiffness operator K is continuous, real, symmetric from V to V', and we have for a real positive constant μ :

$$\forall\ v \in V\ ,\ (Kv,v)_{V',V} \geqslant \mu\ ||v||_V^2\ , \tag{18}$$

For a.a. x fixed in Ω, we introduce the linear observation $Q_x : H \rightarrow \mathbb{C}$ such that, $\forall\ v \in H$:

$$q_x = Q_x v = \sum_{j=1}^{3} \alpha_j(x)\ v_j(x) \tag{19}$$

where $x \rightarrow \alpha(x) = (\alpha_1,\alpha_2,\alpha_3)$ is a given a.e continuous $\mathbb{R}^3$-valued function defined on $\bar{\Omega}$. Then if $u \in W^H$, the observation $q_x = Q_x u \in W^{\mathbb{C}}$ and the FT $\hat{q}_x = Q_x\ \hat{u} \in W^{\mathbb{C}}$.

Frequency response operator

We give the extension of (3) for the infinite dimension case. We have the following result, proved in (ref.47,32).

Under the previous assumptions, $\forall\ \omega \in \mathbb{R}$, the linear operator $-\omega^2 M + i\omega C + K$ in H having the domain Dom K = $\{v \in V,\ Kv \in H\}$ allows as an inverse the compact operator T_ω from H to H :

$$T_\omega = (-\omega^2 M + i\omega C + K)^{-1} \tag{20}$$

Definition of the MF vibrations

As we indicated in the introduction, we are going to study MF vibration by

frequency bands. Let $B_n = [-\Delta\omega/2 + \Omega_n, \Delta\omega/2 + \Omega_n]$ be a MF band with a central frequency Ω_n and a bandwidth $\Delta\omega$: We let $\Omega_0 = 0$ and $B_0 = [-\Delta\omega/2, \Delta\omega/2]$.
To characterize the MF vibrations of the elastic medium in the band B_n, we consider the excitation class of the function $t,x \longrightarrow F_n(t,x)$ belong to $W^H_{B_n}$ and such that :

$$F_n = f_n \otimes \Psi \;; \qquad f_n \in W^{\mathbb{C}}_{B_n} \;;\; \Psi \in H \tag{21}$$

For instance, by using $f_0 \in W^{\mathbb{C}}_{B_0}$ to denote the function :

$$f_0(t) = (\pi t)^{-1} \sin(t\Delta\omega/2) \tag{22}$$

we see that $f_n(t) = f_0(t)\exp(i\Omega_n t) \in W^{\mathbb{C}}_{B_n}$, and is such that : $\hat{f}_n(\omega) = \mathbb{1}_{B_n}(\omega)$; where $\mathbb{1}_{B_n}(\omega) = 1$ if $\omega \in B_n$ and $= 0$ if $\omega \notin B_n$.
For any given $F_n = f_n \otimes \Psi \in W^H_{B_n}$ we define the MF vibration u_n in the band B_n such that its FT $\hat{u}_n$ verifies :

$$\omega \longrightarrow \hat{u}_n(\omega) = T_\omega \hat{F}_n(\omega) = \hat{f}_n(\omega)\,(T_\omega \Psi) \tag{23}$$

where T_ω is the frequency operator defined by eq. (20).
As $\hat{f}_n \in W^{\mathbb{C}}$ has compact support B_n, we can easily verify that the functions.

$$u_n,\ \partial_t u_n,\ \partial^2_t u_n \in W^H_{B_n} \tag{24}$$

The vibration u_n defined by (23) is interpreted as the "forced" solution of :

$$M\,\partial^2_t u_n + C\,\partial_t u_n + K\,u_n = f_n \otimes \Psi \tag{25}$$

where $u_n \in W^V_{B_n} \subset W^H_{B_n}$. In the same way, if $q_x = Q_x u_n$ is an observation of the system, we have for a.a. x fixed in Ω :

$$\forall\, l \in \{0,1,2.\},\ q^{(l)}_x = \frac{d^{(l)}}{dt^l} q_x \in W^{\mathbb{C}}_{B_n} \tag{26}$$

We now introduce some energy quantities that will be used to characterize the vibratory state of the mechanical system in the band B_n.

<u>Global energetic characteristics in B_n</u> : Let $u_{n,j} \in W^H_{B_n}$ be the MF vibration due to the excitation $F_{n,j} = f_{n,j} \otimes \Psi_j \in W^H_{B_n}$ and $q^{(l)}_{x,j} = Q_x\, \partial^l_t u_{n,j} \in W^{\mathbb{C}}_{B_n}$ the associated observation for $l \in \{0,1,2\}$.

Then the energy of interaction of $q^{(l)}_{x,j}$ and $q^{(l)}_{x',j'}$ is defined by :

$$E^n_{q^{(l)}_{x,j}\, q^{(l)}_{x',j'}}(\Psi_j, \Psi_{j'}) = (q^{(l)}_{x,j}, q^{(l)}_{x',j'})_{W^{\mathbb{C}}_{B_n}} \tag{27}$$

and the energy of the observation $q^{(l)}_{x,j}$ is defined by :

$$E^n_{q^{(l)}_{x,j}}(\Psi_j) = E^n_{q^{(l)}_{x,j}\,q^{(l)}_{x,j}}(\Psi_j,\Psi_j) = ||q^{(l)}_{x,j}||^2_{W^{\mathbb{C}}_{B_n}} \tag{28}$$

Let Ω_0 be a subset of Ω. The spatial propagation of the energy of the observation $q^{(l)}_j$ in Ω_0 is found by studying the function $x \rightarrow E^n_{q^{(l)}_{x,l}}(\Psi_j) : \Omega_o \rightarrow \mathbb{R}^+$.

Frequency energy characteristic in B_n : Using the relations (19,23,24), we obtain for $l \in \{0,1,2\}$:

$$\hat{q}^{(l)}_{x,j}(\omega) = H_{j,l,x}(\omega)\,\hat{f}_{n,j}(\omega) \text{ with } H_{j,l,x}(\omega) = (i\omega)^l\, Q_x\, T_\omega \Psi_j \tag{29}$$

The function $\omega \rightarrow H_{j,l,x}(\omega)$ is the $\mathbb{C}$-valued frequency response function defined on $\mathbb{R}$, of the convolution linear filter with input f_n and output $q^{(l)}_{x,j}$, for x, l and Ψ_j fixed.
Let $\{B_{n,k}\}_k$ be a finite sequence of intervals of B_n such that $B_n = \cup_k B_{n,k}$, $\cap_k B_{n,k} = \emptyset$. Then the energy of $q^{(l)}_{x,j}$ in the band $B_{n,k}$ is given by :

$$E^{n,k}_{q^{(l)}_{x,j}}(\Psi_j) = \frac{1}{2\pi}\int_{B_{n,k}} |\hat{q}^{(l)}_{x,j}(\omega)|^2\, d\omega \tag{30}$$

The frequency response function expressed in terms of energy is found by studying the function $k \rightarrow E^{n,k}_{q^{(l)}_{x,j}}(\Psi_j)$.

Cases of general deterministic and random excitations

In this paragraph we examine how certain general deterministic and random excitations can be studied once the MF characteristics are defined, for the particular excitation of the form $F_n = f_n \otimes \Psi$.

Case of a general deterministic excitation on a band B_n : In the general case $F_n \in W^H_{B_n}$. Let $\{\Psi_j\}$, $j \in \mathbb{N}^*$ be an orthonormal basis of the Hilbert space H. Then, using the results of ref.45, we can write :

$$F_n = \sum_{j=1}^{\infty} f_{n,j} \otimes \Psi_j \text{ with } f_{n,j} = (F_n, \Psi_j)_H \in W^{\mathbb{C}}_{B_n} \tag{31}$$

Let u_n be the MF vibration due to the excitation F_n and $q^{(l)}_x$ an associated observation. Using (27) and (28) we see that the energy of $q^{(l)}_x$ is given by :

$$E^{n}_{q_x^{(1)}}(F_n) = \sum_{j,j'=1}^{\infty} E^{n}_{q_{x,j}^{(1)}\, q_{x,j'}^{(1)}}(\Psi_j, \Psi_{j'}) \tag{32}$$

Case of a particular random excitation on band B_n : Let $\{\zeta_n(t),\ t \in \mathbb{R}\}$ be a $\mathbb{C}$-valued second order, continuous, centered, stationary random process, defined on $(\tilde{\Omega}, \mathcal{C}, P)$. Let $\mathcal{E}$ be the mathematical expectation. We assume that its power spectral measure has a density $\mu_{\zeta_n}(d\omega) = S_{\zeta_n}(\omega)\, d\omega$ such that 1) $\mathrm{Supp}_\omega\, S_{\zeta_n} = B_n$ and 2) $S_{\zeta_n} \in L^1(\mathbb{R}, \mathbb{R}^+)$. Under these assumptions, the autocorrelation function of the process $\zeta_n(t)$ is written as :

$$R_{\zeta_n}(\tau) = \mathcal{E}(\zeta_n(t+\tau)\overline{\zeta_n(t)}) = \int_{B_n} e^{i\omega\tau} S_{\zeta_n}(\omega)\, d\omega \tag{33}$$

and there exists (see ref.32) a function $f_n \in W^{\mathbb{C}}_{B_n}$ such that :

$$S_{\zeta_n}(\omega) = \frac{1}{2\pi} |\hat{f}_n(\omega)|^2, \qquad \omega \in \mathbb{R} \tag{34}$$

We consider, on the MF band B_n, a random field $\{\mathbb{F}_n(x,t), x \in \Omega, t \in \mathbb{R}\}$ such that : $\mathbb{F}_n = \zeta_n \otimes \Psi$ where Ψ is a given function in H. Then the observation $x,t \rightarrow q_x^{(1)}(t)$ is a random field denoted by $\{\xi^{(1)}(x,t), x \in \Omega, t \in \mathbb{R}\}$. Then, using the previous results, we can easily verify that $\xi^{(1)}(x,t)$ is a $\mathbb{C}$-valued second order, continuous, centered, stationary in time random field, such that its cross correlation function is written as :

$$R_{\xi^{(1)}}(x,x',\tau) = \mathcal{E}(\xi^{(1)}(x,t+\tau)\, \overline{\xi^{(1)}(x',t)}) = \int_{B_n} e^{i\omega\tau} S_{\xi^{(1)}}(x,x',\omega)\, d\omega \tag{35}$$

and such that the cross spectral density function is written as :

$$S_{\xi^{(1)}}(x,x',\omega) = \frac{1}{2\pi} \hat{q}_x^{(1)}(\omega)\, \overline{\hat{q}_{x'}^{(1)}(\omega)} \tag{36}$$

where $q_x^{(1)} \in W^{\mathbb{C}}_{B_n}$ is the deterministic MF observation associated with the MF vibration $u_n \in W^{H}_{B_n}$, due to the MF deterministic excitation $F_n = f_n \otimes \Psi \in W^{H}_{B_n}$.

From the relations (35,36,27), we obtain :

$$R_{\xi^{(1)}}(x,x',0) = (q_x^{(1)}, q_{x'}^{(1)})_{W_n^{\mathbb{C}}} = E^{n}_{q_x^{(1)}\, q_{x'}^{(1)}}(\Psi,\Psi) \tag{37}$$

$$\mathcal{E}(|\xi^{(1)}(x,t)|^2) = E^{n}_{q_x^{(1)}}(\Psi) = \| q_x^{(1)} \|^2_{W^{\mathbb{C}}_{B_n}} \tag{38}$$

Consequently, we see that the second order characteristics of the random field $\xi^{(1)}(x,t)$ can be computed directly from the deterministic MF characteristics.

Case of a general random excitation in a band B_n : Using the results established previously, one can solve the MF problem for an excitation F_n modelized by a second order, continuous, centered, stationary in time random field $\{\mathbb{F}_n(x,t), x \in \Omega, t \in \mathbb{R}\}$. We have developed a mathematical and numerical method for studying this case, but these developments are too long to be put forth in the present paper.

Case of a broad medium frequency band : Let us assume that the energy of the deterministic or random excitation is in a broad band I. We shall see in the next section that the numerical method is based on the use of narrow MF bands B_n. We can consider some sufficiently narrow MF bands B_n such that $I = \cup_n B_n$, $\cap_n B_n = \emptyset$, and, as the problem is linear, we shall build the solution on I by studying the MF solution for the bands B_n.

ACTUAL CONSTRUCTION OF THE MEDIUM FREQUENCY SOLUTION

In this section we develop an appropriate numerical method for computing directly the quantities (27,28,30) that the MF response of the mechanical system. For a given band B_n, the time part of the excitation is $f_n \in W^{\mathbb{C}}_{B_n}$. In the MF domain, we have by definition $\Omega_n/\Delta\omega >> 1$. Then, the function $t \rightarrow f_n(t)$ presents two scales of time : a large scale $\tau_L = 2\pi/\Delta\omega$ and a small scale $\tau_n = 2\pi/\Omega_n$, because $\tau_n/\tau_L << 1$. Consequently the numerical method consists of using a multiple scale technique for the time parameter. The small scale is directly treated in the frequency domain. Then, no approximation will be introduced for the rapid time variations. The large scale will be treated numerically in the time domain with a step-by-step implicit time integration method. Then the time numerical approximation will only carry on the low time variations.

Low frequency equation associated with the MF problem

Let $f_{n,j}$ be in $W^{\mathbb{C}}_{B_n}$ and let $f_{o,j} \in W^{\mathbb{C}}_{B_o}$ be the function such that :

$$f_{o,j}(t) = f_{n,j}(t)\exp(-i\Omega_n t) \tag{39}$$

Then $f_{o,j}$ is an LF signal associated with the MF signal $f_{n,j}$, because $\text{supp}_\omega \hat{f}_{o,j}(\omega) = B_o = [-\Delta\omega/2, \Delta\omega/2]$. Its time scale is $\tau_L = 2\pi/\Delta\omega$. The MF vibration $u_{n,j} \in W^H_{B_n}$ defined by (23) due to the MF excitation $f_{n,j} \otimes \Psi_j \in W^H_{B_n}$ is searched for in the form :

$$u_{n,j}(t) = u_{o,j}(t)\exp(i\Omega_n t) \tag{40}$$

where $u_{o,j} \in W^H_{B_o}$ is the associated LF vibration defined by its FT :
$\hat{u}_{o,j}(\omega) = \hat{f}_{o,j}(\omega)\,(T_{\omega+\Omega_n}\Psi_j)$ with $\mathrm{supp}_\omega\, \hat{u}_{o,j}(\omega) = B_o$.

For $l \in \{0,1,2\}$ we define the function $v^o_{l,j} \in W^H_{B_o}$ such that :

$$v^o_{o,j} = u_{o,j}$$

$$v^o_{1,j} = \partial_t u_{o,j} + i\Omega_n u_{o,j} \tag{41}$$

$$v^o_{2,j} = \partial^2_t u_{o,j} + 2i\Omega_n \partial_t u_{o,j} - \Omega^2_n u_{o,j}$$

Then, we have for $l \in \{0,1,2\}$:

$$\partial^l_t u_{n,j}(t) = v^o_{l,j}(t)\exp(i\Omega_n t) \tag{42}$$

Let $q^{(l)}_{x,j} = Q_x \partial^l_t u_{n,j}$ be an observation. We obtain :

$$q^{(l)}_{x,j}(t) = Q_x v^o_{l,j}(t)\exp(i\Omega_n t) \tag{43}$$

$$M\partial^2_t u_{o,j} + (C+2i\Omega_n M)\partial_t u_{o,j} + (K+i\Omega_n C-\Omega^2_n M)u_{o,j} = \Psi_j \otimes f_{o,j} \tag{44}$$

by substituting (40) into (25), we see that $u_{o,j}$ is the "forced" solution.

The equation (44) is the LF equation for the MF problem.

Expression of the MF characteristics as a function of the LF quantities

The sampling theorem (ref.17,46,47) yields the following results :

The energy of interaction defined by (27) is written as :

$$E^n_{q^{(l)}_{x,j}\,q^{(l)}_{x',j'}}(\Psi_j,\Psi_{j'}) = \tau_L \sum_{m\in\mathbb{Z}} (Q_x v^o_{l,j}(m\tau_L))\overline{(Q_{x'} v^o_{l,j'}(m\tau_L))} \tag{45}$$

In particular the energy of the observation $q^{(l)}_{x,j}$ defined by (28) is given by :

$$E^n_{q^{(l)}_{x,j}}(\Psi_j) = \tau_L \sum_{m\in\mathbb{Z}} |Q_x v^o_{l,j}(m\tau_L)|^2 \tag{46}$$

The FT of the observation $q^{(l)}_{x,j}$ is such that, for a.a $\omega \in \mathbb{R}$:

$$\hat{q}^{(l)}_{x,j}(\omega) = \tau_L \mathbb{1}_{B_n}(\omega) \sum_{m\in\mathbb{Z}} Q_x v^o_{l,j}(m\tau_L)\exp\{-im\tau_L(\omega-\Omega_n)\} \tag{47}$$

Formula (47) can be used to calculate the frequency response function in terms of energy, defined by (30). Let us not that the series on the right side of equation (47) are convergent in $W^{\mathbb{C}}$. Equations (45) to (47) can also be used to determine all the MF quantities of interest directly, with only the sampled time "forced" solution u_o of the associated LF equation (44), taking equation (41) into account.

Numerical analysis

As the mechanical structures considered can be anisotropic and of any geometry, only finite dimension approximations of these operators can be obtained to represent them. Naturally, to approach the operators M, C and K, we will use the usual method of finite elements (ref.3,4,18,28,37,59). For the MF domain on generally obtains a large number of DOF and the only problem with using this method is selecting the fineness of the grid in the various parts of the structure.

Let us note that all of the approximation procedures for time intergration of equation (44) can be analyzed by not introducing the finite dimension approximation of operators M, C and K.

As the time integration of equation (44) cannot be carried out numerically over all of $\mathbb{R}$, it will be carried out over a bounded time interval $[t_I, t_S]$ where $t_I<0$ and $t_S>0$. To optimize the cost of numerical calculation, $|t_S-t_I|$ must be as small as possible. The selection of t_I is thus related to the asymptotic behaviour of the function f_o for $t \to -\infty$. But since $f_o \in W^{\mathbb{C}}_{B_o}$, we know that for any fixed positive ε as small as we want, $\exists\, t_I$ such that $\int_{-\infty}^{t_I} |f_o(t)|^2 dt < \varepsilon$.

In the same way we can make the energy not taken into account in the interval $(t_S, +\infty)$ as small as we want. But the choice of t_S is also related to the dynamics of the system governed by the structural damping. Consequently t_S must also be chosen so that

$$\int_{t_S}^{+\infty} ||u_o||_H^2 \, dt < +\infty.$$

This is possible, because $u_o \in W^H_{B_o}$.

Practically, t_S must be determined by energy budget considerations. In summary, the proposed numerical method consists of :

1 - calculating the $f_{o,j}$ function associated with $f_{n,j}$ by eq. (39) and to determining t_I,

2 - building, by the finite element method, the finite approximation of the M, C and K operators,

3 - solving, by a numerical method, the LF equation (44) for $t \in [t_I, t_S]$ with the initial conditions $u_o(t_I) = \partial_t u_o(t_I) = 0$. This numerical method must be based on an implicit step-by-step time integration scheme, unconditionally stable (for instance, the Newmark scheme). The integration time step $\Delta t = \tau_L/m_T$ must be such that $m_T > 1$,

4 - computing the MF quantities defined by eq. (45) to (47), with (41) and with the numerical solution $u_o(m\tau_L)$ sampled in time which has been obtained. The choice of t_I and t_S directly gives the truncations to be carried out on the sums $\sum m \in \mathbb{Z}$.

Implantation of the method in a finite element program

Let [M], [C] and [K] be the (m x m) matrices of the m-finite approximation of the M, C and K operators, obtained by the finite element method. Let us introduce the (m x m) matrices :

$$[\mathcal{C}_n] = [C] + 2i\Omega_n[M] \; ; \qquad [\mathcal{K}_n] = [K] + i\Omega_n[C] - \Omega_n^2[M] \tag{48}$$

Then, the space approximation of (44) gives :

$$[M]\,\ddot{U}_{o,j} + [\mathcal{C}_n]\,\dot{U}_{o,j} + [\mathcal{K}_n]\,U_{o,j} = F_{o,j} \tag{49}$$

where $[\mathcal{C}_n]$ and $[\mathcal{K}_n]$ are complex, symmetric matrices which have the same band structure as the matrices [M], [C] and [K].

Consequently, incorporating the proposed method in a general finite element program does not modify its structure, as long as the program can take into account consistent mass and damping matrices and can solve an equation of the type (49) with an implicit step-by-step time integration scheme. The program simply has to be adapted for the complex values. Furthermore, the substructuring technique is very easy to use.

At ONERA, we have developed this method in the ADINA program (ref.2) and we have a scalar version on a CYBER 170/750 computer, and two optimized vectorial versions : one one a CRAY 1 and the other on a CYBER 205. The method has been checked by comparing it with some analytical results to the numerical results 1) for simple structures (plates, shell cylinders, etc) (ref.47,48), 2) for real structures (three dimensional stiffened shell cylinders in different configurations) (ref.49,39).

MF LOCAL EFFECTS IN 3D STIFFENED SHELL CYLINDERS

In this section we give an example of the local effects in the linear dynamic analysis of a structure in the medium frequency range. The structure is a slender steel circular shell cylinder having 56 inside transverse stiffeners, a floor, two bottoms and internal subsystems. This structure is a three dimensional linear elastic medium with an axisymmetric part. The geometry of this structure is defined in Fig. 1 (the internal subsystems are not represented in this figure).

The calculations were made with the MF method developed in this paper, and using the ADINA-ONERA program (ref.48) on a CRAY 1 computer.

The calculated results, which are given in the figures, are taken from study (ref.49). The experimental data, from ONERA, are taken from study (ref.30).

Model by the finite element method

This structure is a 3-D elastic medium with an axisymmetric part. It is possible to mesh all the structure with 3-D finite elements ; but for a global model having a fixed number of DOF , the approximation fits better if the axisymmetric part is meshed with axisymmetric finite elements. We used this latter approach.

We then meshed the axisymmetric part with axisymmetric solid finite elements for some circumferential numbers $N = \{0,1,2,....,12,13\}$. (The displacement field u in cylindrical coordinates is developed in Fourier series : $\vec{u}(r,\Theta,z) = \sum_{N=-13}^{13} \vec{u}_N(r,z)e^{iN\Theta}$.
We therefore implanted in ADINA-ONERA some special finite elements which do not exist in the ADINA version (ref.2). The 3-D part of the structure was meshed with 3-D finite elements (e.g. plate elements for the floor). The axisymmetric part was assembled with the 3-D part using constraint equations which the ADINA-program automatically takes into account. The final model contains 30000 degrees of freedom and is described with 14 substructures and a master structure (we used the substructuring technic for the MF dynamic analysis). Consistent mass matrices are used.

Medium frequency calculations

The dynamic calculations are made for the global frequency band B = [0-3600] Hertz. This band is partioned into 36 MF bands B_n. For each band B_n, the bandwith is 100 Hertz. The spatial part of the excitation is a concentrated force located in the central plane of symmetry, with an amplitude of 1 Newton. (see Fig. 2). The time part $f_n(t)$ is the function such that $\hat{f}_n(\omega) = \mathbb{1}_{B_n}(\omega)$.

The generated damping matrix is $[C\ ] = 2\ \xi\ \Omega_n\ [M]$ with $\xi = 0.006$ and Ω_n the central angular frequency of the band B_n. The observations of the mechanical system are the radial accelerations :

1) at certain points on the shell located on the inside shell generatrix line (this line belongs to the plane containing the force), and

2) at certain points of the shell located in the three crowns M, N and P (see Fig. 2). The frequency response functions expressed in terms of energy are calculated by the formulae (30) and (47), and the energy of the radial acceleration in each band B_n is computed by the formula (46).

Numerical results obtained and experimental comparisons

The fundamental eigenfrequency of vibration of this global system is around 80 Hertz.

We give only partial results of the complete analysis, as the available space here is limited. The experimental data concern only the 2000-36000 Hertz frequency band. It is for this reason that the experimental points are not plotted on Fig. 4.

Fig. 4.a and b show that, in the LF domain, the vibratory energy propagates broadly through the structure. As the frequency increases, reaching the MF domain, Figs. 5 to 6 show that the vibratory energy is partially or completely localized around of the applied force. A detailed examination of these figures shows that the phenomena of the spatial propagation in this structure are relatively complicated.

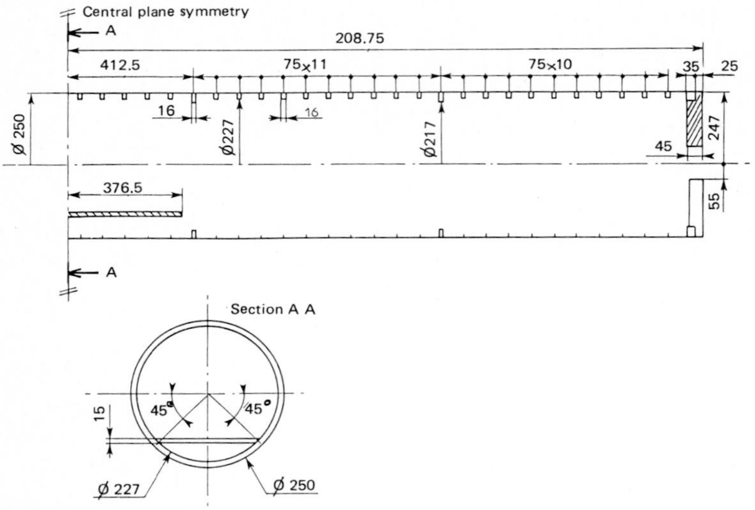

Fig. 1. Definition of the structural geometry

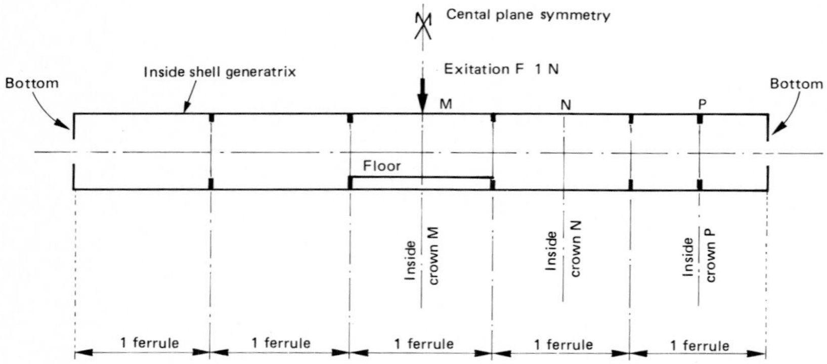

Fig. 2. Definition 1) of the spatial excitation (concentrated force), 2) of the observation points (inside crowns M, N, P and inside generative line of the shell)

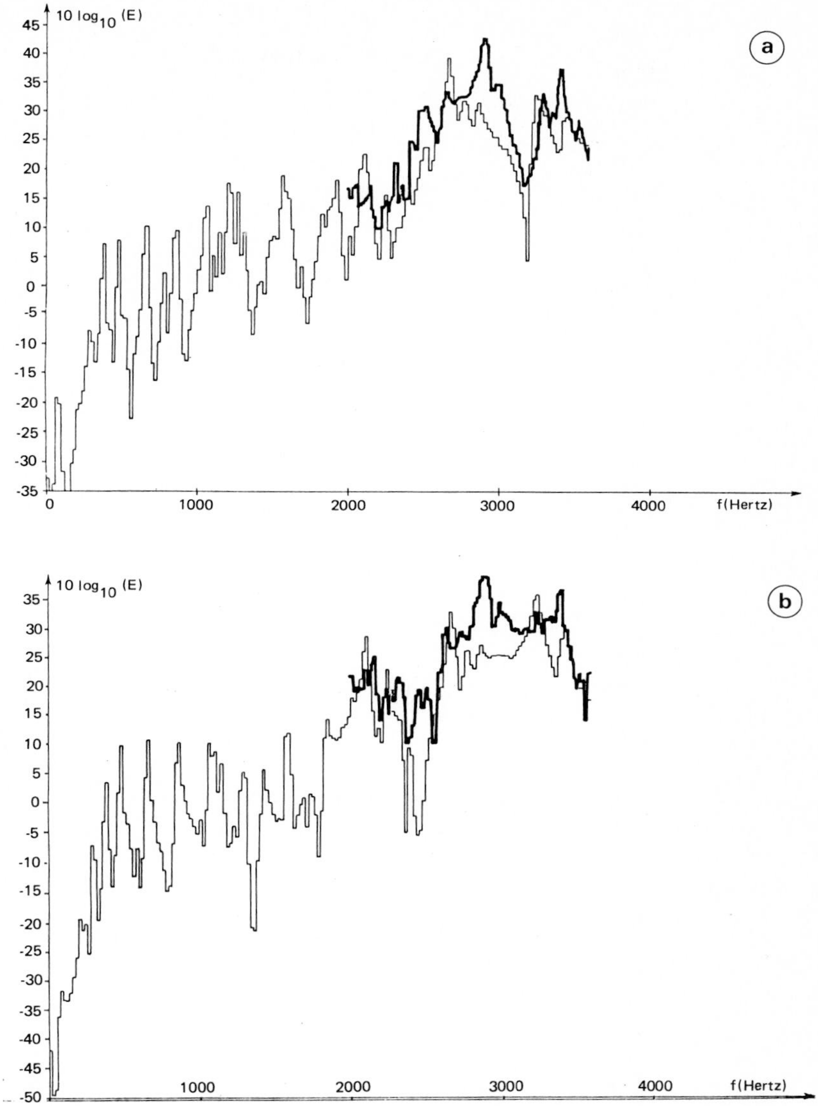

Fig. 3. Frequency response function expressed in terms of energy. (a)-(b) : E = Energy of the radial acceleration integrated by frequency band of 20 Hertz, at two observation points on the shell.

_____ Experimental

_____ Calculation

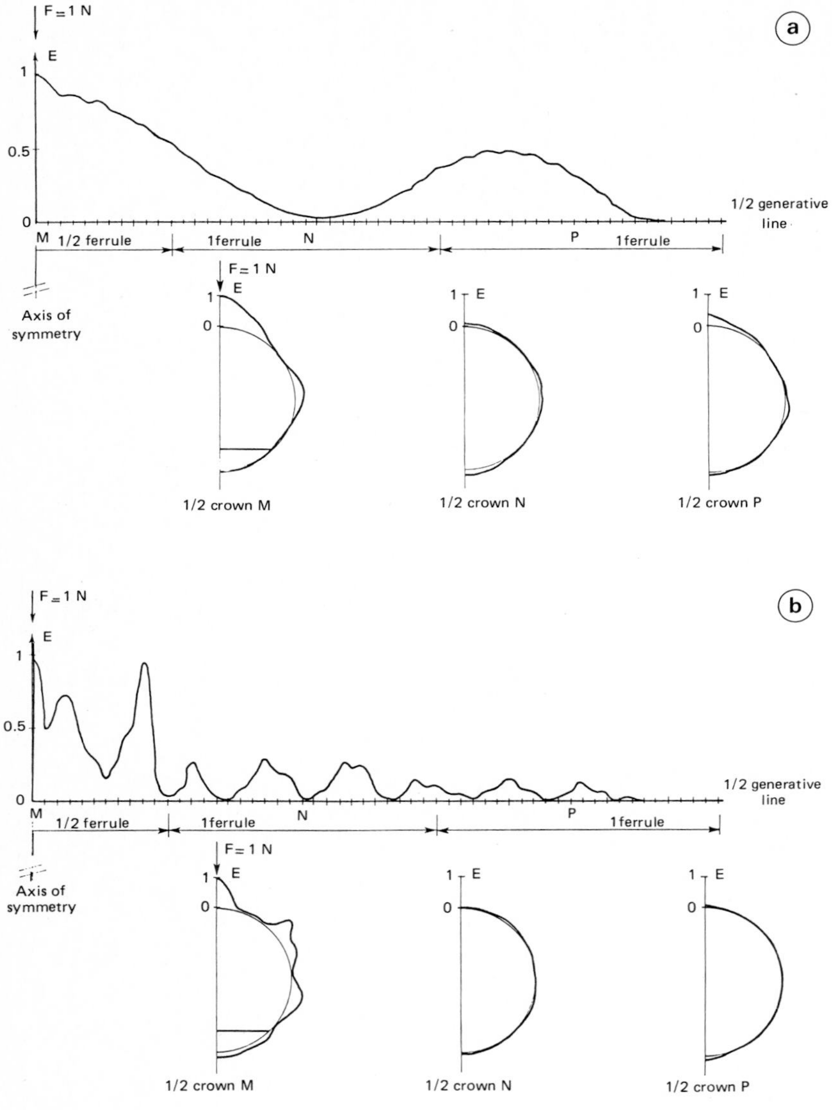

Fig. 4. Spatial propagation of the energy of the radial acceleration E
—— Calculation
(a) Frequency band [300- 400] Hertz, E_{max} = 10.8
(b) Frequency band [1300-1400] Hertz, E_{max} = 79.8

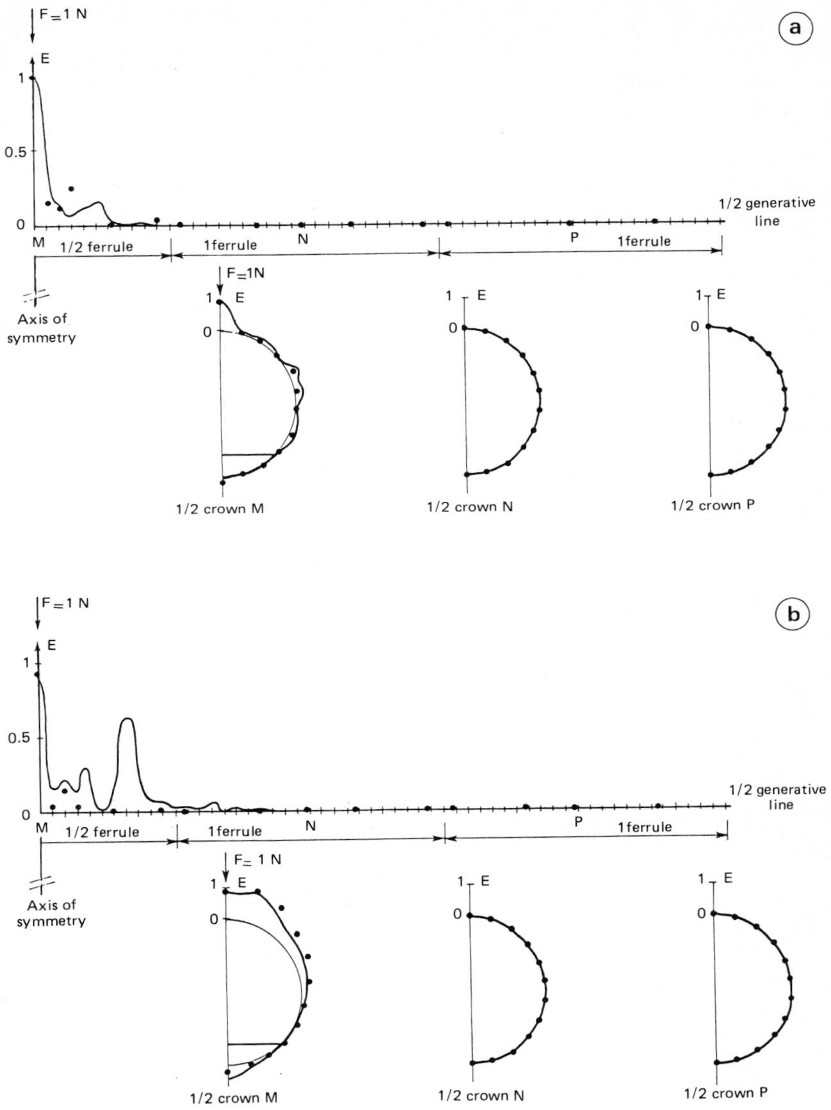

Fig. 5. Spatial propagation of the energy of the radial acceleration E
_____ Calculation
.... Experimental
(a) Frequency band [2000-2100] Hertz, E_{max} = 2932
(b) Frequency band [2700-2800] Hertz, E_{max} = 6062

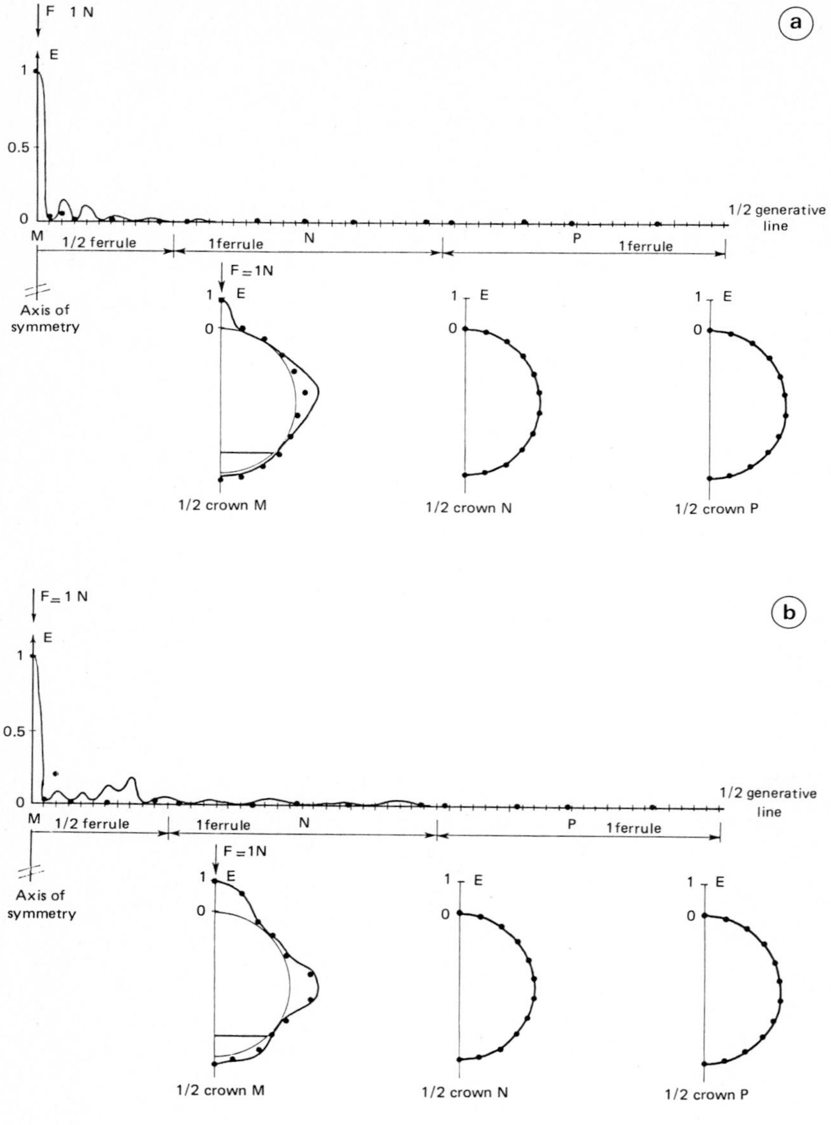

Fig. 6. Spatial propagation of the energy of the radial acceleration E
_____ Calculation
.... Experimental
(a) Frequency band [3100-3200] Hertz, E_{max} = 3897
(b) Frequency band [3400-3500] Hertz, E_{max} = 3845

CONCLUSIONS

We have developed a method for analyzing the dynamics of structures in the medium frequency range, good for some general heterogeneous anisotropic 3-D structures. From a mechanical point of view it seems more theorics need to be developed to forecast the local effects. For the slender structure (bounded or unbounded), several works exist or are currently under development. However, for arbitrary nonslender, bounded structures the problem seems difficult for the MF range.

REFERENCES

1 N.D. Acheback, Wave propagation in elastic solids, Third printing, North-Holland, Amsterdam, 1980.

2 K.J. Bathe, ADINA - A finite element program for Automatic Dynamic Incremental Nonlinear Analysis, Report 82448-1, Acoustics and Vibration Laboratory, Department of Mechanical Engineering, Massachussets Institute of Technology, Cambride, Massach., 1975 (rec. Dec. 1978).

3 K.J. Bathe, Finite element procedures in engineering analysis", Prentice-Hall, Inc. Englewood CLiffs, J.J., 1982.

4 K.J. Bathe, E.L. Wilson, Numerical Methods in Finite Element Analysis, Prentice-Hall, Inc., Englewood Cliffs, N.J., 1976.

5 A. Bensoussan, J.L. Lions, G. Papanicolaou, Asymptotic Analysis for periodic structures, North-Holland, Amsterdam, 1978.

6 R.W. Clough, J. Penzien, Dynamic of Structures, McGraw Hill, New York, 1975.

7 L. Collatz, The numerical treatment of differential equations, Springer-Verlag, New York 1966.

8 R.D. Coodk, Concepts and applications of finite element analysis John Wiley and Sons, New York, 1953.

9 R. Courant, D. Hilbert, Methods of mathematical Physics, John Wiley and Sons, New York, 1953 .

10 G.P. Destefano, Causes of instabilities in Numerical Integration Techniques, Int. Journal of Comp. Math., vol 2, pp. 123-142, 1968

11 J. Dieudonne, Elements d'analyse, Gauthier-Villars, Paris

12 J.L. Doob, Stochastic processes, John Wiley and Sons, New York, 1967.

13 N. Dunford, J.T. Schwartz, Linear operators I, Interscience, New York, 1967.

14 G. Duvaut, J.L. Lions, Les inéquations en mécanique et en physique, Dunod, Paris, 1972.

15 S.J. Fenves, N. Perrone, A.R. Robinson, Schnobrich, Numerical and computer methods in structural mechanics, Academic Press, New York, 1973.

16 G.E. Forsythe, W.R. Wasow, Finite difference methods for partial differential equations, John Wiley and Sons, Inc., New York, 1960.

17 L.E. Franks, Signal theory, Prentice-Hall, Inc., Englewood Cliffs, N.J., 1969.

18 R.H. Gallagher, Finite element analysis fundamentals, Prentice-Hall, Inc., Englewood Cliffs, N.J., 1975.

19 P. Germain, Mécanique des milieux continus, Masson, Paris, 1973.

20 Ph. Gibert, Les modes locaux et globaux en mécanique des vibrations des corps élancés, RT-ONERA, n° 38/3454 RY 081 R, Châtillon, France, oct. 1984.

21 P. Henrici, Error propagation for difference methods, John Wiley and Sons, New York, 1963.

22 H.M. Hilber, T.J.R. Hughes, R.L. Taylor, Improved numerical dissipation for time integration algorithms in structural mechanics, Intern. Journal of Earthquake Eng. and Struc. Dyn., vol. 5, pp. 283-292, 1977.

23 E. Hille, R.S. Phillips, Functional analysis and semigroups, Amer. Math. Soc. Coll. Publ. 31, 1948.

24 S. Hon, Review of modal synthesis techniques and a new approach, Shock and vibration bulletin, vol. 40, n° 4, 1969.

25 T.J.R. Hughes, W.K. Liu, Implicit-Explicit finite elements in transient analysis : stability theory. Journal of Applied Mechanics, vol. 45, June 1978.

26 W.C. Hurty, J.D. Collins, G.C. Hart, Dynamic nalysis of large structures by modal synthesis techniques, Computers and Structures, vol. 1, pp. 535-563, Dec. 1971.

27 W.C. Hurty, M.F. Rubinstein, Dynamics of structures, Prentice-Hall, Inc., Englewood Cliffs, J.J., 1964.

28 J.F. Imbert, Analyse des structures par éléments finis, Ecole Nationale Supérieure de l'Aéronautique et de l'espace, SUPAERO, CEPAD, Paris, 1979.

29 E. Issaacson, H.B. Keller, Analysis of numerical methods, John Wiley and Sons, New York, 1966.

30 A. Jouan, A. Morvan, L. Guillaumie, Experimentation élasto-acoustique de la poute tubulaire en basse et moyenne fréquence au lac de Castillon, RT-ONERA n° 37/3454 RY 082-447 R, Châtillon, France, août 1984.

31 T. Kato, Pertubation theory for linear operators, Springer-Verlag, New York, 1966.

32 P. Krée, C. Soize, Mécanique Aléatoire - Vibrations non linéaires, turbulence, séismes, houle, fatigue", Dunod, Paris, 1983 (English version : Random Mechanic, D. Reidel publishing company, Dordrecht/Boston/ Landcaster/Tokyo, 1985).

33 J.L. Lions, E. Magenes, Problèmes aux limites non homogènes et applications, volumes 1 et 2, Dunod, Paris, 1968.

34 R.H. Lyon, Statistical energy analysis of dynamical systems, Cambridge, Massachussets, the MIT Press, 1975.

35 N.M. Newmark, A method of computation for structural dynamics, Proc. ASCE, vol. 85, n° EM3, 1959.

36 R.E. Nickell, Direct integration in structural dynamics, ASCE, Journal of Eng. Mech. Divis., vol. 99, pp. 303-317, 1973.

37 J.T. Oden, J.N. Reddy, An introduction to the mathematical theory of finite elements, John Wiley and Sons, Inc., New York, 1976.

38 A. Papoulis, Signal Analysis, McGraw-Hill, New York, 1977.

39 G. Piazzoli, Rayonnement acoustique des coques, RTS-ONERA n° 29/3454 RY 060 à 062 R, Châtillon, France, Dec. 1983.

40 P.A. Raviart, P. Faure, Cours d'analyse numérique, Ecole Polytechnique, Paris, 1976.

41 M. Roseau, Asymptotic wave theory, North-Holland, Amsterdam, 1976.

42 M.F. Rubinstein, Structural Systems, Statics, Dynamics and Stability, Prentice-Hall, Inc. Englewood Cliffs, N.J., 1970.

43 E. Sanchez-Palencia, Non homogeneous media and vibration theory, Springer-Verlag, Berlin, 1980.

44 L. Schwartz, Analyse Hilbertienne, Hermann, Paris 1979.

45 L. Schwartz, Produits tensoriels topologiques, Séminaire 1953/1954/Faculté des Sciences de Paris.

46 C. Soize, Elements mathématiques de la théorie déterministe et aléatoire du signal, Ecole Nationale Supérieure des Techniques Avancées, ENSTA, Paris, 1983.

47 C. Soize, Medium Frequency linear vibrations of anisotropic elastic structures", Journal Recherche Aérospatiale, (5) ONERA, Châtillon, France, 1982.

48 C. Soize, J.M. David, A. Desanti, P.M. Hutin, Rayonnement acoustique des coques dans le domaine moyenne fréquence et extension du programme ADINA en couplage fluide structure", RT-ONERA, n° 27/3454 RY 063 R, Châtillon, France, déc. 1983.

49 C. Soize, J.M. David, A. Desanti, P.M. Hutin, Etude dynamique moyenne fréquence de la poutre tubulaire, à sec et en immersion, dans plusieurs configurations et comparaisons expérimentales, Tomes I et II, RT-ONERA n° 34/3454 RY 081, Châtillon, France, mai 1984.

50 G. Strang, G.J. Fix, An analysis of the finite element method, Prentice-Hall, Inc. Englewood Cliffs, N.J., 1973.

51 S. Timoshenko, J.N. Goodier, Theory of elasticity, McGraw-Hill Book Company, New York, 1951.

52 K.N. Tong, Theory of Mechanical Vibration, John Wiley and Sons, New York, 1960.

53 R. Valid, La mécanique des milieux continus et le calcul des structures", Editions Eyrolles, Paris, 1977.

54 K. Wahizu, Variational methods in elasticity and plasticity, Pergamon Press, Inc., Elmsford, New York, 1967.

55 J.R. Whiteman (ed.), The mathematics of finite elements and applications, Academic Press, Inc., Ltd., London, 1973.

56 J.H. Wilkinson, Rounding errors in algebraic processes, Prentice-Hall, Inc., Englewood Cliffs, N.J., 1962.

57 E.L. Wilson, K.J. Bathe, W.P. Doherty, Direct solution of large systems of linear equations". Computers and Structures, vol. 4, pp. 363-372, 1974.

58 K. Yosida, Functional Analysis, Third edition, Springer-Verlag, Berlin, 1971.

59 O.C. Zienkiewicz, The finite element method in engineering science, McGraw-Hill Book Company, New York, 1971.

60 K.N. Tong, Theory of mechanical vibration, John Wiley and Sons, New York, 1960.

CHAPTER 4:

GENERAL NUMERICAL APPROACHES OF LOCAL EFFECTS

IMPLEMENTATION OF LOCAL EFFECTS INTO CONVENTIONAL AND NON CONVENTIONAL FINITE ELEMENT FORMULATIONS

J. JIROUSEK
Swiss Federal Institute of Technology, CH-1015 Lausanne (Switzerland)

ABSTRACT

The first part of this contribution is concerned with the problem of including load dependent stress singularities (such as e.g. those due to concentrated loads) into conventional assumed displacement elements. It is shown that such singularities may be conveniently handled in the form of appropriate initial strain and stresses, each of which may be accounted for following standard finite element technology. A numerical example illustrates the efficiency of the method which does not require any mesh refinement in the vicinity of the singularities.

The second part of this contribution presents the concept of the so-called large finite elements (LFEs). This concept, which may be viewed as a finite element form of the Trefftz's method (ref.1), is based on using parametric displacement fields satisfying, *a priori* , the governing differential problem equations. Any local solution representing a stress singularity or stress concentration may be used as the LFE expansion basis. Boundary conditions and interelement continuity are implicitly imposed by making use of a simple stationary principle and an auxiliary compatible interelement displacement field. The excellent efficiency of the approach is demonstrated by several examples.

IMPLEMENTATION OF LOAD DEPENDENT LOCAL EFFECTS INTO CONVENTIONAL ASSUMED DISPLACEMENT ELEMENTS

Introduction

It is well known that conventionally formulated finite elements with regular displacement fields are inadequate for problems presenting stress singularity. This drawback is particularly awkward for load dependent singularities since local refinement of the finite element (FE) mesh complicates considerably the data preparation and is hardly practical in cases where the location of the singularities (due e.g. to concentrated loads) changes from one loading case to another.

Although the inclusion of singularities in existing assumed displacement elements has considerably attracted the attention of research workers, their efforts have generally been directed towards geometry dependent stress singularities (cracks, angular corners...). In contrast with such methods which generally result in appropriate modifications of the element stiffness matrix, the method presented hereafter acts solely on the loading terms while the stiffness matrix remains unaltered.

Principle of the approach

As shown in Fig.1, the solution $\mathbf{u}=[u\ v]^t$ of a given plane elasticity problem (t designates matrix transposition) including a stress singularity (concentrated loads) may be conveniently represented as a sum of two parts :

- a singular part $\hat{\mathbf{u}}$ which is known, e.g. the elasticity solution for a semi-infinite plane summarized in Fig.2 (see also ref.2),
- and a unknown regular part $\tilde{\mathbf{u}}$.

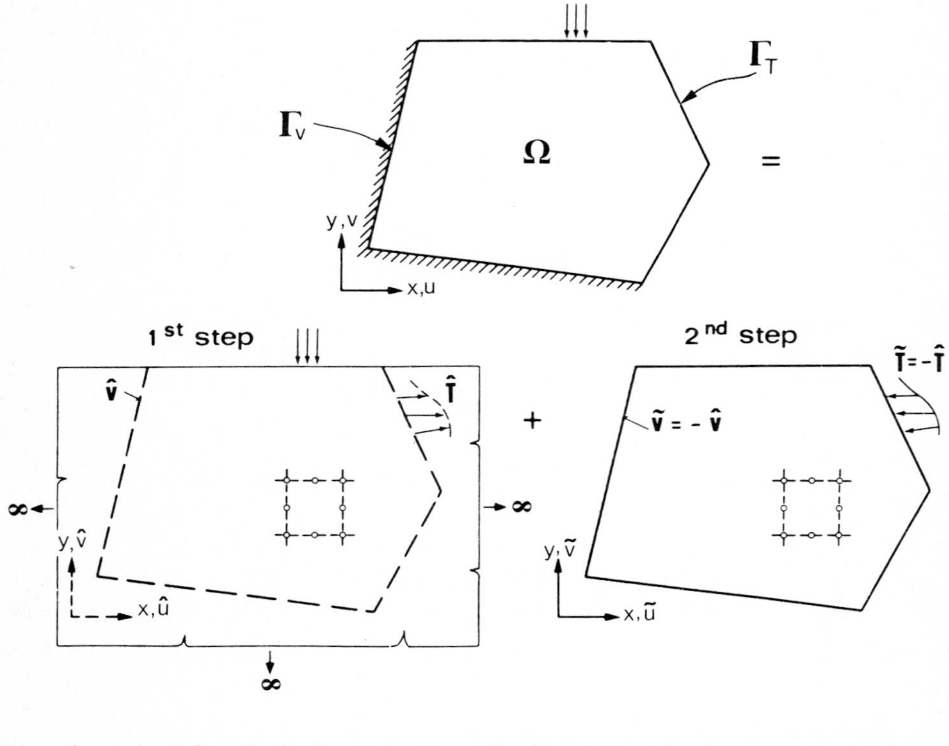

Fig. 1. Principle of the two step singularities computations; an academic approach too tedious for practical use.

In the past, this idea has occasionally been applied to some comparatively simple problems in conjunction with the finite difference method. In such cases this was used to find $\tilde{\mathbf{u}}$ such that the undesired boundary tractions $\hat{\mathbf{T}}$ on the

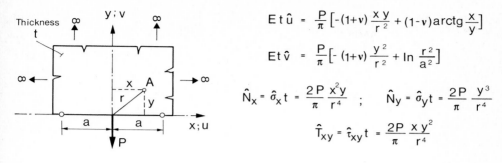

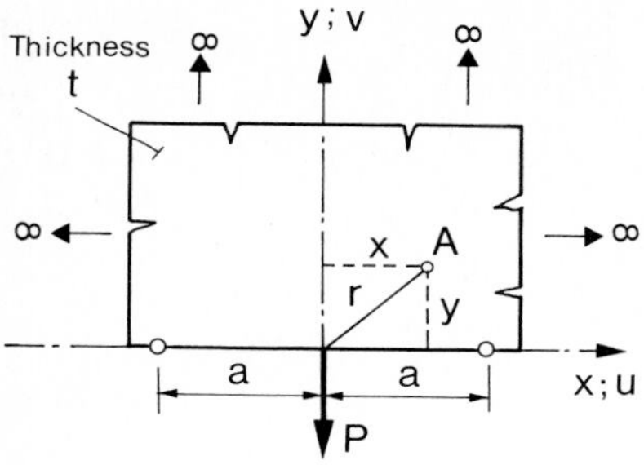

$$Et\hat{u} = \frac{P}{\pi}\left[-(1+\nu)\frac{xy}{r^2} + (1-\nu)\,\mathrm{arctg}\,\frac{x}{y}\right]$$

$$Et\hat{v} = \frac{P}{\pi}\left[-(1+\nu)\frac{y^2}{r^2} + \ln\frac{r^2}{a^2}\right]$$

$$\hat{N}_x = \hat{\sigma}_x t = \frac{2P}{\pi}\frac{x^2 y}{r^4} \quad ; \quad \hat{N}_y = \hat{\sigma}_y t = \frac{2P}{\pi}\frac{y^3}{r^4}$$

$$\hat{T}_{xy} = \hat{\tau}_{xy} t = \frac{2P}{\pi}\frac{x y^2}{r^4}$$

Fig. 2. Example of a known singular solution for elastic semi-infinite plane.

portion Γ_T and boundary displacements $\hat{\mathbf{v}}$ on the portion Γ_V of the boundary were properly compensated and, consequently, the resulting displacement field

$$\mathbf{u} = \hat{\mathbf{u}} + \tilde{\mathbf{u}} \tag{1}$$

satisfied, as well as possible, the prescribed boundary conditions on the whole Γ.

Although the possibility of applying the finite element method to the second step of this two step solution is obvious, it is apparent that the idea as outlined above, can hardly be applied to the complex problems encountered in reality. Clearly, the time necessary to prepare all the input data for an FE program, such as equivalent nodal loads or imposed displacements for all boundary nodes, and the subsequent manual superposition of the results, is discouraging. The method briefly outlined below (see also ref.3) removes this difficulty; it combines the two steps into a single one and enables the resulting solution, $\mathbf{u} = \hat{\mathbf{u}} + \tilde{\mathbf{u}}$, to be obtained directly from the finite element process.

In the sequence we assume that the customary force-displacement relationship is defined for any element e as

$$\mathbf{s}_e = \overline{\mathbf{s}}_e + \mathbf{K}_e \mathbf{d}_e \, , \tag{2a}$$

where $\mathbf{s}_e$ and $\mathbf{d}_e$ are respectively the vector of equivalent nodal forces and the vector of generalized nodal displacement and $\mathbf{K}_e$ is the stiffness matrix of the element. The displacement independent part $\overline{\mathbf{s}}_e$ of the vector $\mathbf{s}_e$ in (2a) is due to the initial deformations and/or distributed loads applied to the element. A similar relationship for the element assembly will be written as

$$\mathbf{s}_a = \bar{\mathbf{s}}_a + \mathbf{K}_a\mathbf{d}_a \ . \tag{2b}$$

For any element let $\hat{\mathbf{d}}_e$ be the known vector of generalized nodal displacement derived from the known singular field $\hat{\mathbf{u}}$ and assume the following initial strains ${}^o\boldsymbol{\varepsilon}$ and stresses ${}^o\boldsymbol{\sigma}$, defined as

$$^o\boldsymbol{\varepsilon} = \mathbf{B}_e\hat{\mathbf{d}}_e \quad \text{and} \quad {}^o\boldsymbol{\sigma} = \hat{\boldsymbol{\sigma}} \ , \tag{3a,b}$$

where $\mathbf{B}_e$ is the customary strain-displacement matrix of the element. Note that whereas the initial stresses are derived directly from $\hat{\mathbf{u}}$, the initial deformation ${}^o\boldsymbol{\varepsilon}$ differs from $\hat{\boldsymbol{\varepsilon}}$. Using (3a) and following standard finite element technology we obtain, for $\bar{\mathbf{s}}_e$

$$\bar{\mathbf{s}}_e = -\int_{\Omega_e} \mathbf{B}_e^t\mathbf{D}\ {}^o\boldsymbol{\varepsilon}d\Omega + \int_{\Omega_e} \mathbf{B}_e^t\ {}^o\boldsymbol{\sigma}d\Omega - \int_{\Gamma_{Te}} \mathbf{N}_e^t\bar{\mathbf{T}}d\Gamma = -\int_{\Omega_e} \mathbf{B}_e^t\mathbf{D}\mathbf{B}_e d\Omega\hat{\mathbf{d}}_e + \int_{\Omega_e} \mathbf{B}_e^t\hat{\boldsymbol{\sigma}}d\Omega - \int_{\Gamma_{Te}} \mathbf{N}_e^t\bar{\mathbf{T}}d\Gamma \ ,$$

where $\mathbf{N}_e$ stands for the matrix of base functions associated with the assumed displacement field, $\mathbf{u}_e = \mathbf{N}_e\mathbf{d}_e$, $\mathbf{D}$ is the stress-strain matrix and Γ_{Te} is the part of the element boundary belonging to Γ_T. Here the first righthand integral is the customary expression of the element stiffness matrix and the second integral may be replaced, by virtue of the virtual work principle, by

$$\int_{\Gamma_e} \mathbf{N}_e^t\hat{\mathbf{T}}d\Gamma \ .$$

Thus

$$\bar{\mathbf{s}}_e = -\mathbf{K}_e\hat{\mathbf{d}}_e - \int_{\Gamma_{Te}} \mathbf{N}_e^t(\bar{\mathbf{T}}-\hat{\mathbf{T}})d\Gamma + \int_{\Gamma_e-\Gamma_{Te}} \mathbf{N}_e^t\hat{\mathbf{T}}d\Gamma \ . \tag{4}$$

The last righthand integral in (4) may, obviously, be omitted. Indeed, the contributions of any two adjacent elements cancel each other along the common portion of the interelement boundary and the boundary tractions on Γ_V have no effect on the solution. If the element assembly is now performed and the vector $\mathbf{s}_a$ of equivalent nodal forces is set to zero, one obtains

$$\mathbf{K}_a\mathbf{d}_a = -\bar{\mathbf{s}}_a = \mathbf{K}_a\hat{\mathbf{d}}_a + \int_{\Gamma_T} \mathbf{N}_a^t(\bar{\mathbf{T}}-\hat{\mathbf{T}})d\Gamma \ .$$

Setting $\bar{\mathbf{T}} = \hat{\mathbf{T}}$ and extending Γ_T to the whole boundary, $\Gamma_T = \Gamma$, leads, consequently, to the identity

$$\mathbf{d}_a = \hat{\mathbf{d}}_a \ .$$

In addition, if one applies the standard stress-strain relationship with the initial stresses (3b), one finds that

$$\boldsymbol{\sigma} = {}^{0}\boldsymbol{\sigma} + \mathbf{D}(\boldsymbol{\varepsilon} - {}^{0}\boldsymbol{\varepsilon}) = \hat{\boldsymbol{\sigma}} + \mathbf{D}\mathbf{B}_e(\mathbf{d}_e - \hat{\mathbf{d}}_e) = \hat{\boldsymbol{\sigma}} \ .$$

The solution obtained is, obviously, the finite element equivalent to step 1 of the two step approach.

Now, to match the actual boundary conditions specified on Γ, the sub-vectors $\mathbf{d}_N$ of the nodal displacements on the portion Γ_v of the boundary will simply be given the prescribed values,

$$\mathbf{d}_N = \overline{\mathbf{d}}_N \ .$$

Thus, if the distributed loads $\overline{\mathbf{f}}$ of the element subdomain in Ω_e are included, the vector $\overline{\mathbf{s}}_e$ may finally be defined as

$$\overline{\mathbf{s}}_e = -\int_{\Omega_e} \mathbf{N}_e^t \overline{\mathbf{f}} d\Omega - \int_{\Gamma_{Te}} \mathbf{N}_e^t (\overline{\mathbf{T}} - \hat{\mathbf{T}}) d\Gamma - \mathbf{K}_e \hat{\mathbf{d}}_e \ . \tag{5}$$

It is interesting to observe that the method presented may, in theory, be generalized to geometry dependent stress singularities, where $\hat{\mathbf{u}}$ is a sum of products of M known singular functions by the corresponding number of undetermined singularity coefficients, which may be considered as additional non-nodal degrees of freedom. The singularity will now be accounted for by acting on the element stiffness matrices rather than the vectors of equivalent nodal forces $\overline{\mathbf{s}}_e$.

Numerical example

To study its efficiency, the method was implemented in the family of isoparametric plane stress elements of the library of the general FE structural analysis program SAFE (ref.4). The singular functions used by the element subroutine are of a more general form than those given in Fig.2 : the boundary need not be parallel to the x-axis and the load of any direction may be uniformly distributed over a given portion of the boundary (the point load is a highly unrealistic assumption in engineering praxis).

The accuracy of the solution was tested on a series of examples one of which is shown in Fig.3. The first solution of this example taken from ref. 5 was

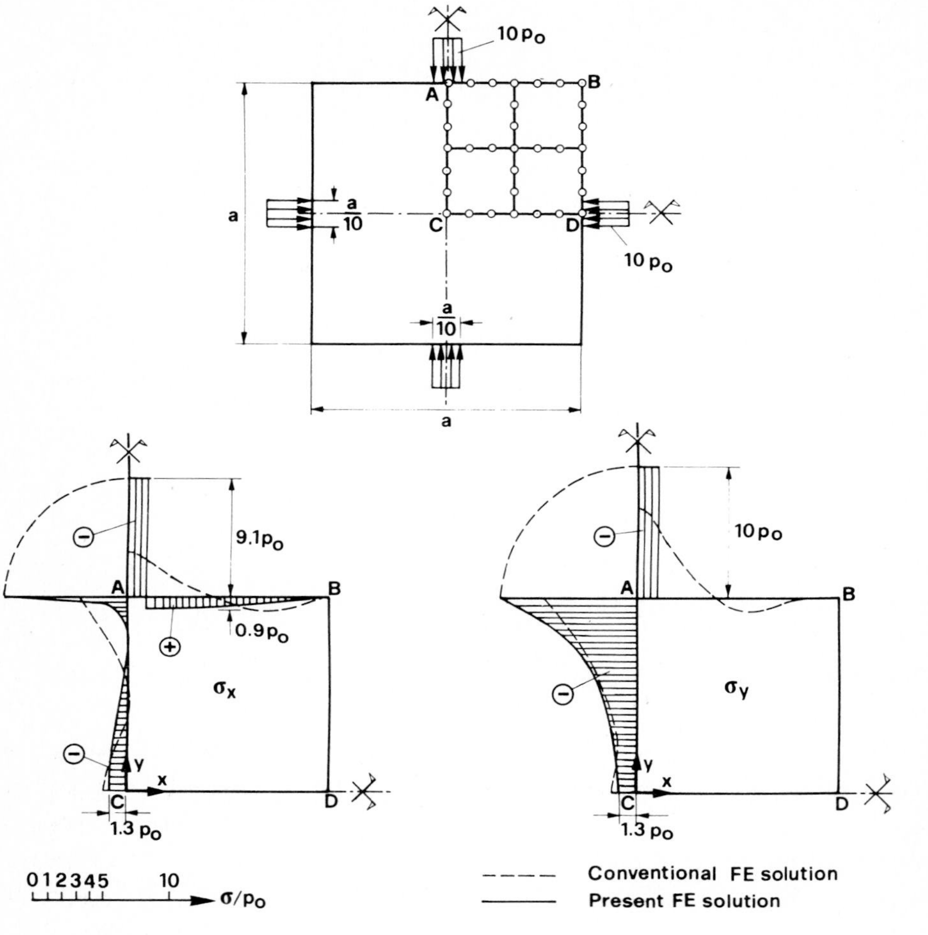

Fig. 3. Symmetrically compressed square plate. Normal stresses along AB and AC.

performed with a rough mesh of 2x2 isoparametric cubic elements over a symmetric quadrant of the plate (Fig.3). In addition, Table 1 shows some numerical results obtained for two different FE meshes and compares them to the published solution.

In the above example, the singular part of the solution extends over the whole region Ω which, clearly, may be unpractical if Ω includes for example two parts Ω_A and Ω_B of different thickness or material. However, the solution is still possible if one limits for example to Ω_A the singular part of the solution due to loads applied to Ω_A while considering, formally, the portion $\Gamma_A \cap \Gamma_B$ of its boundary as a part of Γ_{TA}. As consequence, the solution obtained

TABLE 1

Comparison of some results for the symmetrically compressed plate from Fig. 3.

Quantity	Point	2x2 cubic		4x4 quadratic		Ref. 5
		conventional	present	conventional	present	
$\sigma_x : p_o$	A	-3.47	- 9.10	- 6.19	-9.14	-
$\sigma_y : p_o$	A	-6.82	-10.02	-10.14	-9.99	-10.000
$\sigma_y : p_o$	C	-1.63	- 1.31	- 1.21	-1.26	- 1.285

presents along $\Gamma_A \cap \Gamma_B$ some incompatibility. This incompatibility equals to

$$\Delta\mathbf{u} = \mathbf{N}\hat{\mathbf{d}}_A - \hat{\mathbf{u}}_A$$

and is due to the fact that the continuity between $\hat{\mathbf{u}}_A$ and the corresponding part of the displacements $\mathbf{u}_B$ is now achieved only at nodes. However, this does not seem to severly affect the accuracccy of the solution. Thus e.g. if in the symmetrically compressed plate example a 4x4 mesh of quadratic elements is used and the singular solution $\hat{\mathbf{u}}$ is limited to the first and last elements, the results of Table 1 are modified as follows :

point A...$\sigma_x:p_o$ = -9.99, $\sigma_y:p_o$ = -10.00; point C...$\sigma_y:p_o$ = -1.29.

To conclude, note that the method presented may also be implemented in conventional isoparametric plate bending element subroutines with a view to obtaining accurate moments under loads uniformly distributed on a small specified portion (circular or rectangular) of the middle surface.

METHOD OF LARGE FINITE ELEMENTS

Introduction

The basic idea of the method of large finite elements (LFEs) is to found the finite element formulation on displacement functions that satisfy, *a priori*, the governing differential problem equations (Trefftz's method) rather than the essential boundary conditions and the interelement continuity (Rayleigh-Ritz). Apart from ref. 6 to 12 where this principle is used as a basis for formulation of all FE covering the region, the Trefftz's method, when ocasionally applied in the FE context, seems to have been essentially limited to a particular subregion. In refs. 13 and 14 for example a large internal portion of a plate or shell solved by this method is combined in the boundary region with conventional FEs. A general method for linking a particular subregion treated by a boundary (Trefftz's) method with a conventional FE field is given in ref.15. Moreover, most of the hybrid superelements for crack-type or sharp V-notch singularities (refs. 16, 17 and others) may be viewed as Trefftz's-type elements. More recently, special Trefftz's-type plane elasticity elements for stress singularities and stress concentrations (holes) have also been presented in ref. 18, but once again the basic idea is that, apart from the subregion covered by such special elements, standard finite elements will be used for the remaining part of the region.

In addition to obvious theoretical reasons against mixing FE models, the basing of the whole FE field on a unique theoretical concept as advocated by the LFE method (see e.g. Fig.4) is supported by the following practical arguments :

- the concept of special superelements is not suited to dealing with load dependent singularities (their location may change from one loading case to another);
- using the LFEs rather than the conventional ones may be computationally more efficient even if the problem does not present stress singularities or stress concentrations (ref.6);

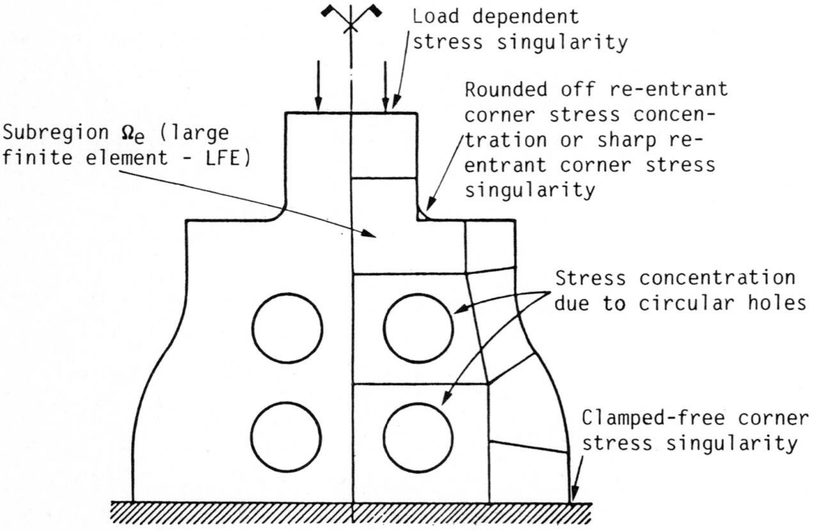

Fig. 4. Typical division of a region into large finite elements (LFEs).

- the mesh definition is much simpler (Fig.4);
- when using graphic output facilities, full advantage may be taken of the knowledge of the analytic solution within each LFE and of the practically negligible discontinuities of both displacements and stresses at interfaces (combining different FE models may severly compromise the possibilities of graphical post-processing).

Various LFE formulations differ essentially in the method used to restore the interelement continuity and to fulfil the boundary conditions. Clearly, this may be achieved in many differents ways (collocation, least square error...). Unfortunately (see e.g. refs. 7,9,10), some LFEs cannot be implemented in the FE libraries of existing FE programs since the way of obtaining the system of simultaneous equations for the undetermined coefficients of all LFEs fails to comply with the standard FE assembly rules of the direct stiffness method. To overcome this important drawback and obtain for the LFEs the customary force-displacement relationship (9), the interelement continuity should be enforced by means of an auxiliary independent interelement field in a way similar to that used in various hybrid models. Although both the theoretical basis of this type of LFE and some studies related to its practical efficiency have already been published some time ago (refs. 6,8,11), the method is not largely known and for the reader's convenience its formulation is briefly summarized in the following section.

Theoretical formulation

Consider a boundary problem stated as follows: Let the internal equilibrium of an elastic continuum occupying the region Ω bounded by $\Gamma = \partial\Omega$ be expressed in terms of displacements by a system of governing differential equations

$$\mathbf{Lu} = \overline{\mathbf{f}} \quad \text{on} \quad \Omega , \tag{6}$$

where $\mathbf{L}$ is a differential operator matrix, $\mathbf{u}$ is a vector of generalized displacements and $\overline{\mathbf{f}}$ is a conjugated vector of generalized body forces. The boundary conditions are defined in terms of the generalized boundary displacements $\mathbf{v} = \mathbf{v}(\mathbf{u})$ and/or the conjugated generalized boundary tractions by

$$\mathbf{v} = \overline{\mathbf{v}} \quad \text{on} \quad \Gamma_V \qquad \text{and} \qquad \mathbf{T} = \overline{\mathbf{T}} \quad \text{on} \quad \Gamma_T , \tag{6a,b}$$

where $\overline{\mathbf{v}}$ and $\overline{\mathbf{T}}$ are the prescribed quantities and $\Gamma_V + \Gamma_T = \partial\Omega$.

The basic idea of the LFEs method is to subdivide (Fig.4)* the region Ω into a small number of subregions Ω_e (large finite elements - LFEs) and assume on each Ω_e an independent field of generalized displacements $\mathbf{u}_e$ expressed in terms of a particular integral $\overset{o}{\mathbf{u}}_e$ and a set of appropriate homogeneous solutions $\boldsymbol{\phi}_1,\ldots\boldsymbol{\phi}_m$ to (6) :

$$\mathbf{u}_e = \overset{o}{\mathbf{u}}_e + \sum_{i=1}^{m} \boldsymbol{\phi}_i \mathbf{a}_i = \overset{o}{\mathbf{u}}_e + \boldsymbol{\Phi}_e \mathbf{a}_e \ , \tag{7}$$

where $\mathbf{a}_e$ is a vector of undetermined coefficients and the known coordinate functions $\overset{o}{\mathbf{u}}_e$ and $\boldsymbol{\phi}_i$ are such that

$$\mathbf{L}\overset{o}{\mathbf{u}}_e = \overline{\mathbf{f}} \quad \text{and} \quad \mathbf{L}\boldsymbol{\phi}_i = \mathbf{0} \quad \text{on} \quad \Omega_e \ . \tag{7a,b}$$

It is important to point out that $\overset{o}{\mathbf{u}}$ may be a "global" function, i.e. a function extended either over the whole Ω or, at least, over several LFEs.

The goal is now to calculate the undetermined coefficients $\mathbf{a}_e$ of all LFEs so that the boundary conditions (6a,b) are matched and the interelement continuity nuity restored, as best as possible, in some conveniently defined sense. If the LFEs are to be implemented to an existing FE program based on the standard FE assembly process of the direct stiffness method, the simplest way to enforce the interelement continuity is based on using an independent interelement boundary displacement field $\tilde{\mathbf{v}}_e$. If

$$\tilde{\mathbf{v}}_e = \mathbf{N}_e \mathbf{d}_e \quad \text{on} \quad \Gamma_e \ , \tag{8}$$

where $\mathbf{d}_e$ are nodal displacement parameters and where the interpolation functions of the matrix $\mathbf{N}_e$ are such that if the corresponding nodal parameters of the adjacent LFEs are matched, $\tilde{\mathbf{v}}$ is the same for the two adjacent elements over their common boundary, such formulation leads in the most straightforward fashion to the conventional force-displacement relationship

$$\mathbf{s}_e = \overline{\mathbf{s}}_e + \mathbf{K}_e \mathbf{d}_e \ . \tag{9}$$

In (9) $\mathbf{K}_e$ is the stiffness matrix of the element and $\overline{\mathbf{s}}_e$ stands for that part of the vector of fictitious nodal forces $\mathbf{s}_e$ that is independent of $\mathbf{d}_e$.

---

* For expository purposes, Figs. 4 and 5 represent simple plane elasticity situations. However, we wish to point out that all considerations apply to any linearly elastic continuum (plate in bending, shell, etc.).

The simplest variational formulation based on $\tilde{\mathbf{v}}_e$ assumes that the boundary $\partial\Omega_e$ of a particular LFE (Fig.5) is represented as a sum of four distinct portions,

$$\partial\Omega_e = \Gamma_{Se} + \Gamma_{ve} + \Gamma_{Te} + \Gamma_{Ie} = \Gamma_e + \Gamma_{Se} , \tag{10}$$

defined as follows :

- Γ_{Se} , portion of $\partial\Omega_e$ on which the prescribed boundary conditions are satisfied *a priori* (this is the case when the approximating functions are derived from a known local solution in the vicinity of a stress singularity or stress concentration);
- Γ_{ve} and Γ_{Te} , portion of the remaining part, $\partial\Omega_e - \Gamma_{Se}$, of the LFE boundary on which either displacements ($\mathbf{v} = \overline{\mathbf{v}}$) or boundary tractions ($\mathbf{T} = \overline{\mathbf{T}}$) are prescribed;
- Γ_{Ie} , interelement portion of $\partial\Omega_e$;
- $\Gamma_e = \partial\Omega_e - \Gamma_{Se}$ (to simplify notation).

Note that the independent field $\tilde{\mathbf{v}}$ is not defined at Γ_{Se} but only on Γ_e .

In the sequence, we consider the following proposition :

$$J(\mathbf{u},\tilde{\mathbf{v}}) = \sum_e \left[-\frac{1}{2}\int_{\Omega_e} \overline{\mathbf{f}}^t \mathbf{u}_e d\Omega - \frac{1}{2}\int_{\partial\Omega_e} \mathbf{T}_e^t \mathbf{v}_e d\Gamma + \int_{\Gamma_e} \tilde{\mathbf{v}}_e^t \mathbf{T}_e d\Gamma - \int_{\Gamma_{Te}} \tilde{\mathbf{v}}_e^t \overline{\mathbf{T}}_e d\Gamma \right] = \text{stationary}, \tag{11}$$

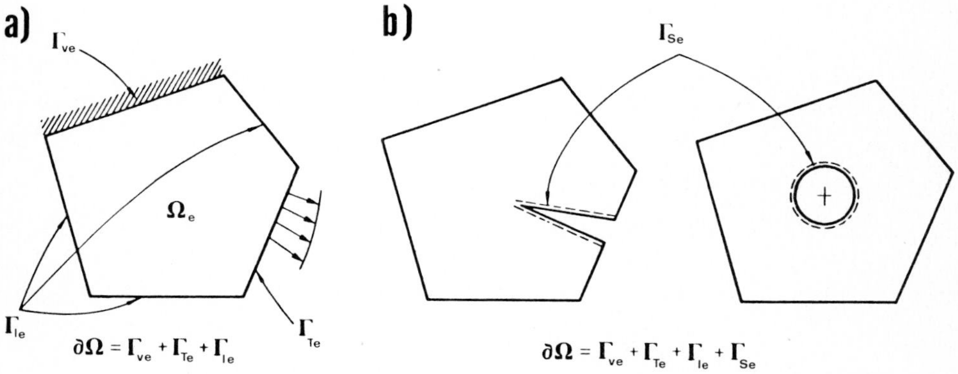

Fig. 5. Regular (a) and stress singularity or stress concentration (b) types of LFEs. Note that boundary conditions along the Γ_{Se} portion of the LFE boundary are satisfied by the approximating functions *a priori* and that all element types are produced by a single LFE routine associated with a library of optional approximating functions.

where the sum $\sum_e$ extends over all subregions (LFE) of Ω. The independent fields $\mathbf{u}_e$ and $\tilde{\mathbf{v}}_e$ of any particular LFE are subjected to the following subsidiary conditions :

a) the field $\mathbf{u}_e$ verifies the governing differential equations of the problem

$$\mathbf{L}\mathbf{u}_e = \bar{\mathbf{f}} \quad \text{on} \quad \Omega_e \tag{11a}$$

and, if relevant (Fig.5b), some boundary conditions

$$\mathbf{v}(\mathbf{u}_e) = \mathbf{0} \quad \text{and/or} \quad \mathbf{T}(\mathbf{u}_e) = \bar{\mathbf{T}} \quad \text{on appropriate portions of } \Gamma_{Se} \;; \tag{11b}$$

b) the boundary field $\tilde{\mathbf{v}}_e$ satisfies the kinematic boundary conditions

$$\tilde{\mathbf{v}} = \bar{\mathbf{v}} \quad \text{on} \quad \Gamma_{ve} \tag{11c}$$

and the kinematic interelement continuity requirements.

Now, if the variation of $J(\mathbf{u},\tilde{\mathbf{v}})$ with respect to $\mathbf{u}$ and $\tilde{\mathbf{v}}$ is performed and one observes that, by virtue of Betti's reciprocity,

$$\int_{\Omega_e} \delta\mathbf{u}_e^t\, \bar{\mathbf{f}}\, d\Omega = \int_{\partial\Omega_e} (\delta\mathbf{T}_e^t \mathbf{v}_e - \delta\mathbf{v}_e^t \mathbf{T}_e)\, d\Gamma$$

one obtains

$$\delta J(\mathbf{u},\tilde{\mathbf{v}}) = \sum_e \left[-\int_{\Gamma_e} \delta\mathbf{T}_e^t(\mathbf{v}_e - \tilde{\mathbf{v}}_e)\, d\Gamma + \int_{\Gamma_{Te}} \delta\tilde{\mathbf{v}}_e^t(\mathbf{T}_e - \bar{\mathbf{T}}_e)\, d\Gamma + \int_{\Gamma_{Ie}} \delta\tilde{\mathbf{v}}_e^t \mathbf{T}_e\, d\Gamma \right] . \tag{12}$$

Clearly, the stationary conditions associated with $\delta J(\mathbf{u},\tilde{\mathbf{v}}) = 0$ are the statical boundary conditions, $\mathbf{T}_e = \bar{\mathbf{T}}$ at Γ_{Te}, and the interelement continuity of displacements $\mathbf{v}$ and the conjugated tractions $\mathbf{T}$. Consequently, the stationary principle (11) implicitly imposes all the missing conditions and, as such, constitutes a convenient basis for development of the LFE matrices $\bar{\mathbf{s}}_e$ and $\mathbf{K}_e$.

Examination of the expression (12) of $\delta J(\mathbf{u},\tilde{\mathbf{v}})$ shows that the solution becomes singular if any of the functions $\boldsymbol{\phi}_i$ describes a rigid-body-motion mode of displacement and thus leads to a vanishing boundary traction term. Therefore, special care should be taken to form the matrix $\boldsymbol{\Phi} = [\boldsymbol{\phi}_1, \ldots \boldsymbol{\phi}_m]$ as a complete set of linearly independent solutions associated with non-vanishing strains. Note that once the solution of the LFE assembly has been performed, the lacking rigid-body modes may again be reintroduced in the internal fields $\mathbf{u}_e$ and their undetermined coefficients calculated by requiring e.g. the least squares adjustment of $\mathbf{v}_e$ and $\tilde{\mathbf{v}}_e$ at all nodes of any particular LFE.

Development of the LFE matrices and program implementation

From Eq. (7) it is easy to derive :

$$\mathbf{v}_e = \overset{o}{\mathbf{v}}_e + \mathbf{\Psi}_e \mathbf{a}_e \quad \text{and} \quad \mathbf{T}_e = \overset{o}{\mathbf{T}}_e + \mathbf{\Theta}_e \mathbf{a}_e \tag{13a,b}$$

where $\overset{o}{\mathbf{v}}_e$, $\overset{o}{\mathbf{T}}_e$ correspond to $\overset{o}{\mathbf{u}}_e$ (particular part of $\mathbf{u}_e$) and $\mathbf{\Psi}_e$, $\mathbf{\Theta}_e$ to $\mathbf{\Phi}_e$ (homogeneous part of $\mathbf{u}_e$). Observing that the variation of J with respect to $\mathbf{u}$ must separately vanish for any LFE enables to express the undetermined LFE coefficients $\mathbf{a}_e$ in terms of the nodal parameters $\mathbf{d}_e$:

$$\mathbf{a}_e = \mathbf{A}_e^{-1} \overset{o}{\mathbf{g}}_e + \mathbf{A}_e^{-1} \mathbf{G}_e \mathbf{d}_e \ , \tag{14}$$

where $\mathbf{A}_e^{-1}$ is inverse of the symmetric matrix

$$\mathbf{A}_e = \int_{\Gamma_e} \mathbf{\Theta}_e^t \mathbf{\Psi}_e d\Gamma = \frac{1}{2} \int_{\Gamma_e} (\mathbf{\Theta}_e^t \mathbf{\Psi}_e + \mathbf{\Psi}_e^t \mathbf{\Theta}_e) d\Gamma \tag{14a}$$

and

$$\overset{o}{\mathbf{g}}_e = - \int_{\Gamma_e} \mathbf{\Theta}_e^t \overset{o}{\mathbf{v}}_e d\Gamma \ , \quad \mathbf{G}_e = \int_{\Gamma_e} \mathbf{\Theta}_e^t \mathbf{N}_e d\Gamma \ . \tag{14b}$$

Performing the variation of J with respect to $\tilde{\mathbf{v}}$ and setting

$$\overset{o}{\mathbf{h}}_e = \int_{\Gamma_e} \mathbf{N}_e^t \overset{o}{\mathbf{T}}_e d\Gamma - \int_{\Gamma_{Te}} \mathbf{N}_e^t \bar{\mathbf{T}} \, d\Gamma \tag{15}$$

finally yields :

$$\bar{\mathbf{s}}_e = \overset{o}{\mathbf{h}}_e + \mathbf{G}_e^t \mathbf{A}_e^{-1} \overset{o}{\mathbf{g}}_e \quad \text{and} \quad \mathbf{K}_e = \mathbf{G}_e^t \mathbf{A}_e^{-1} \mathbf{G}_e \ . \tag{16}$$

The evaluation of the LFE matrices $\bar{\mathbf{s}}_e$ and $\mathbf{K}_e$ only calls for integration along the element boundaries (preceding section) which makes it possible, in two dimensional problems, to generate arbitrary polygonal or even curved sided elements. From the method of matching the governing differential problem equations it is obvious that any local solution for singularity or stress concentration may be straightforwardly used as the expansion bases for the particular LFE which contains the singularity or stress concentration. As suggested in Figs. 4 and 5, this enables the formulation of, for example, elements including angular corners, sharp V-shaped notches, cracks, circular holes, etc. Whilst the standard regular displacement field for the plate in bending, for example, may be easily generated using biharmonic polynomials

$$\phi_{i+1}=r_0^2\mathrm{Re}z_0^k\ ,\quad \phi_{i+2}=r_0^2\mathrm{Im}z_0^k\ ,\quad \phi_{i+3}=\mathrm{Re}z_0^{k+2}\ ,\quad \phi_{i+4}=\mathrm{Im}z_0^{k+2}\ ,\qquad (k=0,1,2,\ldots)$$

where x_0 and y_0 are local coordinates with origin at the center of gravity of the LFE, $r_0^2 = x_0^2 + y_0^2$ and the complex variable $z_0 = x_0 + iy_0$, the displacement field for a perforated or a singular corner LFE may be conveniently represented using the known local solutions in the vicinity of a circular hole and the Williams eigensolutions for the given apex angle respectively (the latter are tabulated for different boundary conditions e.g. in ref.19). The load dependent term $\overset{o}{\mathbf{u}} = \overset{o}{w}$ for a regular plate bending LFE and a uniform load $\bar{p}$ may be set e.g. to

$$\overset{o}{w} = \frac{\bar{p}}{8D}\, x^2y^2 \quad \text{where} \quad D = \frac{Et^3}{12(1-\nu^2)} \quad \text{stands for plate stiffness.}$$

For a concentrated load $\bar{P}$ at x_p , y_p , the corresponding $\overset{o}{w}$ is equal to

$$\overset{o}{w} = \frac{\bar{P}}{16\pi D}\, r^2 \ell n r^2 \quad \text{with} \quad r^2 = (x-x_p)^2 + (y-y_p)^2 \ .$$

It is essential to point out that this load term should be used as a global function extending over the nearest regular LFEs (their choice should be performed automatically by the LFE subroutine). This not only helps in improving the accuracy (the term $\overset{o}{w}$ being already continuous over the interelement boundary) but also simplifies the input data since the program user need not trouble to identify the elements concerned by the applied concentrated loads.

For the classical bending again, the conjugated vectors $\mathbf{T}_e$ and $\mathbf{v}_e$ may be readily defined e.g. as

$$\mathbf{T}_e^t = [Q_n,\ M_n,\ M_{nt}] \quad \text{and} \quad \mathbf{v}_e^t = [w,\ -\frac{\partial w}{\partial n}\ ,\ -\frac{\partial w}{\partial t}\ ]\ .$$

The LFE formulation for other elasticity problems is quite similar (see e.g. ref.12).

Since the LFE formulation for different situations remains essentially unchanged, except for the choice of the expansion set, it is possible to develop for each class of problems (plane elasticity, plate bending, etc.) a single LFE subroutine and provide it with a library of optional sets of trial functions, which may be selected by the user according to the situation encountered, by appropriately specifying a single control parameter.

Figure 6 presents the LFE families implemented within the general purpose program SAFE (ref.4). The elements have an optional number of nodes. The 6 DOF

ELEMENT TYPE	NODAL DOF	Standard elements	Circular hole stress concentration elements		Angular singularity elements
Plane elasticity elements	u, v $\varepsilon_x, \varepsilon_y$ γ_{xy}, ω_{xy} or u, v σ_x, σ_y τ_{xy}, ω_{xy}	1)			2) $\vartheta \geq 0$
Plate bending elements	w, w_x, w_y w_{xx}, w_{xy}, w_{yy}	1)			2) $\vartheta \geq 0$
Elements for cross-sectional properties and stress analysis of beams	ψ, ψ_x, ψ_y (ψ= warping function)				$\vartheta \geq 0$

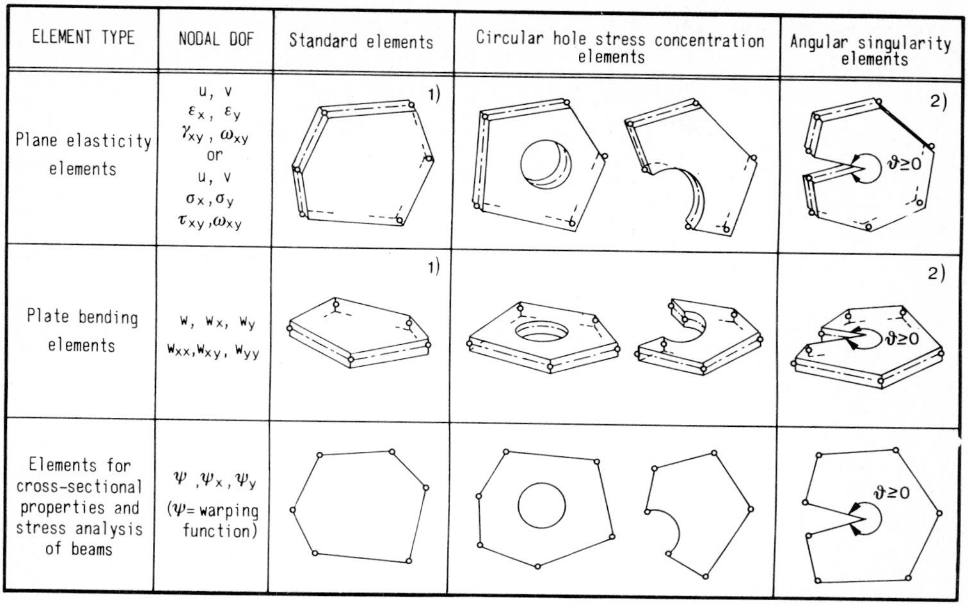

1) Takes into account local effects of concentrated loads
2) Various boundary conditions along the two sides adjacent to singular corner

Fig. 6. Families of LFEs available in the general purpose program SAFE.

used for the first two LFE families enable to represent the independent boundary displacement $\tilde{w}$ and the normal slope $\partial\tilde{w}/\partial n$ along a side of a plate bending LFE by Hermitian polynomials of order 2 and 1 or to use the first order Hermitian interpolation of the normal and tangential components of the independent boundary displacement in the case of a plane elasticity problem. Obtaining a 3 DOF/node version, if preferred, is a simple matter of appropriately changing the interpolation functions N_e at the LFE boundary. An interesting but not yet sufficiently explored possibility to improve the interpolation at the LFE boundary, yet avoid the "overcompatibility" due to the higher order DOF, is the combination of lower order Hermitian polynomials with an optional number of "bubble" functions. Any number of such functions between two Hermitian nodes may be associated, formally, with a single mid-side node. The interested reader may refer to ref.11 for more details.

Assessment of the efficiency and examples

The excellent capacity of LFEs to deal with various singular situations may

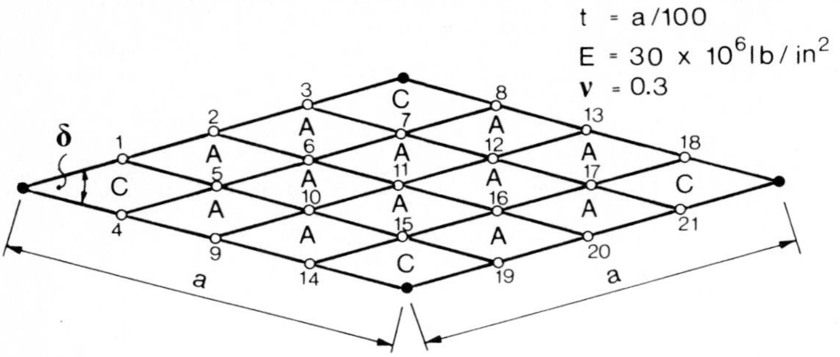

Fig. 7. Typical LFE mesh for skew plate study: A. Regular LFE; C. LFE with singular corner singularity. o assembly nodes with 6 DOF; ● auxiliary nodes without DOF.

be demonstrated on the example of a skew plate. The results shown in Table 2 represent the LFE reply to the skew plate challenge recently proposed by Finite Element News (ref.20). The percentage error has been calculated with respect to the series solution by Morley (ref.21). As may be seen, already the simplest 2x2 LFE mesh produces for any skew angle δ very accurate results whereas the majority of the 33 plate bending elements reported in the Final Report (ref.22) appear as unduly sensitive to the skew angle δ and present large errors despite very fine meshes used (14x14!).

Some other examples of the LFE's efficiency are presented in Figs. 8 to 10. The contour plots (Figs. 8 and 10) show that the interelement continuity upon

TABLE 2

Convergence study for uniformly loaded simply supported skew plate. Uniform meshes of LFEs include 4 singular LFEs for angular corners (see e.g. Fig.7).

Mesh over complete plate	2 x 2			4 x 4			8 x 8		
Percentage error in displacement and principal moments at centre (C)	w_C	M_{maxC}	M_{minC}	w_C	M_{maxC}	M_{minC}	w_C	M_{maxC}	M_{minC}
$\delta = 90^o$	0	1.0	1.0	0	-0.4	-0.4	0	0	0
$\delta = 80^o$	0	0.8	1.8	0	-0.4	-0.2	0	0	0.2
$\delta = 60^o$	0	0.7	3.6	0	-0.2	-0.3	0	0	0
$\delta = 40^o$	0	-3.9	-6.7	0	-0.7	-0.6	0	-0.4	0
$\delta = 30^o$	-0.1	0	0.9	0.1	-1.0	0.9	0	-0.5	0

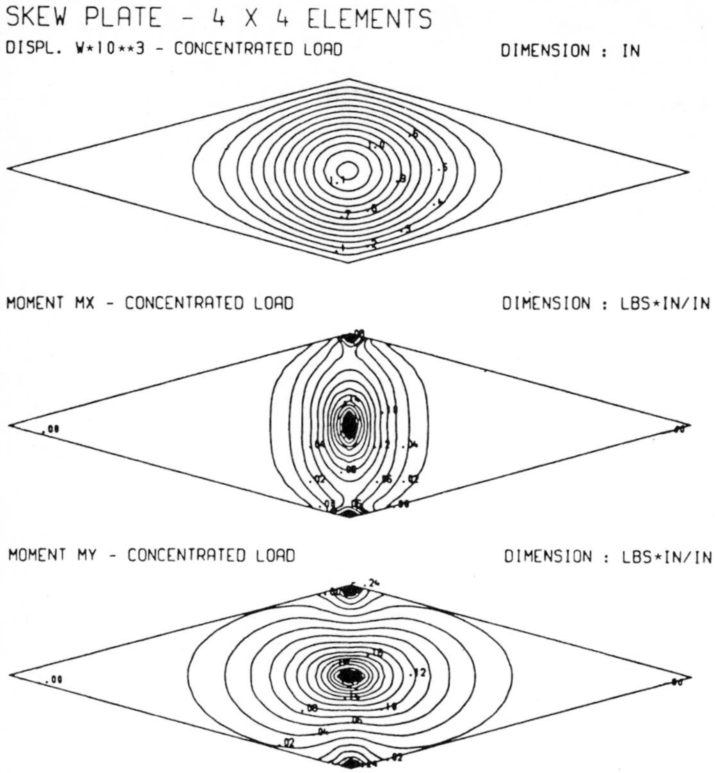

Fig. 8. Simply supported skew plate from Fig.7 (a = 1 in) under concentrated load (P = 1 lb).

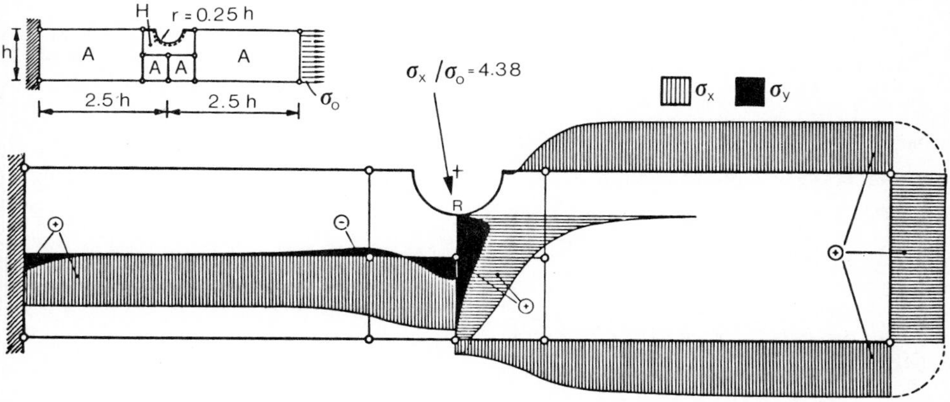

Fig. 9. Results for tension of a notched steel plate (ν=0.3): A. Regular LFE; H. LFE with circular hole functions. 60 simultaneous equations.

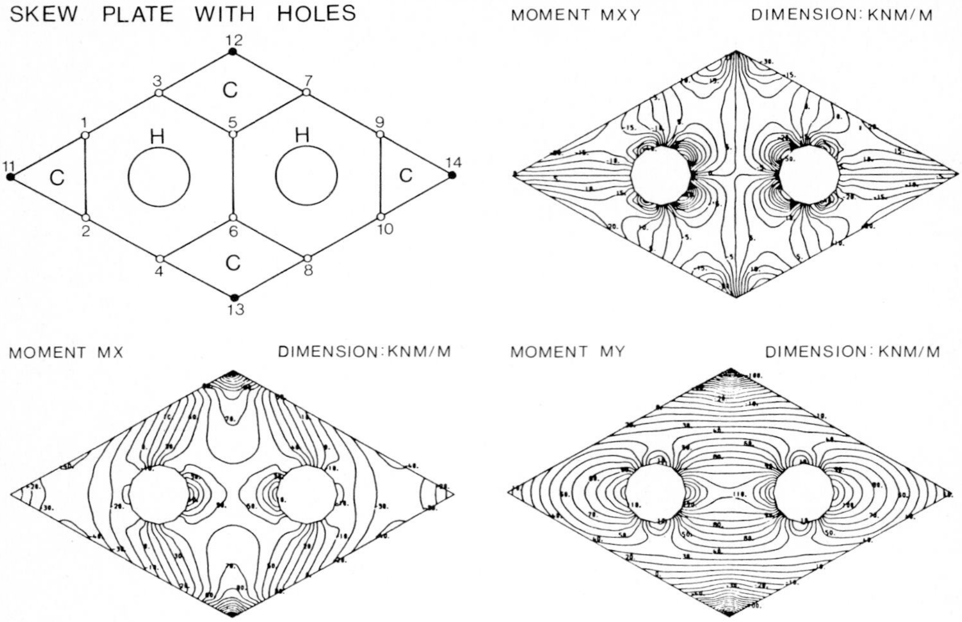

Fig. 10. Results for simply supported uniformly loaded skew plate with holes: C. LFE with singular corner singularity; H. LFE with circular hole functions. o assembly nodes with 6 DOF; ● auxiliary nodes without DOF. 10 nodes x 6 DOF = 60 simultaneous equations.

which bears the only one approximation used is remarkably well satisfied not only be kinematic quantities but also by moments. Note that to allow obtaining good graphics with only few elements, the LFE subroutine evaluates results for optional auxiliary grids of up to 25 internal output points for a regular LFE (and more for a singular or stress concentration LFE) - see ref.23 for more information.

The notched steel plate problem from Fig.9 has also been solved by using a very fine mesh of 700 conventional quadratic isoparametric elements totalising 4522 DOF (see ref.11) and produced at the root of the notch normal stresses $\sigma_{xR} = 4.21\ \sigma_0$ and $\sigma_{yR} = 0.10\ \sigma_0$. The undue radial stress of $0.10\ \sigma_0$ indicates that this solution cannot be considered as sufficiently accurate.

Conclusions

The LFEs method attempts to combine the flexibility of conventional FEs with the accuracy and high convergence rate associated with the Trefftz's method. The method is related to both the so-called boundary solution procedures and the hybrid methods.

All practical experiences with LFEs thus far have been extremely encouraging. The LFEs method seems to be ideally suited to involved elasticity problems presenting stress singularities and stress concentrations. Indeed, the overall cost of the analysis is frequently reduced by a factor of more than 20 as compared to conventional FE solutions. The method seems to be also particularly well adapted to implementation on low capacity computers since both the computer time and the central memory requirements are kept rather low.

Whether the use of local solutions as expansion basis for a LFE will always lead to a convergent solution is a theoretical problem not yet solved. For the time being, the numerical experimentation remains the basic method for testing the LFE adequacy.

REFERENCES

1 E. Trefftz, Ein Gegenstück zum Ritzschen Verfahren, Verh. des 2. Int. Kongr. für Technische Mechanik, Zürich, 1926, pp. 131-137.
2 S.P. Timoshenko and J.N. Goodier, Theory of Elasticty, 3rd edition, McGraw-Hill, 1970.
3 J. Jirousek, A contribution to finite element and associated techniques for analysis of problems with stress singularities, Proceedings of the 2nd International Congress G.A.M.N.I., Dunod, Paris, 1980, vol. 2, pp. 719-729.
4 J. Jirousek, Manuel de l'utilisateur du programme SAFE (Structural Analysis by Finite Elements), IREM - Swiss Federal Institute of Technology, Lausanne, provisional edition, 1984, 260 pp.
5 Zb. Drahonovsky, Contribution to solution of some bi-dimensional elasticity problems (in czech). Proceedings of the 1st International Symposium on the Theory of Structures in Smolenice, Slovak Akademy of Sciences, Bratislava, 1959, pp. 249-261.
6 J. Jirousek and N. Leon, A powerful finite element for plate bending, Comp. Meth. Appl. Mech. Engng., 12 (1977), pp. 77-96.
7 M.D. Tolley, Grands éléments finis singuliers, Ph.D. thesis, Université Libre de Bruxelles, 1977.
8 J. Jirousek, Basis for development of large finite elements locally satisfying all field equations. Comp. Meth. Appl. Mech. Engng., 14 (1978), pp. 65-92.
9 M.D. Tolley, Application de la méthode des grands éléments finis à la résolution de l'équation biharmonique avec conditions de bord discontinues, C.R. Acad. Sci. Paris, t. 287, série A (1978), pp. 875-878.
10 J. Descloux and M.D. Tolley, Approximation of the Poisson problem and of the eigenvalue problem for the Laplace operator by the method of large singular elements, Dept. of Math., Swiss Federal Institute of Technology, Lausanne and Fac. Appl. Sci., Université Libre de Bruxelles, 1980.
11 J. Jirousek and P. Teodorescu, Large finite element method for the solution of problems in the theory of elasticity, Computers and Structures, 15 (1982), pp. 575-587.
12 P. Teodorescu, Grands éléments finis "GEF" pour l'élasticité plane, Ph.D. thesis No. 462, IREM - Swiss Federal Institute of Technology, Lausanne, 1982, 231 pp.

13 E. Stein, Die Kombination des modifizierten Trefftzschen Verfahrens mit der Methode der finiten Elemente, Finite Elemente in der Statik, Ernst, Berlin, 1973, pp. 172-185.

14 G. Ruoff, Die praktische Berechnung der Kupplungsmatrizen bei der Kombination der Trefftzschen Methode und der Methode der finiten Elemente bei flachen Schalen, Finite Elemente in der Statik, Ernst, Berlin, 1973, pp. 242-259.

15 O.C. Zienkiewicz, D.W. Kelly and P. Bettess, The coupling of finite element method and boundary solution procedures, Int. J. Num. Meth. Engng., 11 (1977), pp. 355-375.

16 Pin Tong, P.H. Pian and S.L. Lasry, A hybrid-element approach to crack problems in plane elasticity, Int. J. Num. Meth. Engng., 7 (1973), pp. 297-308.

17 K.Y. Lin and Pin Tong, Singular finite elements for the fracture analysis of V-notched plate, Int. J. Num. Meth. Engng., 15 (1980), pp. 1343-1354.

18 R. Piltner, Spezielle finite Elemente mit Löchern, Ecken und Rissen unter Verwendung von analytischen Teillösungen, Fortschritt-Berichte der VDI Zeitschriften, Reihe 1, Nr. 96, VDI-Verlag, Düsseldorf, 1982, 277 pp.

19 M.L. Williams, Surface stress singularities resulting from various boundary conditions in angular corners of plate in bending, Proceedings of the 1st U.S. National Congress of Applied Mechanics, 1952.

20 J. Robinson, Skew effects - FEM User Project No. 2 - Morley's simply supported skew plate problem, Finite Element News, 3 (1984), pp. 14-15.

21 L.S.D. Morley, Skew plates and structures, Pergamon Press, London, 1963.

22 J. Robinson, An evaluation of skew sensitivity of thirty-three plate bending elements in nineteen FEM systems, Robinson and Associates, Wimborne, Dorset BH21 6NB, England, 31 pp.

23 J. Jirousek, A. Bouberguig and F. Frey, An efficient post-processing approach based on a probabilistic concept, Proceedings of the NUMETA 85 International Conference of the University of Swansea, A.A. Balkema, Rotterdam/Boston, pp. 723-732.

SPECIAL FINITE ELEMENTS FOR AN APPROPRIATE TREATMENT OF LOCAL EFFECTS

R. Piltner

Ruhr-Universität Bochum, Institut für Mechanik, Bochum, West Germany

ABSTRACT

Using two different variational formulations some special finite elements as for example elements with circular and elliptic holes, with internal and external cracks have been developed and tested numerically. These elements are problem adapted as the trial functions satisfy not only the governing differential equations but also some boundary conditions on such influential boundary portions as crack or hole surfaces. So the local character of a solution is taken into account appropriately. Even problems with singularities as crack and sharp corner problems for example can be treated numerically in an effective manner. In both methods used one obtains symmetric element matrices which can be evaluated via simple numerical integrations along the element boundaries. The special problem-adapted finite elements can be coupled with conventional elements. The discussion of the special elements includes the variational formulations and element coupling procedures as well as the construction of special trial functions for 2- and 3-dimensional elasticity problems.

1 INTRODUCTION

For the numerical treatment of partial differential equations the finite-element-method became an effective tool. Unfortunately there exist problems which involve numerical difficulties caused by local effects. In these problems the unknown functions are increasing or decreasing strongly in a part of the solution domain and under circumstances they have singularities in their derivatives. Examples are stress concentration problems in domains with holes, cracks and sharp corners. To gain a suitable problem treatment it is sensible to analyse the local solution character of a critical part of the domain and to use it within the numerical computations. For this purpose one looks for sequences of functions which satisfy a priori the governing differential equation and some influential boundary conditions. If we use such function terms for the trial functions we obtain problem-adapted finite elements.

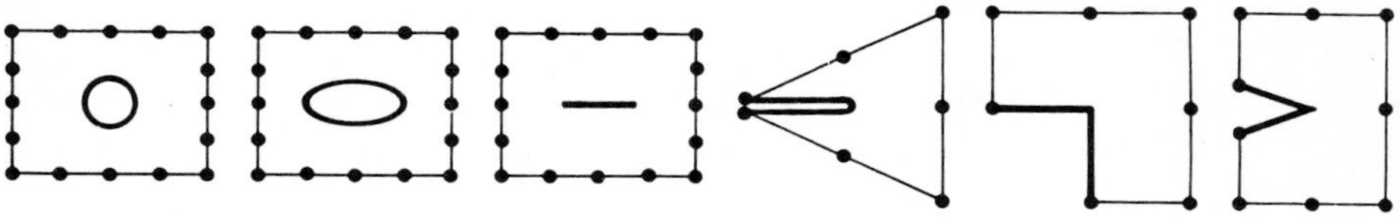

Figure 1. Special problem-adapted finite elements

The trial functions for the special elements in figure 1 satisfy the differential equation in each element domain and the boundary conditions on the surfaces of the holes, on the surfaces of the internal and external crack, on the boundary portion of the sharp corner and the notch. Beside the special problem adapted finite elements figured one can use the standard elements.

For the construction of special finite elements two main points have to be considered, namely

I. proper variational formulations, including the according element coupling procedures and
II. systematic ways of constructing special trial functions.

2 VARIATIONAL FORMULATIONS AND ELEMENT COUPLING PROCEDURES

Concerning the problem formulations two different variational formulations have been used for the numerical treatment of the partial differential equation (Navier-equations in matrix form)

$$\underline{D}^T \underline{E}\, \underline{D}\, \underline{u} + \underline{\bar{F}} = \underline{0} \qquad \text{in } V$$

with given displacement and traction boundary conditions

$$\underline{u} = \underline{\bar{u}} \qquad \text{on } O_u$$

$$\underline{n}\, \underline{E}\, \underline{D}\, \underline{u} = \underline{T} = \underline{\bar{T}} \qquad \text{on } O_\sigma$$

To get displacement elements the variational functionals given in figure 2 have been used. Figure 2 shows schematicly the basic procedures of the two methods which both lead to stiffness matrices.

The following notations have been used in the basic equations of the two methods:

$\underline{u} = \underline{u}_h + \underline{u}_p$
$\quad = \underline{\Phi}_h \underline{c} + \underline{u}_p$ — special trial functions for the displacement field, containing a homogeneous solution part $\underline{u}_h$ and a particular solution part $\underline{u}_p$. $\underline{u}$ satisfy the governing differential equations. $\underline{c}$ = vector of free parameters.

$\underline{T} = \underline{T}_h + \underline{T}_p$
$\quad = \underline{\Psi}_h \underline{c} + \underline{T}_p$ — element boundary tractions, which are related to the displacements through $\underline{T} = \underline{n}\, \underline{E}\, \underline{D}\, \underline{u}$, where $\underline{n}$ is the matrix of unit normals on the boundary,

$\underline{\tilde{u}} = \underline{\tilde{\Phi}}\, \underline{q}$ — chosen boundary displacements of an element, where
$\underline{q}$ = vector of element node displacements
$\underline{\tilde{\Phi}}$ = matrix, containing 1-dim. shape functions for the 2-dim. case or 2-dim. shape functions for the 3-dim. case.

$O^i = O^i_u + O^i_c + O^i_r$ — boundary of element i

O^i_u — boundary portion with prescribed displacements $\underline{u} = \underline{\bar{u}}$

O^i_c — boundary portion with prescribed tractions $\underline{T} = \underline{\bar{T}}$

O^i_r — inter-element boundary

method I : pure displacement method, using a separately minimized least square functional J^i

$$\Pi = \sum_i \Pi^i$$

$$\Pi^i = \int_{V^i} \left[\tfrac{1}{2}(\underline{u}^T \underline{D}^T)\underline{E}(\underline{D}\underline{u}) - \underline{u}^T \underline{\bar{F}}\right] dV^i - \int_{O^i_\sigma} \underline{u}^T \underline{\bar{T}}\, dO^i$$

← substitution of $\underline{c}$

$$J^i = \int_{O^i_u + O^i_r} (\underline{\tilde{u}} - \underline{u})^T (\underline{\tilde{u}} - \underline{u})\, dO^i = \text{minimum for element } i$$

⇓

$\dfrac{\partial J^i}{\partial \underline{c}^T} = \underline{0}$ yields a relationship between free parameters $\underline{c}$ and element nodal values $\underline{q}$: $\underline{c} = \underline{G}\,\underline{q} + \underline{g}$

for special elements: Π^i can be expressed by boundary integrals

method II: hybrid displacement method (based on an extended functional for the potential energy)

$$\Pi = \sum_i \Pi^i_H$$

$$\Pi^i_H = \Pi^i + \int_{O^i_u + O^i_r} \underline{T}^T (\underline{\tilde{u}} - \underline{u})\, dO^i$$

← substitution of $\underline{c}$

⇓

$\dfrac{\partial \Pi^i_H}{\partial \underline{c}^T} = \underline{0}$ yields a relationship between free parameters $\underline{c}$ and element nodal values $\underline{q}$: $\underline{c} = \underline{G}\underline{q} + \underline{g}$

Figure 2. Comparison of two methods leading to symmetric stiffness matrices

To ensure the interelement compatibility in both methods an optimal relationship between free parameters $\underline{c}$ of the trial functions and the element nodal values $\underline{q}$ is required. In matrix notation the general relationship between free parameters and nodal values can be expressed through

$$\underline{c} = \underline{G}\,\underline{q} + \underline{g} \quad .$$

So an optimal computation of the matrix $\underline{G}$ and the vector $\underline{q}$ in the sense of the chosen method is the basis for each element coupling procedure. The principle of the coupling procedures can be explained with the help of figure 3.

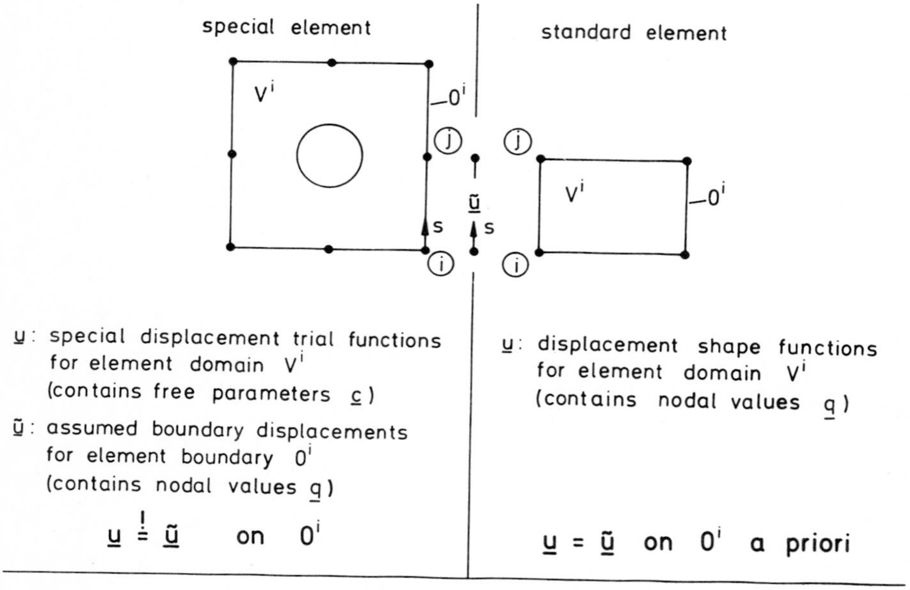

linear $\tilde{\underline{u}}$ between (i) and (j):

$$\tilde{\underline{u}} = \begin{bmatrix} \left(1-\frac{s}{s_{ij}}\right) & \frac{s}{s_{ij}} & 0 & 0 \\ 0 & 0 & \left(1-\frac{s}{s_{ij}}\right) & \frac{s}{s_{ij}} \end{bmatrix} \begin{bmatrix} u_i \\ u_j \\ v_i \\ v_j \end{bmatrix} , \left(0 \leq s \geq s_{ij}\right)$$

Figure 3. Illustration for the element coupling procedures using independently assumed boundary displacements $\tilde{\underline{u}}$

To achieve the interlement compatibility we use in addition to the special trial functions $\underline{u}$ independently assumed boundary displacements $\tilde{\underline{u}}$. The special displacement trial function vector $\underline{u}$ is valid for the element domain V^i and contains the free parameters $\underline{c}$. The vector of assumed boundary displacements $\tilde{\underline{u}}$ for the element boundary 0^i contains the nodal values $\underline{q}$ and is chosen to be linear between between two nodes of a 2-dimensional element or quadratic between 3 nodes according to the neighbouring standard elements. In figure 3 an example for the boundary displacements $\underline{\tilde{u}}$ is given, which vary linearly between node i and node j. For this example $\underline{\tilde{u}}$ contains one-dimensional shape functions and the node displacement values of the according nodes i and j.

In general the assumed functions $\underline{\tilde{u}}$ of a special element are chosen in such a way that they agree with the linear or quadratic boundary displacements of the conventional adjacent elements, for which we use the standard shape functions. For a 2-dim. problem $\underline{\tilde{u}}$ contains 1-dim. shape functions, whereas for a 3-dim. problem $\underline{\tilde{u}}$ contains 2-dim. shape functions. On the boundary of a special element we require that $\underline{u} = \underline{\tilde{u}}$. This matching is realized in the two variational formulations in different ways.

In the first method the matching of the special element trial functions $\underline{u}$ with the boundary displacements $\underline{\tilde{u}}$ is achieved through the minimization of the least square functional J^i. The result of the minimization of J^i is an optimal relationship between free parameters $\underline{c}$ and nodal values $\underline{q}$. This relationship is substituted into the displacement functional portion Π^i for an element i.
As a particularity it should be noticed that the least square expression J^i is not added as a penalty function to the variational functional Π^i but is minimized separately for the special element. So the problem of finding an optimal penalty factor for a penalty term does not arise /1/.
For the standard elements we use continuous linear or quadratic shape functions so that $J^i=0$ a priori. This means that for standard elements nothing has changed.

A second effective possibilty for the coupling of special elements with standard displacement elements is the application of an extended variational functional /2,3,4,5/ (method II in figure 2). The matching of $\underline{u}$ and $\underline{\tilde{u}}$ is done in this method with the extension term. Also this procedure first gives a relationship between free parameters $\underline{c}$ and element nodal values $\underline{q}$. The resulting expression for the vector $\underline{c}$ is then substituted into the extended functional portion Π_H^i.
In the case of continuous standard elements the extension term is zero and so the basic relation for standard displacement elements is included in the extended functional.

As the trial functions for a special element satisfy the governing differential equation we can simplify the expression for Π^i in both methods:
After integrating by parts we obtain a sum of boundary integrals plus a domain integral vanishing for the special trial functions. So if we use functions satisfying the Euler equation we have only boundary integrals in the expression for Π^i.
The idea of a variational formulation with boundary integrals instead of domain integrals was presented by Trefftz /6/ in 1926 and is based on the use of solution series of a treated differential equation.

To get suitable formulas, the special trial function vector $\underline{u}$ is decomposed into a homogeneous solution part $\underline{u}_h$ and a particular solution part $\underline{u}_p$, where only the homogeneous part has free parameters $\underline{c}$. After the decomposition $\underline{u} = \underline{u}_h + \underline{u}_p$ we obtain the following boundary integral form for Π^i :

$$\Pi^i = \int_{O^i} \frac{1}{2} \underline{u}_h^T [\underline{n}\,\underline{E}\,\underline{D}\,\underline{u}_h]\, dO^i + \int_{O^i} \underline{u}_h^T [\underline{n}\,\underline{E}\,\underline{D}\,\underline{u}_p]\, dO^i - \int_{O^i_\sigma} \underline{u}_h^T \underline{\bar{T}}\, dO^i \quad + \text{terms without } \underline{u}_h$$

In both methods mentioned we obtain symmetric stiffness matrices which are evaluated via simple numerical integrations along the element boundaries /1,7/ . There are no integrations necessary along the element boundary portions where boundary conditions are satisfied a priori. Thereby we have the advantage of this procedures that singular stress points do not appear in the numerical calculations, although the singular character of a solution is completely taken into account.
One essential difference between the method with the least square functional and the method with the extended functional is that there must be rigid body terms in method I whereas in method II no rigid body terms are allowed.

3 THE CONSTRUCTION OF TRIAL FUNCTIONS

The condition for the procedures presented is that we can construct series of functions which satisfy the treated differential equation. For the special elements these solution series also have to satisfy the boundary conditions on such influential boundary portions as hole or crack surfaces. In the following chapters some advantages of complex formulations are used for the construction of homogeneous trial functions for plane problems and for 3-dimensional problems.

3.1 PLANE STRESS AND PLANE STRAIN PROBLEMS

To give a particular solution $\underline{u}_p$ we need specifications for the body forces. In the case of constant body forces $\bar{f}_x$, $\bar{f}_y$ the particular solution for the displacements and stresses is given by

$$2\mu u_p = \frac{1}{2}\, a \left(1 - \frac{\lambda}{2(\lambda+\mu)}\right) \bar{f}_x x^2 - \frac{\lambda}{2(\lambda+\mu)}\, b \bar{f}_y xy + \left[\frac{\lambda}{4(\lambda+\mu)}\, a - (1+a)\right] \bar{f}_x y^2$$

$$2\mu v_p = \frac{1}{2}\, b \left(1 - \frac{\lambda}{2(\lambda+\mu)}\right) \bar{f}_y y^2 - \frac{\lambda}{2(\lambda+\mu)}\, a \bar{f}_x xy + \left[\frac{\lambda}{4(\lambda+\mu)}\, b - (1+b)\right] \bar{f}_y x^2$$

$$\sigma_{xx_p} = a \bar{f}_x x$$

$$\sigma_{yy_p} = b \bar{f}_y y$$

$$\tau_{xy_p} = -(1+b) \bar{f}_y x - (1+a) \bar{f}_x y$$

a and b are any real values which are to be chosen.
The homogeneous solution which contains the free parameters can be written in the well known Muskhelishvili-formulation /8/ with a complex variable $z = x + iy$. With the help of two complex functions $\Phi(z)$ and $\Psi(z)$ the displacements and stresses for the homogeneous

solution can be given in the following simple form:

$$2\mu\,(u_h + i v_h) = \varkappa\,\Phi(z) - z\,\overline{\Phi'(z)} - \overline{\Psi(z)}$$

$$\sigma_{xx_h} + i\tau_{xy_h} = \Phi'(z) + \overline{\Phi'(z)} - z\,\overline{\Phi''(z)} - \overline{\Psi'(z)}$$

$$\sigma_{yy_h} - i\tau_{xy_h} = \Phi'(z) + \overline{\Phi'(z)} + z\,\overline{\Phi''(z)} + \overline{\Psi'(z)}$$

The special advantage of this complex formulation is that with any function Φ and Ψ the Navier-equations are automatically satisfied for vanishing body forces.

For finite element computations one can choose the two complex functions in form of complex power series

$$\Phi(z) = \sum_j a_j z^j \,, \quad \Psi(z) = \sum_j b_j z^j$$

with complex coefficients $a_j = \alpha_j + i\,\beta_j$ and $b_j = \gamma_j + i\,\delta_j$. The exponents of the power series depend on the type of the element domain: For a simple connected domain we have only positive exponents whereas for a double connected domain we can also admit negative exponents.

Using such complex power series we obtain the following real trial functions for the homogeneous displacements:

$$2\mu u_h = \sum_{j=-N}^{M} \Big\{\big[\varkappa Re\,[z^j] - jx Re\,[z^{j-1}] - jy\,Im\,[z^{j-1}]\big]\,\alpha_j + \big[-\varkappa Im\,[z^j] + jx Im\,[z^{j-1}] - jy Re\,[z^{j-1}]\big]\,\beta_j - \gamma_j\,Re[z^j] + \delta_j Im[z^j]\Big\}$$

$$2\mu v_h = \sum_{j=-N}^{M} \Big\{\big[\varkappa Im\,[z^j] + jx Im\,[z^{j-1}] - jy Re\,[z^{j-1}]\big]\,\alpha_j + \big[\varkappa Re\,[z^j] + jx Re\,[z^{j-1}] + jy\,Im\,[z^{j-1}]\big]\,\beta_j + \gamma_j\,Im[z^j] + \delta_j Re\,[z^j]\Big\}$$

where $Re\,[z^j] = r^j \cos j\varphi$

$$Im\,[z^j] = r^j \sin j\varphi$$

$$\alpha_o = \beta_o = 0$$

These functions satisfy the governing differential equation and can be used for the evaluation of element stiffness matrices via one of the mentioned boundary procedures of the Trefftz-type.

Now for special elements for the treatment of local effects we need functions which also ensure the satisfaction of influential boundary conditions. To satisfy also boundary conditions we can not assume the two functions Φ and Ψ independently. There must be a relationship between Φ and Ψ to ensure the satisfaction of a boundary condition.

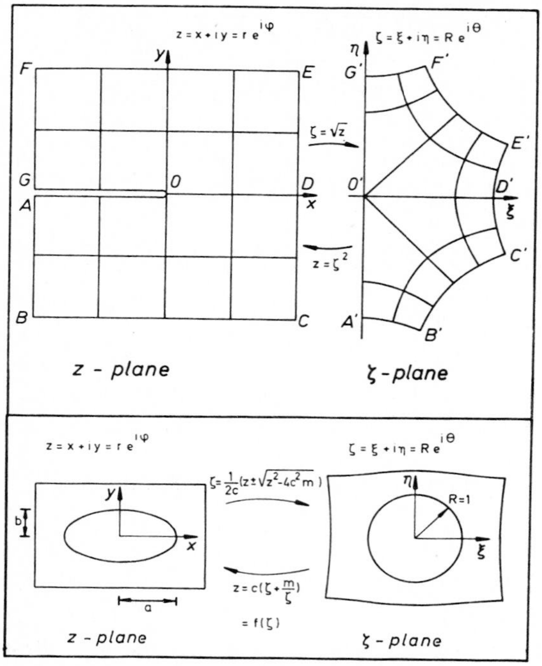

Figure 4. Conformal mapping of a finite element domain

Normally it is very difficult or impossible to construct suitable series of complex functions in the original domain. In these cases a conformal mapping can be very helpful (figure 4). By means of a conformal mapping the domain of a finite element is mappped onto another domain with an altered boundary. The point of such a mapping is that an important boundary portion of an element under consideration can be found after the mapping on a geometrically simpler curve such as the unit circle or the real and imaginary axis.

Using a conformal mapping the functions are assumed in the transformed element domain. This can be compared with the isoparametric element concept where for example the trial functions are not taken in a general quadrilateral domain but in a unit square which is the suitable mapped element domain for that case.
In complex form the boundary conditions for the original domain (z-plane) can be given by

$$\varkappa \Phi(z) - z\overline{\Phi'(z)} - \overline{\Psi(z)} = 2\mu(\bar{u} + i\bar{v}) \quad \text{on } \Gamma_1$$

$$\Phi(z) + z\overline{\Phi'(z)} + \overline{\Psi(z)} = i\int(\bar{T}_x + i\bar{T}_y)ds \quad \text{on } \Gamma_2$$

whereas for the transformed domain (ζ -plane) the boundary conditions are

$$\Psi(\zeta) = x\overline{\Phi} - \overline{f}\,\frac{\dot{\Phi}}{\dot{f}} - 2\mu(\overline{\bar{u} + i\bar{v}}) \qquad on\ \Gamma_1'$$

$$\Psi(\zeta) = -\overline{\Phi} - \overline{f}\,\frac{\dot{\Phi}}{\dot{f}} - i\,\overline{\int(\overline{T}_x + i\overline{T}_y)ds} \qquad on\ \Gamma_2'$$

Here Γ_1 is the boundary portion with prescribed displacements $\underline{u} = \underline{\bar{u}}$ and Γ_2 is the boundary portion with prescribed tractions $\underline{T} = \underline{\bar{T}}$. Γ_1' and Γ_2' are the according boundary portions for the mapped domain.

The boundary equations in the ζ -plane are arranged in such a way that the complex function Ψ is on the left-hand sides. This form has been chosen as we want to assume only the complex function Φ in form of a power series to compute after that the unknown function from a boundary equation under consideration.

So on the right-hand sides of the transformed boundary equations we have only given or chosen quantities:

Φ is chosen as a power series

f is the mapping function

$\bar{u}$, $\bar{v}$, $\bar{T}_x$, $\bar{T}_y$ are prescribed displacements and tractions on the boundary

Now on the left-hand sides we have the analytic function Ψ , but after substituting the terms Φ , f, $\bar{u}$, $\bar{v}$, $\bar{T}_x$ and $\bar{T}_y$ into one of the boundary equations we obtain a function which is not analytic. The complex conjugate terms on the right-hand sides are the reason of this unhappy situation. If we are able to replace all complex conjugate terms by analytic expressions we can find the unknown analytic function Ψ , which in connection with the assumed function Φ ensures the satisfaction of a treated boundary condition.

In the treatment of a problem with a conformal mapping of a boundary onto a unit circle we can use the important relationship

$$\overline{\zeta^{\alpha}} = \zeta^{-\alpha} \qquad on\ \ |\zeta| = 1$$

which is an example for the replacement of conjugate complex function terms. On the unit circle the nonanalytic function $\overline{\zeta^{\alpha}}$ is equal to the analytic function $\zeta^{-\alpha}$. Corresponding relationships exist for mappings onto a half plane where the transformed boundary portion is the real or imaginary axis.

With the chosen function Φ and the computed function Ψ we can get all real trial functions for the displacements and the according stresses.

The mentioned procedure of satisfying a boundary condition after assuming the complex function Φ as a power series in the mapped domain was successful for the treated problems with circular and elliptic holes, with internal and external cracks. In these examples the exponents of the power series for Φ are integer numbers.

Unfortunately we can not always satisfy a boundary condition using only integer exponents. An example for such a situation is the evaluation of solution functions for sharp re-entrant corners with an angle α according to figure 5. In this case we can use real and complex exponents for the assumed power series. For an unloaded corner for example one can construct

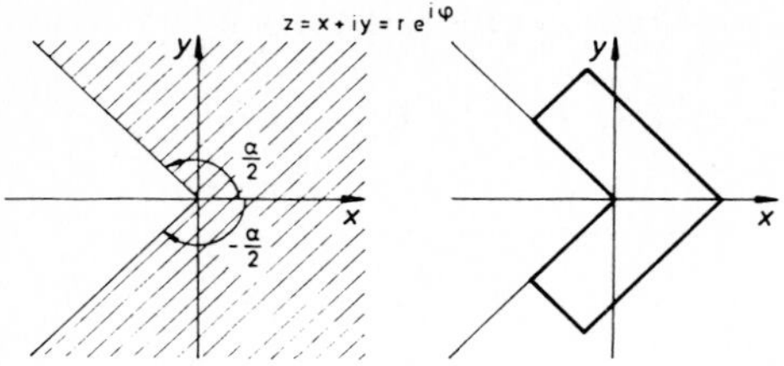

boundary conditions (example):

$$\sigma_{\varphi\varphi} = 0 \qquad \text{on} \quad \varphi = \pm \frac{\alpha}{2}$$

$$\tau_{r\varphi} = 0 \qquad \text{on} \quad \varphi = \pm \frac{\alpha}{2}$$

Figure 5. Domains with sharp re-entrant corners

the following series for Φ and Ψ :

$$\Phi(z) = \sum_{j=0}^{\infty} a_j \, z^{\lambda_j} + \sum_{j=0}^{\infty} \overline{a}_j \, z^{\overline{\lambda}_j}$$

$$+ \sum_{j=0}^{\infty} a_j^* \, z^{\omega_j} - \sum_{j=0}^{\infty} \overline{a}_j^* \, z^{\overline{\omega}_j}$$

$$= \sum_{j=0}^{\infty} \left[\frac{1}{2} \alpha_j (z^{\lambda_j} + z^{\overline{\lambda}_j}) + i \frac{1}{2} \beta_j (z^{\lambda_j} - z^{\overline{\lambda}_j}) \right] +$$

$$+ \sum_{j=0}^{\infty} \left[\frac{1}{2} \alpha_j^* (z^{\omega_j} - z^{\overline{\omega}_j}) + i \frac{1}{2} \beta_j^* (z^{\omega_j} + z^{\overline{\omega}_j}) \right]$$

$$\Psi(z) = - \sum_{j=0}^{\infty} a_j \left[e^{i\lambda_j \alpha} + \lambda_j e^{i\alpha} \right] z^{\lambda_j} - \sum_{j=0}^{\infty} \overline{a}_j \left[e^{i\overline{\lambda}_j \alpha} + \overline{\lambda}_j e^{i\alpha} \right] z^{\overline{\lambda}_j} +$$

$$+ \sum_{j=0}^{\infty} a_j^* \left[e^{i\omega_j \alpha} - \omega_j e^{i\alpha} \right] z^{\omega_j} - \sum_{j=0}^{\infty} \overline{a}_j^* \left[e^{i\overline{\omega}_j \alpha} - \overline{\omega}_j e^{i\alpha} \right] z^{\overline{\omega}_j}$$

where

$$a_j = \frac{1}{2} (\alpha_j + i \beta_j) \qquad \text{and} \qquad a_j^* = \frac{1}{2} (\alpha_j^* + i \beta_j^*)$$

The infinite sequences of the exponents depend on the corner angel α. For a given angle α we can compute the exponents as the roots of the two characteristic equations

$$\sin \alpha \lambda_j + \lambda_j \sin \alpha = 0$$

and

$$\sin \alpha \omega_j - \omega_j \sin \alpha = 0$$

Singular stresses at the corner point are obtained if the real parts of the exponents lie between 0 and 1.

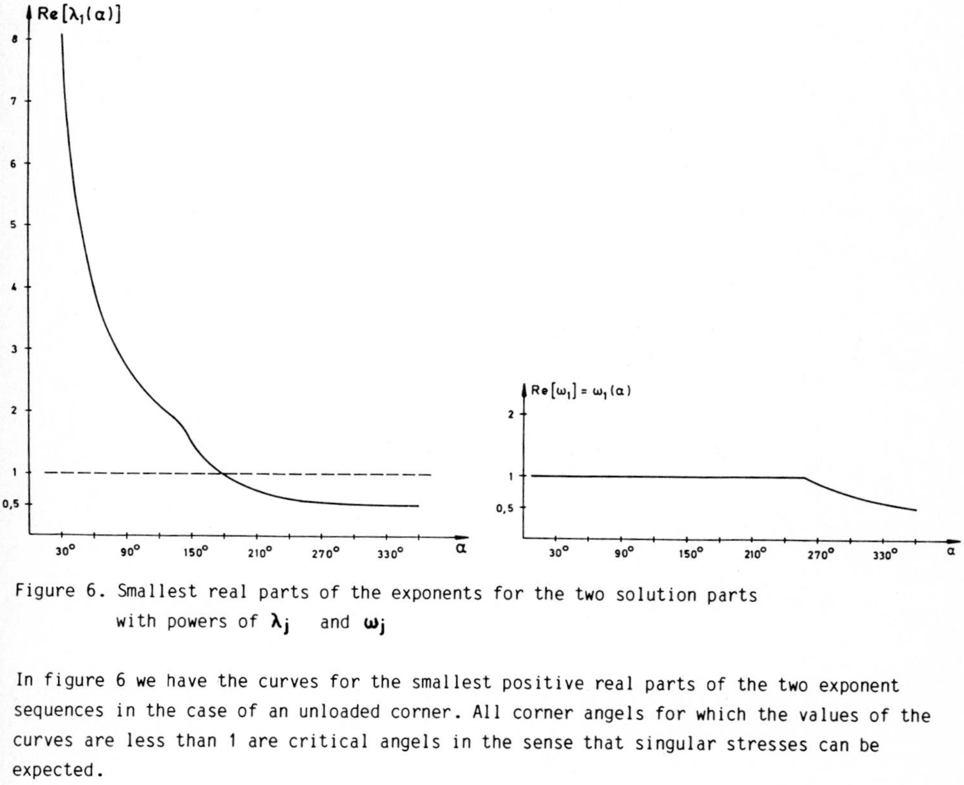

Figure 6. Smallest real parts of the exponents for the two solution parts with powers of λ_j and ω_j

In figure 6 we have the curves for the smallest positive real parts of the two exponent sequences in the case of an unloaded corner. All corner angels for which the values of the curves are less than 1 are critical angels in the sense that singular stresses can be expected.

3.2 PLATE BENDING

The displacement function w for plate bending is decomposed into a homogeneous and a particular solution part also. Analogously to the plane stress and plane strain problems the homogeneous solution can be expressed by help of two complex functions. So for the displacement function w we have

$$w = \frac{1}{2}\left[\bar{z}\,\Phi(z) + z\,\overline{\Phi(z)} + \Psi(z) + \overline{\Psi(z)}\right] + w_p$$

In the case of a constant load function p(x,y) = const. the particular solution is given by

$$w_p = \left\{ax^4 + \frac{1}{8}(1-24\,[a+b])\,x^2y^2 + by^4\right\}\frac{p}{D}$$

where a and b are any real values, which are to be chosen. Of course also for plate bending problems one can make use of the conformal mapping techniques.

3.3 THREE-DIMENSIONAL PROBLEMS

As the use of complex functions turned out to be very helpful for the treatment of plane problems I tried to do something in an analogous way for 3-dimensional problems. My aim was to express the displacements and stresses by help of any complex functions in such a way that the homogeneous Navier-equations and the equilibrium equations, respectively, are automaticly satisfied.

To obtain such a fomulation it is convenient to look at first for solutions of the 3-dimensional biharmonic equation $\Delta\Delta F = 0$, as for plane problems the 2-dimensional biharmonic equation is of great importance.

The biharmonic solution function F is decomposed into 3 real functions according to $F = U + V + W$. Each of the three functions U, V, W is expressed with the help of two complex functions in the following form:

$$U = \frac{1}{2}\left[\bar{\zeta}_1\Phi_1(\zeta_1) + \zeta_1\overline{\Phi_1(\zeta_1)} + \chi_1(\zeta_1) + \overline{\chi_1(\zeta_1)}\right]$$

$$V = \frac{1}{2}\left[\bar{\zeta}_2\Phi_2(\zeta_2) + \zeta_2\overline{\Phi_2(\zeta_2)} + \chi_2(\zeta_2) + \overline{\chi_2(\zeta_2)}\right]$$

$$W = \frac{1}{2}\left[\bar{\zeta}_3\Phi_3(\zeta_3) + \zeta_3\overline{\Phi_3(\zeta_3)} + \chi_3(\zeta_3) + \overline{\chi_3(\zeta_3)}\right]$$

$$\zeta_1 = ix + b_1y + c_1z \quad \text{where} \quad b_1^2 + c_1^2 = 1$$

$$\zeta_2 = a_2x + iy + c_2z \quad \text{where} \quad a_2^2 + c_2^2 = 1$$

$$\zeta_3 = a_3x + b_3y + iz \quad \text{where} \quad a_3^2 + b_3^2 = 1$$

So F is expressed with the help of six complex functions. Three different complex variables have been used for the complex formulation. They contain free parameters and parameter functions, respectively, which only have to satisfy the characteristic equations ($b_1^2 + c_1^2 = 1$ etc.). An example for the parameter functions is

$$b_1 = c_2 = a_3 = \pm\cos t$$

$$c_1 = a_2 = b_3 = \pm\sin t$$

Using the biharmonic functions of the above given complex form one can construct suitable expressions for the stresses satisfying the equilibrium equations. Since the 6 Beltrami-equations are automaticaly satisfied for the stress representation first published in reference /9/, an integration of the stress expressions was possible yielding the relationships for the displacements.
The construction of the 3-dimensional homogeneous solution with the help of the complex functions gave the following stress and displacement representation:

$$\sigma_{xx} = U_{yy} + U_{zz} + V_{yy} + V_{zz} - c_2^2(1-\nu)\Delta V + W_{yy} + W_{zz} - b_3^2(1-\nu)\Delta W$$

$$2\mu u = -U_x + (1-\nu)\int \Delta U\, dx - V_x + (1-\nu)a_2^2\int \Delta V\, dx - W_x + (1-\nu)a_3^2\int \Delta W dx$$

$$\sigma_{yy} = U_{xx} + U_{zz} - c_1^2(1-\nu)\Delta U + V_{xx} + V_{zz} + W_{xx} + W_{zz} - a_3^2(1-\nu)\Delta W$$

$$2\mu v = -U_y + (1-\nu)b_1^2\int \Delta U\, dy - V_y + (1-\nu)\int \Delta V\, dy - W_y + (1-\nu)b_3^2\int \Delta W\, dy$$

$$\sigma_{zz} = U_{xx} + U_{yy} - b_1^2(1-\nu)\Delta U + V_{xx} + V_{yy} - a_2^2(1-\nu)\Delta V + W_{xx} + W_{yy}$$

$$2\mu w = -U_z + (1-\nu)c_1^2\int \Delta U\, dz - V_z + (1-\nu)c_2^2\int \Delta V\, dz - W_z + (1-\nu)\int \Delta W\, dz$$

$$\tau_{xy} = -U_{xy} - V_{xy} - W_{xy} + a_3 b_3 (1-\nu)\Delta W$$

$$\tau_{xz} = -U_{xz} - V_{xz} - W_{xz} + a_2 c_2 (1-\nu)\Delta V$$

$$\tau_{yz} = -U_{yz} - V_{yz} - W_{yz} + b_1 c_1 (1-\nu)\Delta U$$

All derivatives and integrals in these formulas can be expressed in a simple way by the complex functions and their complex derivatives. For example we have

$$U_x = Im\left[\Phi_1 - \bar{\zeta}_1\Phi_1' - \chi_1'\right]$$

and

$$\int \Delta U dx = 4\, Im\left[\Phi_1\right]$$

If one uses parameter functions for the complex variables instead of discret parameter values, one can obtain systematicly various solution functions after choosing the kind of the complex functions and integrating all right-hand sides with respect to the parameter variable t /9,10/. It is hoped that this complex formulation enables us to analyse local effects of 3-dimensional problems and to construct systematicly special trial functions containing the local character of a solution.

4 NUMERICAL EXAMPLES

For the numerical calculations some examples have been chosen for which the exact solutions are known. Here the classical solution for an infinte plate with an elliptic hole under normal tension is considered. Choosing different values for the major axis a and the minor axis b of the elliptic hole, we consider the following three cases:

1. general elliptic hole ($a \neq b$)
2. a circular hole ($a = b$)
3. an internal crack ($a \neq 0$, $b = 0$)

The infinite plate under normal tension is idealized by a finite element model according to figure 7. The special element is coupled with standard displacement elements.

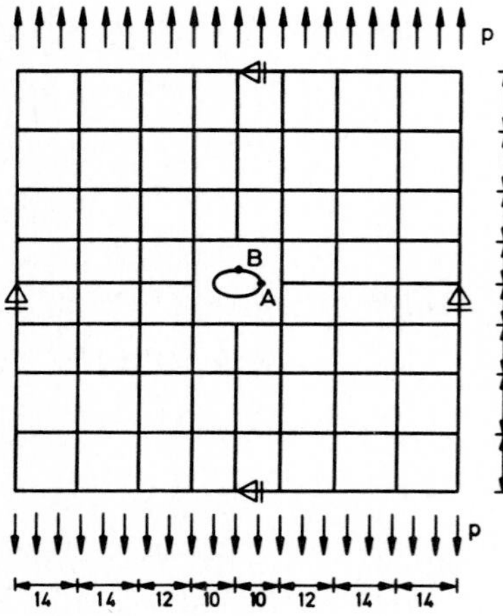

Figure 7. Finite element model for an infinite plate with an elliptic hole with major axis a and minor axis b.

4.1 EXAMPLE 1

In the case of an elliptic hole ($a = 2$, $b = 1$) a special element with quadratic boundary displacements have been chosen. The results for the stress tips contain errors which lie between 0.2 and 0.4 per cent. At point A one obtains $\sigma_{\varphi\varphi}/p = 5.010$ for method I and $\sigma_{\varphi\varphi}/p = 5.011$ for method II. The results for the stress tip at point B are $\sigma_{\varphi\varphi}/p = -1.003$ for method I and $\sigma_{\varphi\varphi}/p = -1.004$ for method II. The exact stress tip values are 5.0 and - 1.0, respectively.

4.2 EXAMPLE 2

The stress results for a plate with a circular hole ($a = b = 2$) agree fairly well with the analytical solution. The exact and the numerical calculated stresses in the domain of the special hole element are plotted for a vertical and a horizontal cut (figure 8). Using elements with quadratic boundary displacements $\underline{\tilde{u}}$ one can scarcely see differences in the results between method I and method II.

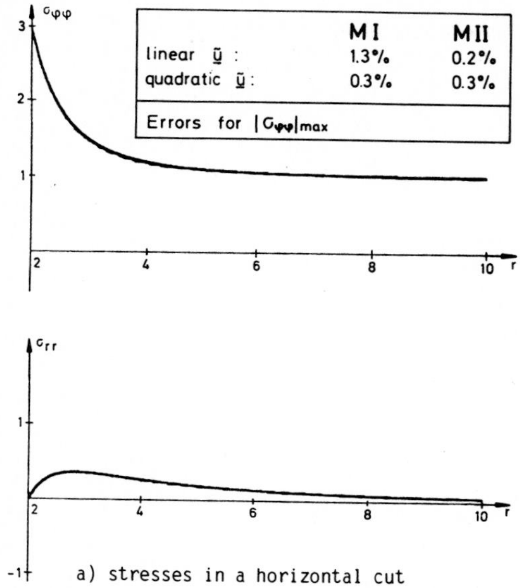

a) stresses in a horizontal cut

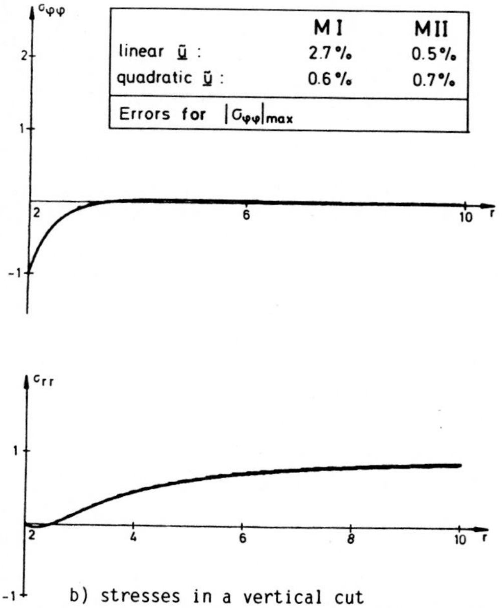

b) stresses in a vertical cut

Figure 8. Stresses in the domain of a special element with a circular hole (The numerically calculated curves in these plots have been obtained with method I)

4.3 EXAMPLE 3

In the case of an internal crack (a = 2, b = 0) we have the exact stress intensity factors K1 = 1.414 and K2 = 0. In the numerical computations with quadratic boundary displacements $\bar{u}$ we obtain K1 = 1.4165 for method I and K1 = 1.4164 for method II. For K2 we obtain values, which are of the order of 10^{-6}. More numerical examples using special problem-adapted finite elements can be found in the references /1,7/.

REFERENCES

1. R. Piltner, "Spezielle finite Elemente mit Löchern, Ecken und Rissen unter Verwendung von analytischen Teillösungen", Doctor thesis, Ruhr-Universität Bochum (1982), VDI-Verlag Düsseldorf, 1982
2. T. H. H. Pian, "Derivation of element stiffness matrices by assumed stress distribution", AIAA J. 2 No. 7 , 1333 - 1336 (1964)
3. P. Tong, T. H. H. Pian, S. Lasry, "A hybrid element approach to crack problems in plane elasticity", Int. J. num. Meth. Engng. 7, 297 - 308 (1973)
4. O. C. Zienkiewicz, D. W. Kelly, P. Bettes, "The coupling of the finite element method and boundary solution procedure", Int. J. for Num. Meth. in Engng., Vol. 11, 355 -375 (1977)
5. O. C. Zienkiewicz, D. W. Kelly, P. Bettes, "Marriage a la mode - the best of both worlds (Finite elements and boundary integrals)", Proc. Int. Symposium on Innovative Numerical Analysis in Applied Engineering Science, Versailles, May 1977, 19 -26 .
   Also Ch. 5 p. 81 -106, Energy Methods in Finite Element Analysis, ed. R. Glowinski, E. Y. Rodin and O. C. Zienkiewicz, J. Wiley & Son, 1979
6. E. Trefftz, "Ein Gegenstück zum Ritzschen Verfahren", 2. Int. Kongr. f. Techn. Mech., Zürich 1926, 131 - 137
7. R. Piltner, "Special finite elements with holes and internal cracks", (to appear in: Int. J. num. Meth. Engng.)
8. N. I. Muskhelishvili, "Some Basic Problems of the Mathematical Theory of Elasticity", Nordhoff, Groningen, Holland, 1953
9. R. Piltner, "Finite Elemente mit Ansätzen im Trefftz'schen Sinne", Proc. of the Conference on "Finite Elemente - Anwendung in der Baupraxis", München, March 1984 Wilhelm Ernst & Sohn, Berlin/München/Düsseldorf
10. R. Piltner, "Die Lösung der Grundgleichungen der Elastizitätstheorie für den räumlichen Fall mit Hilfe von komplexen Funktionen", (in preparation)

ON THE CONSIDERATION OF LOCAL EFFECTS IN THE FINITE ELEMENT ANALYSIS OF LARGE STRUCTURES

W. DIRSCHMID
AUDI AG, P.O.Box 220, D-8070 Ingolstadt, West Germany

INTRODUCTION

Nowadays the finite element method is the most used tool for structural analysis during the design process. The main problem to the user is to set up a mesh which is fine enough for carrying out the stress distribution even in small regions of the structure. Fig. 1 shows the finite element model of a car body. The model consists of 15000 degrees of freedom, and it is suitable very well for calculating the overall stiffness, but it cannot consider the stress distribution.

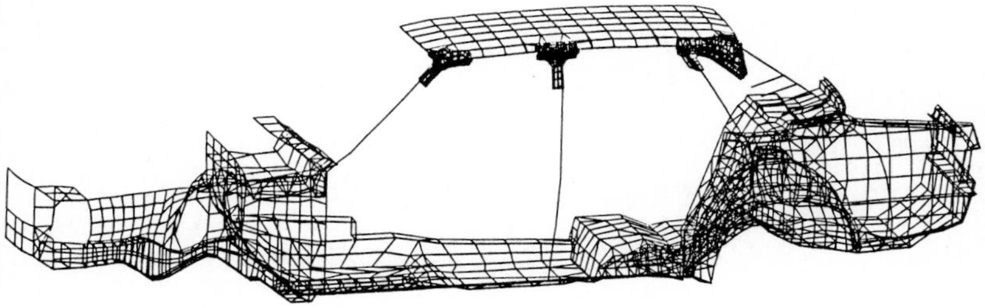

Fig. 1

The number of the degrees of freedom must be increased in order to be able to work out a reliable stiffness distribution. In most cases it is not possible because of the effort of the calculation.

So an Iteration procedure is proposed here which allows to get the solution of a locally refined system. The effort of the calculation depends on the size of the refined region, no matter how big the structure is; thereby local stress distribution can be worked out economically even for large structures.

The algorithm is a combination of an iteration procedure with a direct solution. The mathematical formulation is described thoroughly in ref. 1, here only a rough description will be given.

THEORETICAL FORMULATION

Let us assume the structure of fig. 2 is to be analysed by the finite element method. Usually the displacements of the nodal points are found out

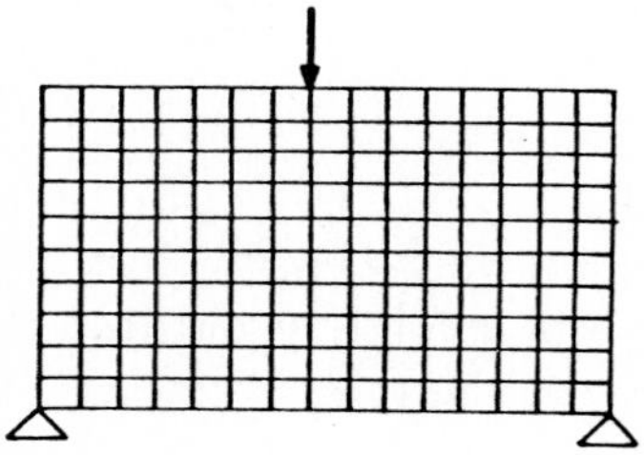

Fig. 2

by the solution of the simultaneous equations

$$[K_T]\{r_T\} = \{R_T\} \tag{1}$$

Where $\{R_T\}$ is the load vector, $[K_T]$ the stiffness matrix and $\{r_T\}$ the displacement vector.

The system is split into several macroelements like superelements. First the interior gridpoints of the macroelements are reduced by static condensation (fig. 3).

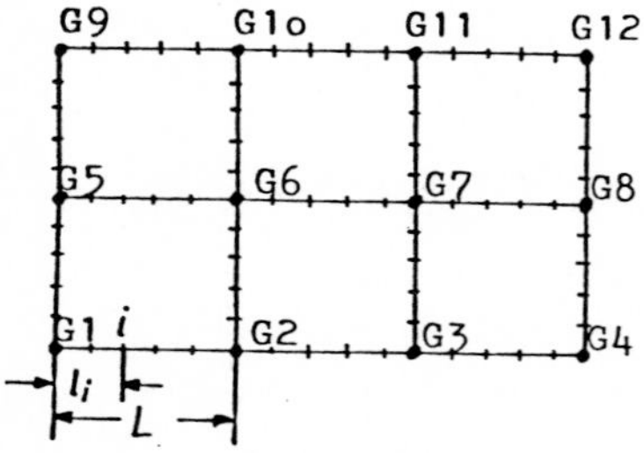

Fig. 3

$$[K]\{r\} = \{R\} \tag{2}$$

The relationship concerns the residual structure; the stiffness matrix [K] is smaller, but the bandwith is larger. So approximatly the effort for carrying out the vector $\{r\}$ is the same as above.

In order to find out the solution economically a further reduction of the unknowns in eq. (2) must be done.

It is distingueshed between (see fig. 3): the points G1, G2, ... on the corner of the macroelements and the points lying on the sides of the macroelements, i.e. the point "i" on the side G1 - G2.

The total displacement of the point "i" is split into a term given by a linear varying displacement state along the edge G1- G2 and into a term $\{b\}$ due to the rest. Using the values L, l_i (fig. 3) the displacement vector $\{r_i\}$ of the point "i" may be written as

$$\{r_i\} = \{r_{G1}\} + (\{r_{G2}\} - \{r_{G1}\})\frac{l_i}{L} + \{b_i\} \tag{3}$$

$$\{r_i\} = \{1 - \frac{l_i}{L} \quad 1\}\begin{Bmatrix} r_{G1} \\ r_{G2} \end{Bmatrix} + \{b_i\} = [a]\begin{Bmatrix} r_{G1} \\ r_{G2} \end{Bmatrix} + \{b\} \tag{4}$$

Analogically, the displacements of the points on all sides may be expressed

$$\{\bar{r}\} = [A]\{r_G\} + \{b\} \tag{5}$$

$\{r_G\}$ being the displacement vector of all corner points

$\{\bar{r}\}$ being the displacement vector of all points lying on the sides (the displacements of the points G1, G2, ... are not included)

$\{b\}$ being the displacements which depart from the linear varying displacement of each side.

Two types of degrees of freedom are established by means of eq. (5): the absolut values of displacements $\{r_G\}$ and the relative ones $\{b_i\}$.

Assuming

$$\{b\}^{(0)} = \{0\} \tag{6}$$

a system is turned out which contains the displacements of the corner points only; there is a linear varying displacement state on the sides of the macroelements. This statement is only valid if the elements of each macroelement have a shpe function leading to a linear varying displacement state along the edges. This is provided by the most elements used in practice. Nevertheless, the method presented here is working even this condition is not fullfiled.

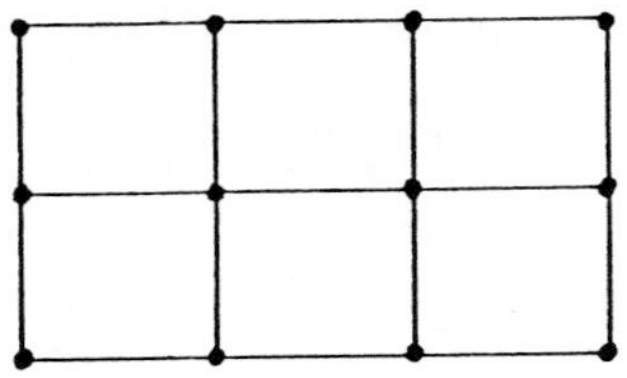

Fig. 4

The system we obtained now is called "basic-system" (fig. 4). The stiffness matrix of each macroelement can be determined by static condensation of the interior points (fig. 3) and by introduction of constraints according to eq. (5). So the "basic-system" behaves like a conventional finite element system with [K] as the stiffness matrix

$$[K_G]\{r_G\}^{(0)} = \{R_G\} \tag{7}$$

$[K_G]$ is a very small matrix compared to $[K_T]$ and $[K]$, respectively. $[K_G]$ is compiled using the stiffness matrices of the macroelements; the big matrices $[K_T]$ and $[K]$ need not to be established.

The solution of eq. (7) gives a first approximation $\{r_G\}^{(0)}$ of $\{r_G\}$. The exact solution will be found by an iteration process which relaxes the constraints of the points on the sides of the macroelements.

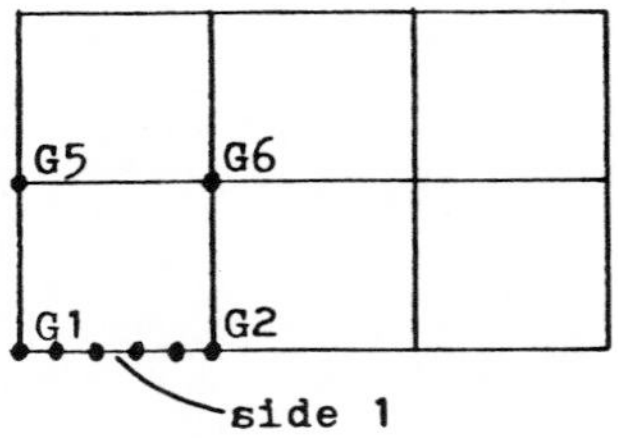

Fig. 5

In the first step of iteration the "basic-system" is enlarged by the points lying on the side 1 (fig. 5). Using the modification method suggested by Roy (ref. 2) the "basic-system" is reduced to the points G1, G2, G5 and G6 and the points of side 1. The reduction leads to a stiffness matrix $[K_1]$ whose size is much smaller than the size of $[K_G]$ and a load vector $\{R_1\}$. The displacements of these points are determined by

$$[K_1]\{r\}^{(1)} = \{R_1\} \tag{8}$$

A vector $\{b_i\}^{(1)}$ is obtained which contains the displacements departed from the linear displacement state of the points on the side G1 - G2

$$\{b_i\}^{(1)} = \{r_{side\ 1}\}^{(1)} - [a]\begin{Bmatrix} r_{G1} \\ r_{G2} \end{Bmatrix} \tag{9}$$

With that the first step is completed.

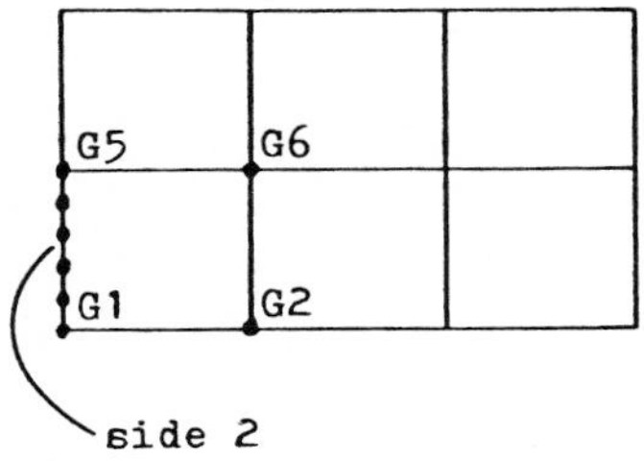

Fig. 6

In the second one another side, i.e. side 2 is perfomed in the same way, see fig. 6. A reduced system consisting of the degrees of freedom of the points G1, G2, G5, and G6 and the points on the side 2 is set up giving a stiffness matrix $[K_2]$ and a load vector $\{R_2\}$. An additional load vector $-\{T\}^{(2)}$ must be introduced which considers the improvements of the previous steps.

$$[K_2]\{r\}^{(2)} = \{R_2\} - \{T\}^{(2)} \tag{10}$$

A vector $\{b_i\}^{(2)}$ is achieved as it is shown before using the solution of eq.(10)

Each side is relaxed in this way turning out a vector $\{b_i\}^{(i)}$. A cycle of the iteration is completed when all sides are considered. After completing a cycle the iteration starts again with the first side giving a new vector $\{b_i\}^{(1)}$. It is important to realize that the iteration vectors consist of the vectors $\{b_i\}^{(i)}$ the values of which are very small compared to the values of $\{r\}$ in eq. (2).

The big problem (eq. (2)) is split into "n" (number of sides) small ones. So a hypermatrix consisting of the matrices $[K_1]$, $[K_2]$ ... $[K_n]$ can be set up.

$$[K_S] = \begin{bmatrix} K_1 & & & & \\ & K_2 & & & \\ & & K_3 & & \\ & & & \cdot & \\ & & & & \cdot \end{bmatrix} \tag{11}$$

and the hypervectors

$$\{R_S\} = \{R_1 \quad R_2 \quad \dots\} \tag{12}$$

$$\{T_S\} = \{T^{(1)} \quad T^{(2)} \quad \dots\} \tag{13}$$

$$\{r_S\} = \{r^{(1)} \quad r^{(2)} \quad \dots\} \tag{14}$$

the relationship concerning all sides is established as

$$[K_S]\{r_S\} = \{R_S\} - \{T_S\} \tag{15}$$

From the hypermatrix $[K_S]$ the hyperfactor $[U_S]$ may be deduced easily

$$[U_S]^T[U_S] = [K_S] \tag{16}$$

Performing one cycle of iteration, a displacement vector is found out

$$\{r_S\} = [U_S]^{-1}[U_S]^{-T}(\{R_S\} - \{T_S\}) \tag{17}$$

The vector $\{T_S\}$ changes during the operation of the solution.

Instead of performing the decomposition of the big matrix $[K]$ (eq. (2)) the Choleski factors of "n" small matrices of $[K_S]$ must be turned out. These matrices are small enough to be factorized in the core of the computer.

The iteration procedure is similiar to the multigrid method. The essential difference is that all macroelements are refined one after another, and each refinement considers the improvements obtained by the previous ones. This is carried out by the vectors $\{T\}$ which are attached on the "basic-system".

Eigenvalue analysis

Assuming M as the mass matrix of the system in fig. 3 the eigenvalue problem may be stated as

$$([K] - \lambda [M])\{x\} = \{0\} \quad (18)$$

It can be shown (ref. 3) that the solution can be found out by means of the iteration process using the hypermatrix $[K_S]$ and a decoupeled mass matrix $[M_S]$ carrying out the following iteration procedure

$$\{x_{i+1}\} = [U_S]^{-T}[M_S][U_S]^{-1}\{x_i\} - \frac{1}{\lambda_i}[U_S]^{-T}\{T_S\} \quad (19)$$

$$\{x_i\}^T[M_S]\{x_i\} = 1 \quad (20)$$

$$\{x_{i+1}\} = [U_S]^{-1}\{x_{i+1}\} \quad (21)$$

Convergence behaviour

As it is described in ref. 1the method is based on the Gauß-Seidel iteration procedure. The essential difference is that the iteration values $\{b\}$ don´t contain rigid body motions of the side. As the values $\{b\}$ are much smaller than the unknowns itselfs the rate of convergence is very high (ref. 3).

The convergence of the iteration process is always guaranteed for a positive definite matrix $[K_S]$.

STRUCTURAL MODIFICATION

The solution process is prformed by means of the matrix $[K_S]$ which consists of "n" decoupeled matrices. So it is easy to get a new matrix $[K_S^m]$ which belongs to a modified structure. The macroelements of the modified parts are recalculated only, and eq. (17) must be performed again. The effort for the computation of a modification is mainly restricted to those parts which are changed.

Using this feature a local refinement of the finite element mesh of very large structures can be done, and a accurate stress distribution can be worked out economically.

This is also applicable to eigenvalue analyses. Material nonlinearities can also be included in the local investigations.

Obviously it is also possible to carry out a geometric modification. Fig. 7 shows as an example a part of a car body; it is the join of an A-pillar and the sill. The "basic-system" coincides with the macro-structure of the mesh generator; the refined mesh is used for the analysis.

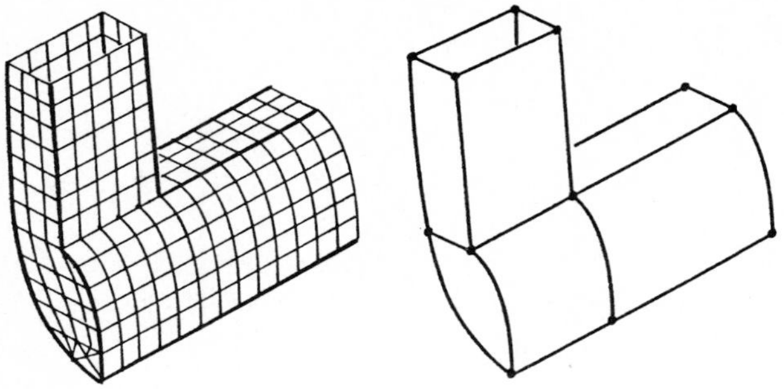

Fig. 7

Now the influence of a whole in the structure is to be calculated. Two macroelements must be recalculated only, no matter how big the structure is (fig. 8).

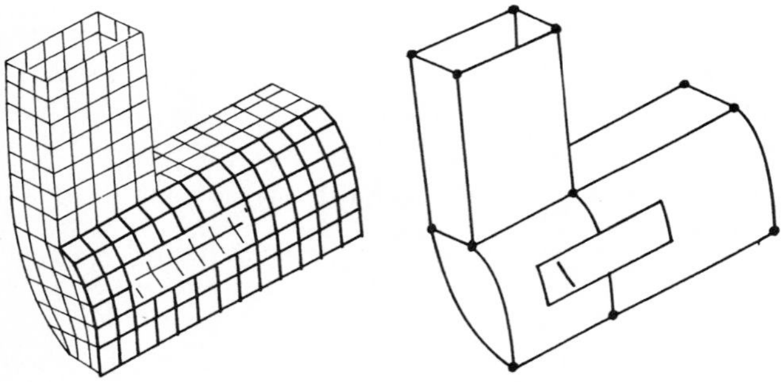

Fig. 8

NUMERICAL EXAMPLE

The convergence rate depends on the stress distribution in the structure. For example, in case of a constant stress distribution the solution of the "basic-system" yields the exact answer of the analysis.

For demonstration a cantilever beam (fig. 9) is chosen which needs very many cycles of iteration because of high stress gradients caused by the bending moment and by the loads which are attached on single points.

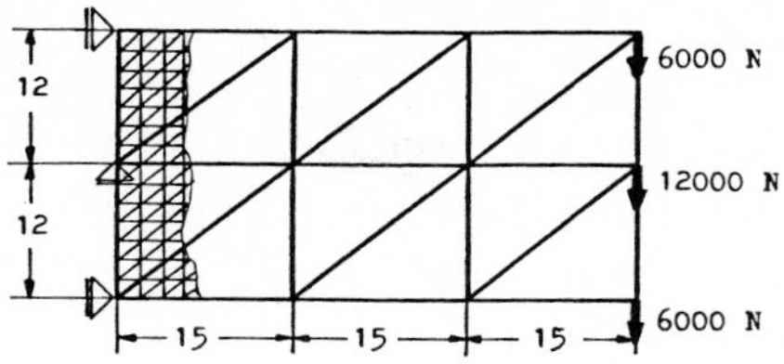

Fig. 9

Fig. 10 shows the stress distributions at the upper edge of the beam turned out by the cycles of iteration and by the direct solution. Fig. 11 shows how the smallest eigenvalue of the beam converges to the exact value.

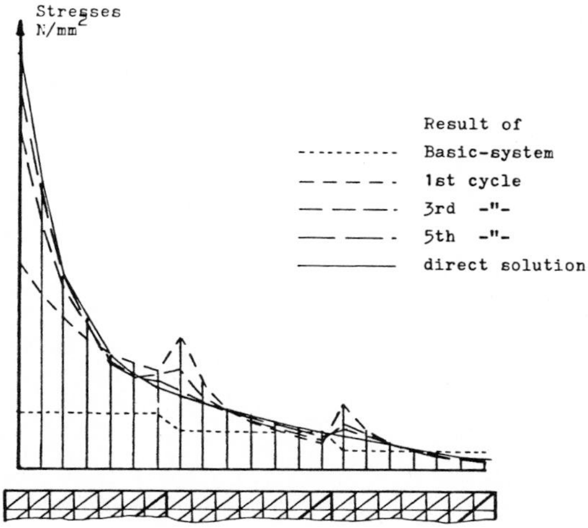

Fig. 10

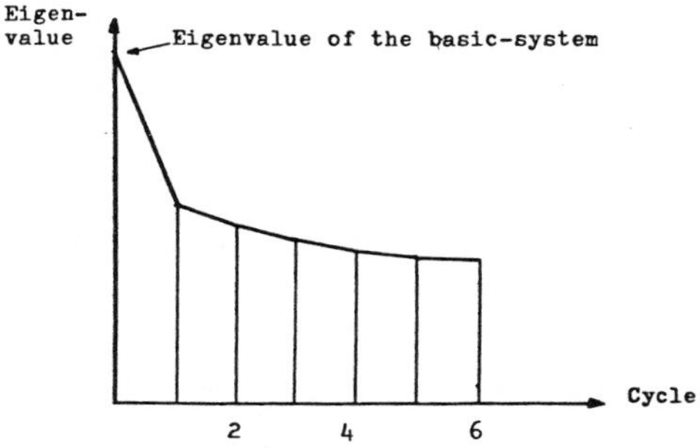

Fig. 11

ESTIMATION OF NUMERICAL EFFORT

An estimation of the number of numerical operations which are required for a computation can be found from the equations of the algorithm explained in ref. 1. A comparison will be given between the number of operations for the iteration process, the direct method and the effort for the modification of the structure. The effort for setting up the stiffness matrices is not included.

As an example a square plate was chosen having about 10000 points. The iteration process is assumed to need 50 cycles for the original run and 10 cycles for the modification. The modification concerns 1 macroelement (of 200).

Number of numerical operations ($.10^8$):

	Iteration	direct	modification
Decomposition	9	62	0,6
For-/back.-substitution	50 x 0,04	0,5	10 x 0,04
Total	11	62,5	1,0

Working space in computer:

Iteration 3 mega words

Direct method 25 mega words

REFERENCES

1 W. Dirschmid, Ein Iterationsverfahren zur Finite-Element Methode in der Elastostatik, doctoral thesis Univ. Stuttgart 1979.

2 John H. Argyris, John R. Roy, General Treatment of Structural Modifications, Journal of the Structural Division, Proceedings of the American Society of Civil Engineers, Febr. 1972

3 W. Dirschmid, An Iteration Procedure for Reducing the Expense of Static, Elastoplastic and Eigenvalue Problems in Finite Element Analyses, Computer Methods in Applied Mechanics and Engineering 35 (1982).

LOCAL EFFECTS OF GEOMETRY VARIATION IN THE ANALYSIS OF STRUCTURES

E. SCHNACK

Institute of Solid Mechanics, Karlsruhe University, Kaiserstr. 12, 75 Karlsruhe (F.R. Germany)

ABSTRACT

When constructing machine parts, the risk of cracks is caused by construction parts with high stress concentrations. To avoid cracks, a possibility exists to reduce stress concentrations by changing the shape of the construction part. This change of shape, however, is only possible as far as the technical conditions of application of the part allows it. On the other hand, the local change of the shape of the construction part must not lead to an increase of the stress values within the crack area. So the task of optimizing the stress concentrations arises under consideration of the given restrictions.

An optimization strategy out of the group of "non-linear programming" can be developed, which is of the class of "dynamic-programming" procedures. The necessary stress fields are produced by the FEM. It can be shown, that the knowledge about the local effects of elastic structures leads to avoiding the computation of gradients, as usually necessary for optimization procedures. The numerical procedure has hitherto been applicable for the optimization of plane as well as of axi-symmetric elastic structures concerning stress concentrations.

INTRODUCTION

To build up the rule for optimization, the continuum must first be defined. A continuum is assumed to be such a body of homogenious linear elastic, isotropic material that under deformation only small displacements are allowed. The domain can be given as plane or axi-symmetric bodies:

$$V \subset \mathbb{R}^n,\ n = 2,3 \quad \text{with the subdomain } V^* \subset V \tag{1}$$

The body is devided into subdomains (see Fig. 1), with boundary $\partial V \cap \partial V^*$ containing the partial boundary Γ, which will be modified so as to minimize the stress concentration on it.

It is shown in Fig. 1, that the body can be under multiple loading, i.e. ∂V_s^μ and ∂V_k^μ are loaded under various traction vectors and displacement vectors. For Γ there exists the relationship:

$$\Gamma \subset \partial V^* \cap \partial V. \tag{2}$$

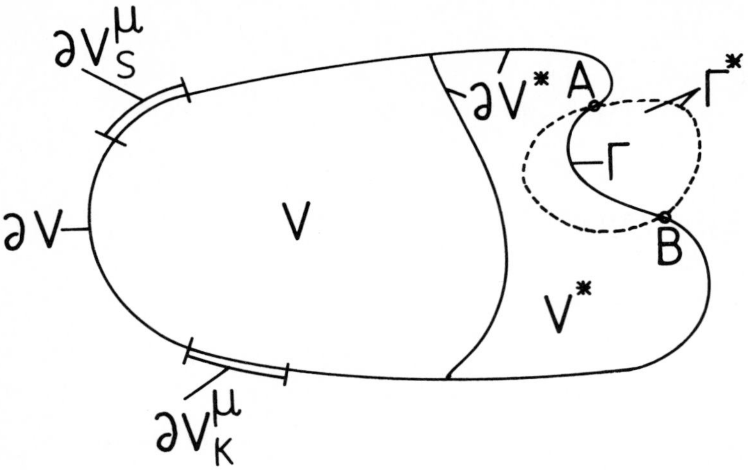

Fig. 1. Sketch of the system

V: domain given

V^*: domain to be optimized

∂V: boundary of V

∂V^*: boundary of V^*

∂V_s^{μ}: static boundary of the μ-th loading case

∂V_k^{μ}: kinematic boundary of the μ-th loading case

Γ^*: variation domain

Γ: boundary to be optimized between A and B

Additionally

$$V^* \equiv V \tag{3}$$

can be allowed, so that the global effective maximum of the stress in V can also be reduced.

If

$$V^* \subset V, \tag{4}$$

it must be guaranteed, that the global effective maximum of the stress $\tilde{\sigma}$, located in domain $V \setminus V^*$, does not raise at any matter.

So the optimizing of the effective stress $\bar{\sigma}_{\mu}$ can be formulated, if M is the number of loading cases, as follows:

$$\text{Min } (\underset{V^*}{\text{Max}}\ \bar{\sigma}_{\mu}) \qquad \mu = 1(1)M \tag{5}$$

$$\Gamma \subset \Gamma^{*} \tag{6}$$

$$\bar{\sigma}_{\mu} - \tilde{\sigma} \leq 0 \in V. \tag{7}$$

The effective stress $\bar{\sigma}_{\mu}$ will be defined according to the octahedral shear-stress hypothesis for the loading case μ (see Fig. 1). It will be assumed, as known from fracture mechanics, that, generally the local effective stress maximum appears at Γ. It leads combined with (5) to:

$$\underset{V^{*}}{\mathrm{Max}}\ \bar{\sigma}_{\mu} = \underset{\partial V^{*}}{\mathrm{Max}}\ \bar{\sigma}_{\mu} = \underset{\Gamma}{\mathrm{Max}}\ \bar{\sigma}_{\mu}. \tag{8}$$

Because in general, analytical solutions are impossible in shape optimization, a discretization becomes necessary (see Fig. 2).

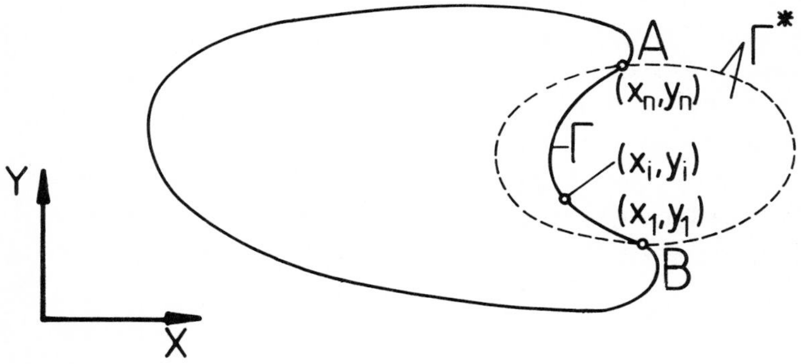

Fig. 2. Definition of the design vector

First, the design vector of Fig. 2 will be defined:

$$\{X\}^{T} = [x_1, y_1, \ldots, x_i, y_i, \ldots, x_n, y_n] \tag{9}$$

then it results:

$$\mathrm{Min}\ (\mathrm{Max}\ [\bar{\sigma}^{1}_{\mu}(\{x\}), \ldots, \bar{\sigma}^{N}_{\mu}\ (\{x\})]) \qquad \mu = 1(1)M \tag{10}$$

$$(x_i, y_i) \in \Gamma^{*} \qquad i = 1(1)N \tag{11}$$

$$\bar{\sigma}^{i}_{\mu} - \tilde{\sigma} \leq 0 \qquad i = 1(1)\bar{N}, \qquad \mu = 1(1)M. \tag{12}$$

N denotes the number of discrete interpolation nodes of Γ and $\bar{N}$ the total number of the discrete interpolation nodes of the total domain V.

For the next, a structure analysis of the strength problem will be made. The following operator-equations must be satisfied:

$\{\varepsilon\} = [C]\{u\} \in V$ (geometric relation between strain $\{\varepsilon\}$ and displacement vector $\{u\}$) (13)

$\{\sigma\} = [D]\{\varepsilon\} \in V$ (material law between stress $\{\sigma\}$ and strain vector $\{\varepsilon\}$) (14)

$[B]\{\sigma\} = \{0\} \in V$ (condition of equilibrium) (15)

$[T]\{\sigma\} = \{\bar{t}_\mu\} \in \partial V_S^\mu$ (static boundary condition, $\bar{t}_\mu$ is the given traction vector) (16)

$\{u\} = \{\bar{u}_\mu\} \in \partial V_K^\mu$ ($\bar{u}_\mu$ is the given boundary displacement vector) (17)

To solve the partial differential equations, the FEM will be used.

As in this presented process a successive reduction of the stress concentration is to be made, the classical FEM-discretization with the method of displacement is precise enough (ref. 1). Therefore the triangular plane element and the triangular torus element with the quadratic function for displacements will be used in the structure analysis for plane and axi-symmetric bodies. For this process the BEM is inadequate, because the stress field in the interior of the domain V will also be needed. That is why the FEM works more efficiently. In the following the most important equations will be described:

$$\{u_e\} = [N_e]\{d_e\} \qquad e = 1,\ldots,E \tag{18}$$

$[N_e]$ represents the form function. $\{d_e\}$ are the displacements of nodes, which are free and still to be generated. E is the number of the elements. The differentiation of equ. (18) yields the strain vector $\{\varepsilon_e\}$:

$$\{\varepsilon_e\} = [B_e]\{d_e\}. \tag{19}$$

The individual stiffness matrix $[k_e]$ will be built up with the Hooke's matrix [D]:

$$[k_e] := \int_{V_e} [B_e]^T [D] \; [B_e] \; dV. \tag{20}$$

The nodal forces can be computed according to:

$$\{F_e\}^T := \int_{\partial V_e} \{\bar{t}_e\}^T [N_e] \; ds. \tag{21}$$

$\{\bar{t}\}$ means the given traction vector of the element. The superposition of all $[k_e]$ to $[K]$ and all $\{F_e\}$ to $\{R\}$ as well as $\{d_e\}$ to $\{r\}$ yields the total stiffness relation for the loading case μ:

$$[K_\mu]\{r_\mu\} = \{R_\mu\} \qquad \text{with } \mu = 1(1)M. \tag{22}$$

The stress fields, which are to be computed according to

$$\{\sigma_e\} = [D]\{\varepsilon_e\} \tag{23}$$

are in general not continuous in V, with:

$$V: = \bigcup_{e=1}^{E} V_e. \tag{24}$$

Therefore the gap value defined by E. Schnack in 1973 (ref. 2) can be used to estimate the accuracy of the FEM structure analysis. The stresses of all elements n_i, which are in contact on node i, will first be computed (see Fig. 3).

$$e = 1(1)n_i \tag{25}$$

and then the arithmetical mean.

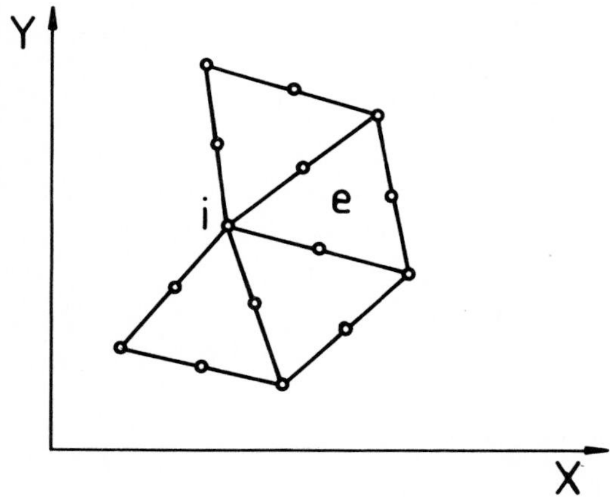

Fig. 3. Node i in element e

The gap is defined as:
"The quotient of the largest stress difference $|\Delta\bar{\sigma}_i|$ of a node in V related to the largest mean of stress $\bar{\sigma}_i$ in V, which has $\bar{N}$ nodes".

$$\text{gap:} = \frac{\underset{i}{\text{Max}}|\Delta\bar{\sigma}_i|}{\underset{i}{\text{Max}}\ \bar{\sigma}_i} 100\% \qquad i = 1(1)\bar{N} \tag{26}$$

FORMULATION OF THE STRATEGY OF OPTIMIZATION WITH LOCAL EFFECTS OF ELASTIC STRUCTURES

Beginning with an allowable start-design-vector $\{x_0\}$, the boundary to be optimized will be modified with an iterative process until the maximum of the effective stress successively decreases in consideration of all the loading cases.

The fundamentals of this process are the theorems of the notch theory, which were presented by E. Schnack in 1976 (ref. 1,3,4,5,6):

- the fade-away-law
- the theory of relieving notch
- the reaction law of the notch effect.

In 1985, U. Spörl (ref. 7) has made a generalization to the effective stress, so that some rules can be summarized as follows:

1. A geometrical disturbance on the boundary to be optimized will cause a fast fade-away of the effective stress $\bar{\sigma}$ around the disturbed place (a special form of the Saint-Venant-principle).
2. The effective stress $\bar{\sigma}$ can be influenced through the variation of the normalized curvature along Γ. The monotonization behaviour of $\bar{\sigma}$ corresponds to that of the normalized curvature.
3. Through raising the minimum of the effective stress, the maximum in the immediate neighbourhood can be decreased (the reaction law of notch effect).

From the theorems 1 to 3 E. Schnack has derived for single (see ref. 1,3,4,5,6) and multiple loadings (see ref. 8) a process of optimization which was later extended to axi-symmetrical problems by U. Spörl (ref. 7). According to the knowledge of latest studies, it can be described as follows:

To reduce the maximum of the effective stress $\bar{\sigma}^j_{max}$ on the j-th iteration step in the optimizing process, a transition function will be derived from the theorems 1 to 3, which has the following structure:

A) The maximum of the effective stress $\bar{\sigma}^j_{max}$ will be reduced through varying the normalized local curvature.
B) The minimum of the effective stress $\bar{\sigma}^j_{min}$ will raise through changing the normalized curvature, which yields additionally a reduction of the maximum of effective stress according to 3.
C) The geometrical restriction $\Gamma \subset \Gamma^*$ will be examined.
D) Roughing of the start contur Γ_j, caused by variation of the local points of maximum and minimum stress values, will be prevented through a smoothing pro-

cess on the residual nodes (see Fig. 5).

It is shown that the geometrical parameter "curvature" represents the controlling parameter.

For solutions of multiple loading cases, the procedure estimates, which points on Γ_j will be chosen in order to reduce $\bar{\sigma}^j_{max}$ and to increase $\bar{\sigma}^j_{min}$ on the j-th iteration step i. According to U. Spörl (ref. 7) for the case of maximum the node

$$i^* \in \{1,\ldots,N\} \quad \text{on } \Gamma_j \tag{27}$$

is chosen, which has the largest effective stress among all loading cases.

For the case of minimum it can occur, that a chosen point for the case of minimum happens to be also the point for the maximum of another loading case. This yields a contradiction in algorithm.

Therefore the maximum under all loading cases will first be found out, and the curvature will be reduced:

$$\Phi_{i^*} := \max \bar{\sigma}^i_\mu \qquad i = 1(1)N, \qquad \mu = 1(1)M \tag{28}$$

To intensify the curvature for a point with minimum stress, the points with the property $\Psi_{\tilde{i}}$ will be chosen:

$$\Psi_{\tilde{i}} := \min_{\mu=1(1)M} \; (\max_{i=1(1)N} \bar{\sigma}^i_\mu) \tag{29}$$

On the j-th step of the iteration will therefore be decided:

<u>j-th iteration step</u>:

<u>reduction of the maximum</u>

change of the position of i^* on Γ_j with $\bar{\sigma}^{i^*}_{\mu^*} = \max_{\mu=1(1)M} \; (\max_{i=1(1)N} \bar{\sigma}^i_\mu)$ (30)

<u>increasement of the minimum</u>

change of the position of $\tilde{i}$ on Γ_j with $\bar{\sigma}^{\tilde{i}}_{\tilde{\mu}} = \min_{\mu=1(1)M} \; (\max_{i=1(1)N} \bar{\sigma}^i_\mu)$ (31)

The local change of the position of the nodes is shown in Fig. 4.

Because the local curvature on the boundary point i is clearly defined with the neighbouring nodes i+1 and i-1, the curvature on i can be altered through changing the co-ordinates of node i, or through changing the position of i-1 and/or i+1, if the position of i is fixed in the domain of variation Γ^*. The new position of node i will be determined with the half angle of i-1, i, i+1. For the shifting of i an auxiliary point i'-1 will be set up, which has the same euclidean distance to i as that from i+1 to i (see Fig. 4).

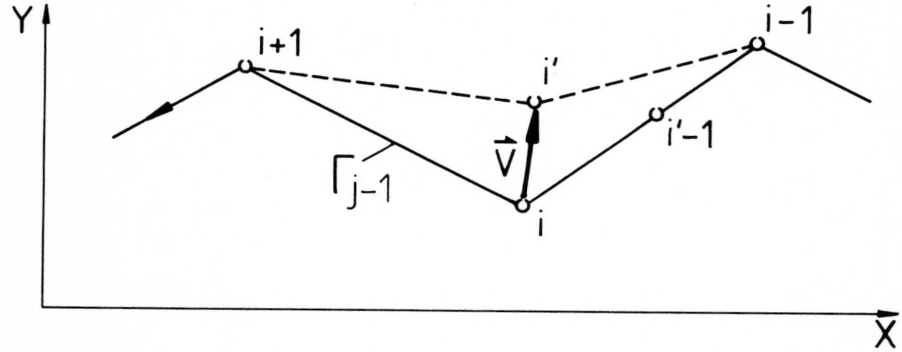

Fig. 4. Local change of the position from i to i' for the alteration of curvature in i
$\vec{V}$: shifting vector
Γ_{j-1}: optimization boundary on the j-1 step

It yields for the shifting of node i on the j-th iteration step

$$\left.\begin{aligned} x_j^i &= x_{j-1}^i + u_j^i \{y_{j-1}^i + \gamma_{j-1}^i (y_{j-1}^{i-1} - y_{j-1}^i) - y_{j-1}^{i+1}\} \\ y_j^i &= y_{j-1}^i + u_j^i \{-[x_{j-1}^i + \gamma_{j-1}^i (x_{j-1}^{i-1} - x_{j-1}^i) - x_{j-1}^{i+1}]\} \end{aligned}\right\} \tag{32}$$

with

$$\gamma_{j-1}^i := \sqrt{\frac{(x_{j-1}^{i+1} - x_{j-1}^i)^2 + (y_{j-1}^{i+1} - y_{j-1}^i)^2}{(x_{j-1}^{i-1} - x_{j-1}^i)^2 + (y_{j-1}^{i-1} - y_{j-1}^i)^2}} \qquad i = 2(1)(N-1). \tag{33}$$

The sign and the amount of the shifting parameter u_j^i are still unknown. For the shifting direction, the local information about the local curvature is necessary: It is necessary to know whether the curvature on i is convex, concave, or zero. This will be determined with the auxiliary parameter ζ_i:

$$\zeta_i := -\frac{\det D}{|\det D|} \qquad \det D := \begin{bmatrix} x_{i-1} & x_i & x_{i+1} \\ y_{i-1} & y_i & y_{i+1} \\ 1 & 1 & 1 \end{bmatrix} \tag{34}$$

with

$$\zeta_i = \begin{cases} -1 \\ 0 \\ +1 \end{cases} \tag{35}$$

Through the superposition of equ. (32) for each node i, the change of the design vector from $\{x_{j-1}\}$ to $\{x_j\}$ on the step j will be given (ref. 7):

$$\begin{bmatrix} x_j^1 \\ \vdots \\ x_j^i \\ \vdots \\ x_j^N \\ \hline y_j^1 \\ \vdots \\ y_j^i \\ \vdots \\ y_j^N \end{bmatrix} = \begin{bmatrix} x_{j-1}^1 \\ \vdots \\ x_{j-1}^i \\ \vdots \\ x_{j-1}^N \\ \hline y_{j-1}^1 \\ \vdots \\ y_{j-1}^i \\ \vdots \\ y_{j-1}^N \end{bmatrix} + \begin{bmatrix} u_j^1 \\ \vdots \\ u_j^i \{y_{j-1}^i + \gamma_{j-1}^i (y_{j-1}^{i-1} - y_{j-1}^i) - y_{j-1}^{i+1}\} \\ \vdots \\ u_j^N \\ \hline u_j^1 \\ \vdots \\ u_j^i \{-[x_{j-1}^i + \gamma_{j-1}^i (x_{j-1}^{i-1} - x_{j-1}^i) - x_{j-1}^{i+1}]\} \\ \vdots \\ u_j^N \end{bmatrix} \tag{36}$$

for

$i = 2(1)N-1$

u_j^i will be combined into the decision vector $\{u_j\}$, and it yields with the transition function $\{f\}$ the following rule of iteration (see Fig. 5):

$$\{x_j\} = \{f(\{x_{j-1}\},\{u_j\})\}. \tag{37}$$

The transition function is now known except the decision vector $\{u_j\}$.

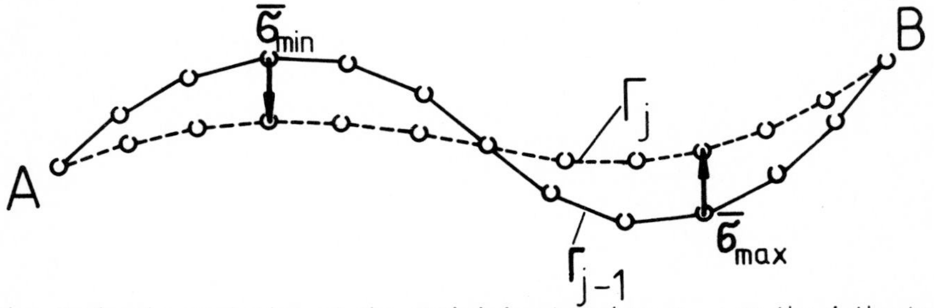

Fig. 5. On the variation of the optimizing boundary Γ_{j-1} on the j-th step

U. Spörl (ref. 7) has shown, that the algorithm, developed by E. Schnack (ref. 1) is suitable for a dynamic program. For that purpose a function g_j is to be defined:

$$g_j = \bar{\sigma}_j^{max} - \bar{\sigma}_{j-1}^{max} \qquad j \in \{1,\dots,l\} \text{ with } \bar{\sigma}_j^{max} := \max_{\substack{i=1(1)N \\ \mu=1(1)M}} \bar{\sigma}_\mu^i \tag{38}$$

and

$$\bar{\sigma}_0^{max} = f(\{x_0\}) \tag{39}$$

the objective function can be described as:

$$\min \sum_{j=1}^{l} g_j(\{X_{j-1}\},\{u_j\}). \tag{40}$$

As shown in equ. (37), the state vector $\{x_j\}$ is a function of the state vector in step j-1 and the decision vector $\{u_j\}$. It will be started with a starting vector $\{x_s\}$:

$$\{x_0\} \equiv \{x_s\}. \tag{41}$$

The possible vector $\{x_j\}$ in the period j is contained in a non-empty state space:

$$\{x_j\} \in \Xi_j \qquad j = 1(1)l. \tag{42}$$

The decision vector $\{u_j\}$ is contained in a non-empty decision space Ω_j:

$$\{u_j\} \in \Omega_j(\{x_{j-1}\}) \qquad j = 1(1)l. \tag{43}$$

The procedure of the dynamic program is shown in Fig. 6.

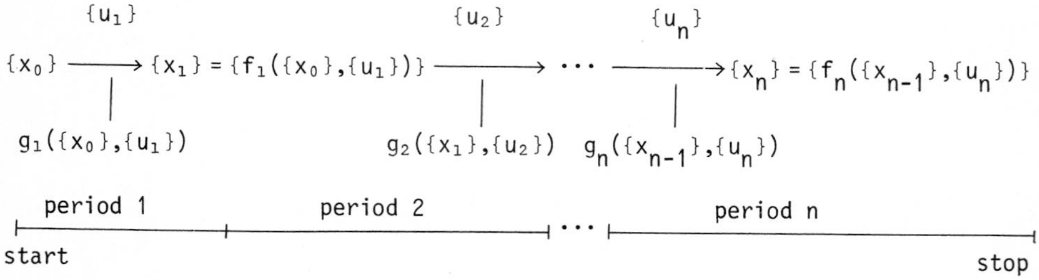

Fig. 6. Graph of the discrete dynamic program

To solve this class of problems the Bellman's functional equation method (ref.9) can be used for computing the decision vector $\{u_j\}$ according to the Bellman's principle of optimality. Because the numerical effort increases exponentially with the dimension of the state vectors (ref. 10), the optimization strategy, established before, will be maintained. The transition condition has been set up. In the following, the decision vectors will be quantitatively handled.

As already shown, the maximum of the effective stress on Γ can be reduced through altering the local curvature on the nodes with the maximal and minimal stress. But with a change of the curvature of these two points the structure could be roughed, so that a smoothing process on Γ is necessary. The amount of the shifting in the direction of the half angle decreases, according to the law of the arithmetical series of each point, on which the local curvature is to be altered, see E. Schnack (ref. 1,3). If u_j^{i*} designates the shifting of the node i*:

$$i* \in \{1,\dots,N\} \quad \text{on } \Gamma_j, \tag{44}$$

with the maximal or minimal effective stress, or their neighbouring nodes in case of unshiftability because of some restrictions, the shifting u_j^i on node i can be computed with

$$i \in \{1,\dots,N\} \setminus \{i*\} \text{ to:} \tag{45}$$

$$u_j^i := \begin{cases} 0 & i=1 \\ u_j^{i*} \quad \xi_i \quad (1 - \frac{i*-i}{i*-1}) & i<i* \\ u_j^{i*} \quad \xi_i \quad (1 - \frac{i-i*}{N-i*}) & i>i* \\ 0 & i=N \end{cases} \qquad i = 2(1)N-1 \tag{46}$$

The auxiliary value ξ_i from equ. (46) can take the value -1, 0, or +1 and determines with ξ_i = -1 v+1 the positive or negative direction to the direction of the bisector of an angle (see Fig. 4). ξ_i with the values -1 v+1 is a function of the variable ζ_i from equ. (34) as well as the stress value (formulation 30 respectively 31).

The node is unshifted for ξ_i = 0, because

- the node remains at the same position by reason of some special restrictions in the iteration,
- otherwise the node would be out of the variation region Γ^*.

It is obvious, that the components of the decision vectors are dependant (see Fig. 5, too). On changing the curvature on the nodes of extreme, the following transition condition will be generated:

$$\begin{bmatrix} x_j^1 \\ \vdots \\ x_j^{i*} \\ \vdots \\ x_j^i \\ \vdots \\ x_j^N \\ \hline y_j^1 \\ \vdots \\ y_j^{i*} \\ \vdots \\ y_j^i \\ \vdots \\ y_j^N \end{bmatrix} = \begin{bmatrix} x_{j-1}^1 \\ \vdots \\ x_{j-1}^{i*} \\ \vdots \\ x_{j-1}^i \\ \vdots \\ x_{j-1}^N \\ \hline y_{j-1}^1 \\ \vdots \\ y_{j-1}^{i*} \\ \vdots \\ y_{j-1}^i \\ \vdots \\ y_{j-1}^N \end{bmatrix} + u_j^{i*} \begin{bmatrix} 0 \\ \vdots \\ \xi_{i*} \quad \{y_{j-1}^{i*} + \gamma_{j-1}^{i*}(y_{j-1}^{i*-1} - y_{j-1}^{i*}) - y_{j-1}^{i*+1}\} \\ \vdots \\ \xi_i (1 - \frac{i-i*}{N-i*}) \{y_{j-1}^{i} + \gamma_{j-1}^{i}(y_{j-1}^{i-1} - y_{j-1}^{i}) - y_{j-1}^{i+1}\} \\ \vdots \\ 0 \\ \hline 0 \\ \vdots \\ \xi_{i*} \quad \{-[x_{j-1}^{i*} + \gamma_{j-1}^{i*}(x_{j-1}^{i*-1} - x_{j-1}^{i*}) - x_{j-1}^{i*+1}]\} \\ \vdots \\ \xi_i (1 - \frac{i-i*}{N-i*}) \{-[x_{j-1}^{i} + \gamma_{j-1}^{i}(x_{j-1}^{i-1} - x_{j-1}^{i}) - x_{j-1}^{i+1}]\} \\ \vdots \\ 0 \end{bmatrix} \tag{47}$$

As shown in equ. (47), the transition condition has the following structure:

$$\{x_j\} = \{x_{j-1}\} + \alpha_j \{S_j\} \quad \text{with} \tag{48}$$

$$\alpha_j \equiv u_j^{i*}. \tag{49}$$

Therefore it is not necessary to determine the complete decision vector $\{u_j\}$, but the parameter of the optimal step distance α_j.

As the numerical process is very CPU-time-consuming, because the FEM must be involved, an optimization of $\{s_j\}$ and α_j according to Zoutendijk (ref. 11) becomes unpractical.

Otherwise a process like that is proved to be not necessary because it results automatically in a reduction of the objective function W, as far as an adequate α_j can be found, because $\{S_j\}$ is known:

$$W(\{x_{j-1}\} + \alpha_j \{s_j\}) < W(\{x_{j-1}\}) \quad \text{with } W(\{x_{j-1}\}) := \bar{\sigma}_{j-1}^{max} \tag{50}$$

According to E. Schnack (ref. 1) α_j will be weighted with the mean side length of all triangular elements on Γ, which is to be optimized, and has a factor:

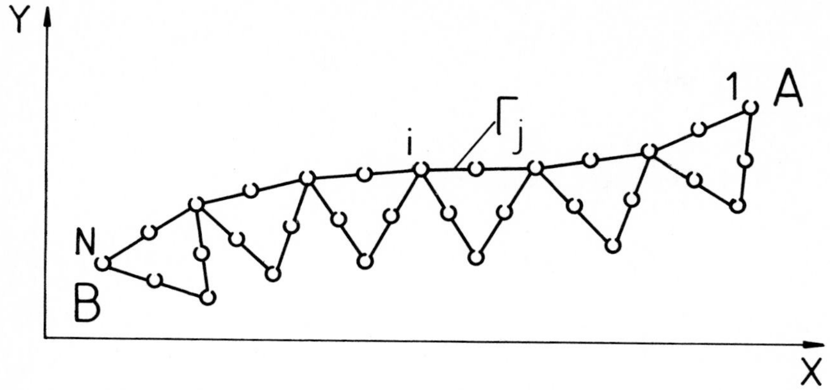

Fig. 7. Computation of the mean side length of the boundary triangle

$$\tau = \frac{1}{N-1} \sum_{i=1}^{N-1} \sqrt{(x_j^i - x_j^{i+1})^2 + (y_j^i - y_j^{i+1})^2} \tag{51}$$

$$\alpha_j := \beta\tau. \tag{52}$$

According to U. Spörl (ref. 7) α has generally the value 1/2, as shown in the test computation. If the objective function is not reduced with an estimated α_j, a new estimate will be made with linear interpolation. In addition to the descent condition the allowableness of $\alpha_j\{S_j\}$ must also be verified. As the variable ξ_i in equ. (47) will be set to zero, on leaving the domain of variation Γ^*, the node continues to be unshifted and remains in the domain. Therefore the allowableness of $\alpha_j\{S_j\}$ is made sure through construction.

In order to control, whether a minimum among the maximum of the effective stress values has been obtained, it is necessary to verify the Kuhn-Tucker-condition. But because a computation of the gradients is necessary, the verification of the Kuhn-Tucker-condition will not be carried out.

Instead of the Kuhn-Tucker-condition two hypothesises will be applied, which were proved to be useful in practice:

I) If there exists in a given region, bounded by Γ^*, between two points A and B a notch surface with constant effective stress $\bar{\sigma}$, the appearing notch stress is minimal.

II) Exisitng in a given region, bounded by Γ^*, between two points A and B no notch surface, over which the distribution of the effective stress $\bar{\sigma}$ is constant, then the appearing effective stress is minimal, if the length of Λ on which $\bar{\sigma}$ is constant, is maximal and on $(\Gamma \setminus \Lambda)$ $\bar{\sigma}$ is not higher than the constant stress $\bar{\sigma}$ of Λ.

STRUCTURE ANALYSIS WITH FEM

In order to prevent the degeneration of a network, a rezoning process has been developed by U. Spörl (ref. 7). It means that not only the boundary of the notch but also all of the points in the variation region should be shifted: For a node k in the shifting region, which is not shifted until now and has m_k neighbouring nodes, two neighbouring nodes

$$i,l \in \{1,\dots,m_k\} \tag{53}$$

with the smallest distance to node k, will be chosen. The sum of the shifting vectors $\{v_i\}$ and $\{v_l\}$, multiplied with the weighing factor γ gives the shifting vector $\{v_k\}$:

$$\{v_k\} = \gamma(\{v_i\} + \{v_l\}) \tag{54}$$

with

$$|\{p_k\} - \{p_i\}| \leq |\{p_k\} - \{p_q\}| \qquad q \in \{1,\dots,m_k\} \tag{55}$$

and

$$|\{p_k\} - \{p_l\}| \leq |\{p_k\} - \{p_q\}| \qquad q \in \{1,\dots,m_k\} \setminus \{i\} \tag{56}$$

$\{p_k\}$ denotes the position vector of node k.

Large CPU-time-consumption may be needed after each iteration step to solve the system of linear equations of FEM for the whole structure with a direct procedure like that of Cholesky:

$$[K_\mu]\{r_\mu\} = \{R_\mu\} \qquad \mu = 1(1)M. \tag{57}$$

As suggested by E. Schnack (ref. 1), it is better to set up an iterative procedure for the calculation of the physical displacement vector $\{r_\mu^j\}$ (see equ. 22).

Considering the stiffness relationship for the starting profile (marked with the index ^{0}):

$$[K_\mu^0]\{r_\mu^0\} = \{R_\mu^0\} \tag{58}$$

one can formulate the ansatz:

$$\{r_\mu^j\} = \{r_\mu^0\} + \{s_\mu\} \tag{59}$$

and this leads to the equation:

$$[K_\mu^0]\{s_\mu\} = ([K_\mu^0] - [K_\mu^j])\ \{r_\mu^0\} + ([K_\mu^0] - [K_\mu^j])\ \{s_\mu\}. \tag{60}$$

From equ. (60) an iteration rule can be derived:

$$\{s_\mu^{k+1}\} = [A]\{s_\mu^k\} + \{a\} \tag{61}$$

with

$$[A]: = [K_\mu^0]^{-1}\ ([K_\mu^0] - [K_\mu^j]) \tag{62}$$

and

$$\{a\}: = [K_\mu^0]^{-1}\ ([K_\mu^0] - [K_\mu^j])\ \{r_\mu^0\}. \tag{63}$$

The convergence will be guaranteed, if for the spectral radius is valid:

$$\rho < 1. \tag{64}$$

This iteration procedure is advantageous, because the total stiffness matrix is to be factorized only by the start:

$$[K_\mu^0] = [L_\mu^0]^T[L_\mu^0]. \tag{65}$$

TEST EXAMPLE

A large plate, alternately loaded in x and y direction, is shown in Fig. 8, on which the optimal form of the hole inside a rectangular variation region was found. The starting profile is the oblong hole (b in Fig. 8). The contour with the symbol c in Fig. 8 is the optimal one in this case. If the stresses σ_1 and σ_2 act on the plate at the same time, the optimal contour will be an ellipse (a in Fig. 8). Fig. 9 shows, that the maximum of the effective stress under multiple loading was able to be reduced by an amount of 25%.

Fig. 10 demonstrates the optimization for the rounding out of a gear root, and Fig. 11 shows, that for such elastic problems the maximum of the effective stress was reduced by 20%, compared to a conventional one.

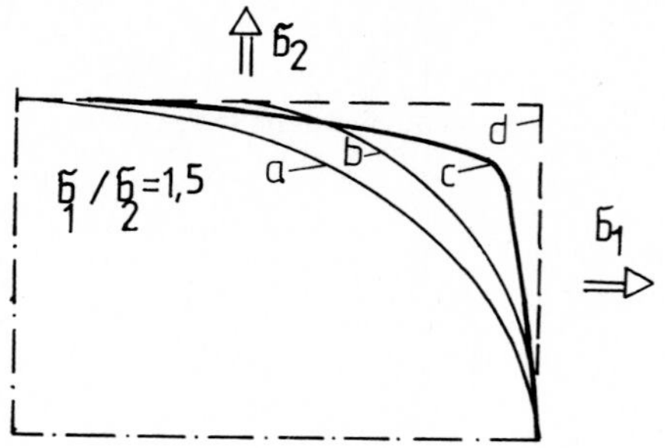

Fig. 8. A large plate with optimized hole contour under biaxial loading

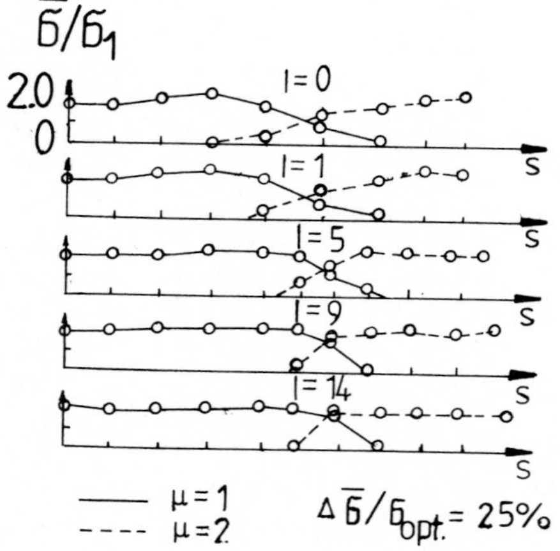

Fig. 9. Curve of convergence for the problem shown in Fig. 8

$\sigma_{opt} := \bar{\sigma}_1^{max}$

$\Delta\bar{\sigma} := \bar{\sigma}_0^{max} - \bar{\sigma}_1^{max}$

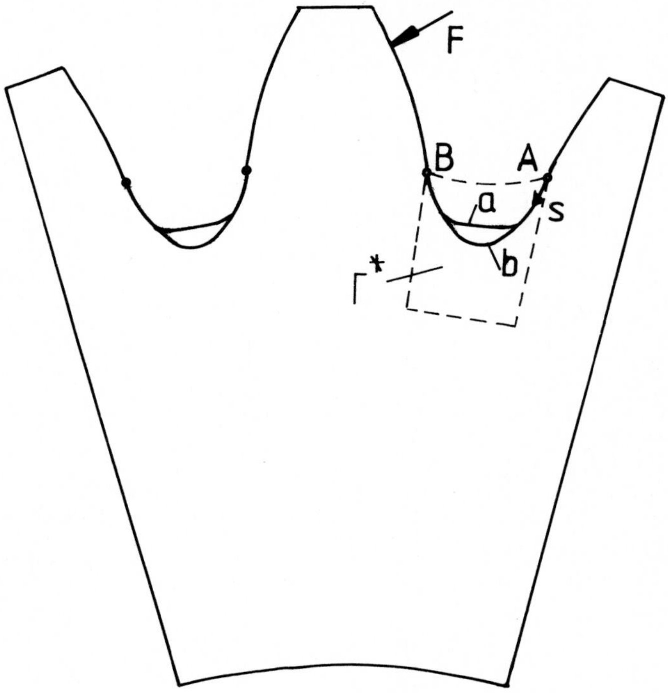

Fig. 10. Comparison between starting profile (a) and optimized profile (b) according to U. Spörl (ref. 7)

A,B   points of notch boundary

Γ*   variation domain

a   starting profile with state vector $\{x_0\}$

b   optimium of Γ, $\bar{\sigma}^{max}$ is a minimum without the stress peak of the singularity, where the single force F is acting

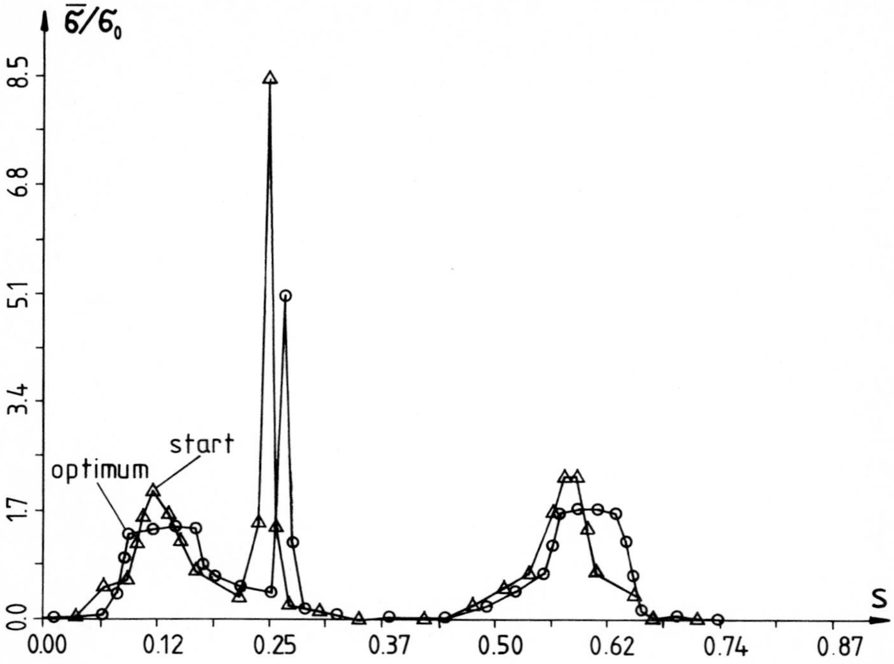

Fig. 11. Stress distribution between points A and B for the rounding out of a gear root (see Fig. 10)

REFERENCES

1 E. Schnack, Ein Iterationsverfahren zur Optimierung von Spannungskonzentrationen, inaugural dissertation, University of Kaiserslautern, 1977.
2 E. Schnack, Beitrag zur Berechnung rotationssymmetrischer Spannungskonzentrationsprobleme mit der FEM, Dr.-Ing. Diss., University of Munich, 1973.
3 E. Schnack, An Optimization Procedure for Stress Concentrations by the Finite-Element-Technique, Int. J. Num. Meth. Engng., Vol. 14, No. 1, 1979, pp.115-124.
4 E. Schnack, Ein Iterationsverfahren zur Optimierung von Kerboberflächen, VDI-Forschungsheft Nr. 589, VDI-Verlag, Düsseldorf, 1978.
5 E. Schnack, Optimal Designing of Notched Structures without Gradient Computation, Proceedings of the Third IFAC Symposium "Control of Distributed Parameter Systems", Toulouse, 29 June-2 July 1982, Pergamon Press 1983, pp. 365-369.
6 E. Schnack, Computer-Simulation of an Experimental Method for Notch-Shape-Optimization, Proceedings of the Int. Symp. "Simulation in Engineering Science", Nantes, 9-11 May 1983, ENSM (Eds.), Nantes, 1983, pp. 311-316.
7 U. Spörl, Spannungsoptimale Auslegung elastischer Strukturen, Dr.-Ing. Diss., University of Karlsruhe, 1985.
8 E. Schnack, Optimierung von Spannungskonzentrationen bei Viellastbeanspruchung ZAMM, Bd. 60, 1980, pp. 151-152.
9 R. Bellman, Dynamic Programming, Princeton University Press, 1957.
10 U. Spörl and E. Schnack, Gradientenfreie Strukturoptimierung mit der FEM, DFG-final report, University of Karlsruhe, 1985.
11 G. Zoutendijk, Methods of Feasible Directions, Elsevier Publ. Comp., Amsterdam, Princeton, 1960.

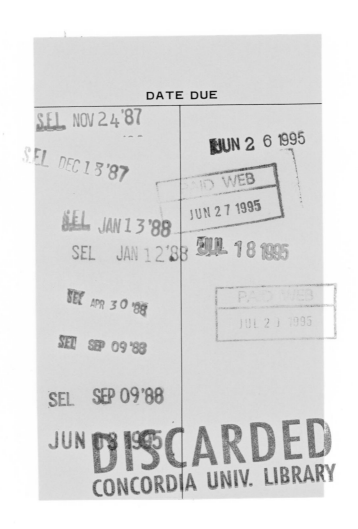